ADVANCES IN CHEMICAL PHYSICS

VOLUME XX

EDITORIAL BOARD

Advances in CHEMICAL PHYSICS

EDITED BY

I. PRIGOGINE

University of Brussels,
Brussels, Belgium

AND

STUART A. RICE

Department of Chemistry
and
The James Franck Institute
The University of Chicago
Chicago, Illinois

VOLUME XX

WILEY-INTERSCIENCE

A DIVISION OF JOHN WILEY AND SONS, INC.
NEW YORK · LONDON · SYDNEY · TORONTO

Library of Congress Catalog Card Number: 58 9935

ISBN 0 471 69925 X

Printed in the United States of America

10 9 8 7 6 5 4 3 2 1

INTRODUCTION

In the last decades, chemical physics has attracted an ever-increasing amount of interest. The variety of problems, such as those of chemical kinetics, molecular physics, molecular spectroscopy, transport processes, thermodynamics, the study of the state of matter, and the variety of experimental methods used, makes the great development of this field understandable. But the consequence of this breadth of subject matter has been the scattering of the relevant literature in a great number of publications.

Despite this variety and the implicit difficulty of exactly defining the topic of chemical physics, there are a certain number of basic problems that concern the properties of individual molecules and atoms as well as the behavior of statistical ensembles of molecules and atoms. This new series is devoted to this group of problems which are characteristic of modern chemical physics.

As a consequence of the enormous growth in the amount of information to be transmitted, the original papers, as published in the leading scientific journals, have of necessity been made as short as is compatible with a minimum of scientific clarity. They have, therefore, become increasingly difficult to follow for anyone who is not an expert in this specific field. In order to alleviate this situation, numerous publications have recently appeared which are devoted to review articles and which contain a more or less critical survey of the literature in a specific field.

An alternative way to improve the situation, however, is to ask an expert to write a comprehensive article in which he explains his view on a subject freely and without limitation of space. The emphasis in this case would be on the personal ideas of the author. This is the approach that has been attempted in this new series. We hope that as a consequence of this approach, the series may become especially stimulating for new research.

Finally, we hope that the style of this series will develop into something more personal and less academic than what has become the standard scientific style. Such a hope, however, is not likely to be completely realized until a certain degree of maturity has been attained—a process which normally requires a few years.

At present, we intend to publish one volume a year, and occasionally several volumes, but this schedule may be revised in the future.

In order to proceed to a more effective coverage of the different aspects of chemical physics, it has seemed appropriate to form an editorial board. I want to express to them my thanks for their cooperation.

I. Prigogine

CONTRIBUTORS TO VOLUME XX

T. Erber, Department of Physics, Illinois Institute of Technology, Chicago, Illinois

H. H. Farrell, Department of Applied Science, Brookhaven National Laboratory, Upton, New York

B. N. Harmon, Department of Physics, Northwestern University, Evanston, Illinois

M. R. Hoare, Department of Physics, Bedford College, London, England

Taro Kihara, Department of Physics, Faculty of Science, University of Tokyo, Tokyo, Japan

John E. Kilpatrick, Department of Chemistry, Rice University, Houston, Texas

H. G. Latal, Department of Physics, Illinois Institute of Technology, Chicago, Illinois

Ian G. Ross, Department of Chemistry, School of General Studies, The Australian National University, Canberra, Australia

G. A. Somorjai, Inorganic Materials Research Division, Lawrence Radiation Laboratory and Department of Chemistry, University of California, Berkeley, California

CONTENTS

ADVANCES IN CHEMICAL PHYSICS

VOLUME XX

MULTIPOLAR INTERACTIONS IN MOLECULAR CRYSTALS

TARO KIHARA

Department of Physics, Faculty of Science, University of Tokyo, Tokyo

CONTENTS

I. INTRODUCTION

The structure of a molecular crystal is governed by the shape of the molecule as well as by the intermolecular forces. The relation between these molecular characteristics and the crystal structure has been clarified to a certain extent by the use of molecular models.[1]

The models simulate magnetically the electric forces between actual molecules. They consist of barium ferrite magnets and manganese-zinc ferrite pieces in such a way that the electric multipole and the electric polarizability of each molecule are represented by the magnetic multipole and the magnetic susceptibility of the model, respectively. Here the manganese-zinc ferrite pieces, whose permeability is greater than 1000, are not permanently magnetized but magnetization is induced in these pieces by the "hard" barium ferrite magnets coming nearby; thus they represent the effect of induced polarization of a molecule. When the shape and the magnetic characteristic of a model are adequately patterned after a given molecule, a structure into which these models are assembled will simulate the actual crystal structure.

The greater part (usually more than 60%) of the cohesive energy of a molecular crystal originates in the van der Waals attraction, which is not very sensitive to the mutual orientation of the molecules. The structure of a molecular crystal, on the other hand, is largely governed by orientation-sensitive multipole interactions; such interactions can be simulated by our molecular models.

The principles upon which our molecular models are based are "simplicity" and "accuracy." The shape of a model is chosen as simple as possible, unessential details of the actual molecular shape being neglected. For example, our model of the benzene molecule is axially symmetric, showing that the deviation of the actual molecular shape from axial symmetry is not essential for predicting the crystal structure. Furthermore, the models are chosen in such a way that a structure formed by the models represents not only the symmetry elements of the crystal but also the cell dimensions as accurately as possible.

II. ELECTRIC MULTIPOLES OF MOLECULES

The power of an atom in a molecule to attract electrons to itself is called the electronegativity. Table I gives a part of the electronegativity scale determined by Pauling.

Fluorine is by far the most electronegative of the atoms. Oxygen has

TABLE I

Electronegativity Values for Some Elements[a]

H 2.1						
Li 1.0	Be 1.5	B 2.0	C 2.5	N 3.0	O 3.5	F 4.0
Na 0.9	Mg 1.2	Al 1.5	Si 1.8	P 2.1	S 2.5	Cl 3.0
			Ge 1.8	As 2.0	Se 2.4	Br 2.8
			Sn 1.8	Sb 1.9	Te 2.1	I 2.5

[a] The values are taken from L. Pauling, *The Nature of the Chemical Bond*, 3rd Ed., Cornell University Press, Ithaca, N.Y., 1960.

the second largest electronegativity. Nitrogen, chlorine, and bromine are also quite electronegative.

The shape of the molecule of 1,2-dichloroethane, $ClCH_2CH_2Cl$, in a crystal is rodlike or like a prolate spheroid (Fig. 1a). Since the chlorine atoms attract the electrons, this rodlike molecule has an electric quadrupole, with negative charge being near the ends and positive charge in the middle.

The shape of the molecule of β-hexachlorocyclohexane, $C_6H_6Cl_6$, is like an oblate spheroid, the six chlorine atoms forming a ring outside the ring of carbon atoms (Fig. 1b). The surface of this molecule is composed of one circular zone of negative charge and two caps of positive charge; thus the molecule has an axially symmetric quadrupole.

The quadrupole of such a "uniaxial" molecule is characterized by the moment Q defined by

$$Q = \frac{1}{2} \int (2z^2 - x^2 - y^2)\rho \, d\tau$$

Here $\rho \, d\tau$ is the electric charge in the volume element $d\tau$ at the point (x, y, z), the z-axis being on the axis of the molecule. According to this

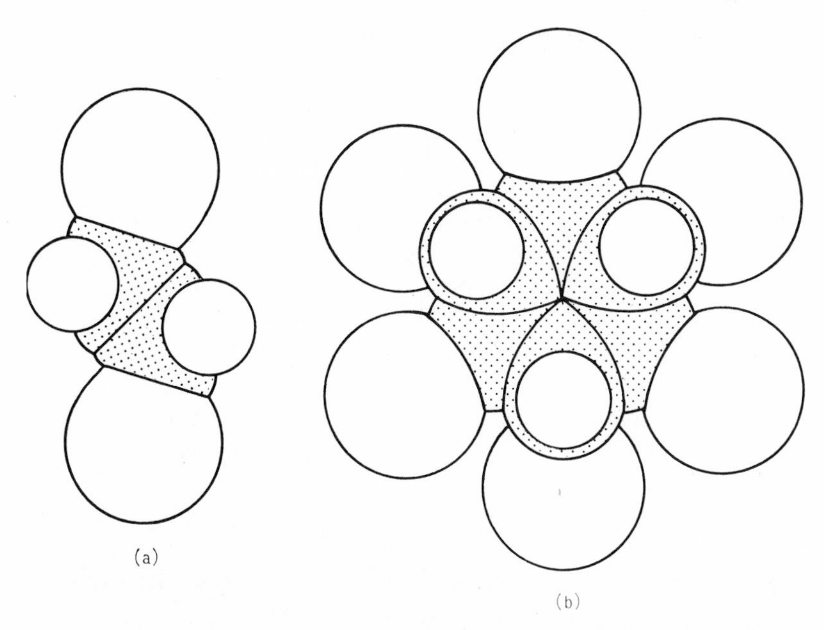

Fig. 1. (a) The molecule of 1,2-dichloroethane. (b) The molecule of β-hexachlorocyclohexane. The carbon atoms are dot-shaded; the large and small spheres are chlorine and hydrogen atoms, respectively.

TABLE II

Quadrupole Moments of Molecules[a]

	CO_2	N_2	C_2H_6	O_2	H_2	F_2	Cl_2
$10^{26}Q$ (esu)	-4.2	-1.5	-0.6	-0.4	$+0.66$	$+0.9$	$+6.1$

[a] Values for CO_2, N_2, C_2H_6, and O_2 are taken from A. D. Buckingham, *Advances in Chemical Physics*, Vol. 12, Wiley-Interscience, New York, 1967. Values for H_2, F_2, and Cl_2 are from D. E. Strogryn and A. D. Strogryn, *Mol. Phys.*, 11, 371 (1966).

definition, the sign of the quadrupole moment of 1,2-dichloroethane is negative, and the sign of the quadrupole moment of β-hexachlorocyclohexane is positive.

The Pauling scale of electronegativity holds quantitatively for single bonds; qualitatively, however, it can also be applied to double and triple bonds. Thus carbon dioxide O=C=O and cyanogen N≡C—C≡N should have quadrupole moments with negative sign.

The molecular quadrupole moments determined recently by Buckingham et al.[2] and those recommended by Strogryn and Strogryn[3] are listed in Table II.

In the nearly spherical molecule of hexamethylenetetramine, $(CH_2)_6N_4$,

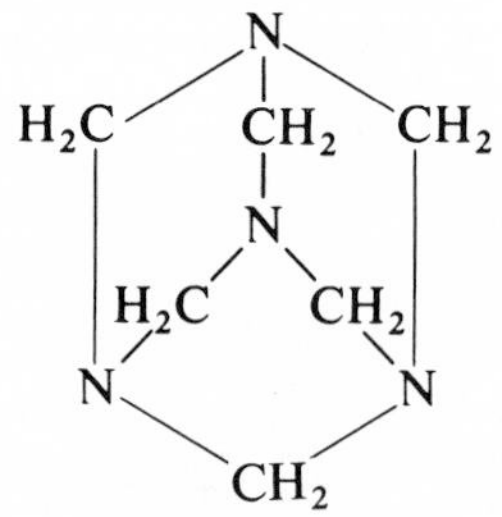

N
H_2C CH_2 CH_2
N
H_2C CH_2
N N
CH_2

four nitrogen atoms form a regular tetrahedron. The surface of this molecule is composed of four poles of negative charge at the positions of nitrogen atoms and four poles of positive charge in between. Thus the molecule has an electric octopole. Similarly, the molecule of silicon tetrafluoride, SiF_4, has an octopole.

UF_6, UCl_6, and WCl_6 are examples of molecules with higher multipoles.

III. THE CRYSTAL STRUCTURE OF CARBON DIOXIDE

The crystal structure of carbon dioxide belongs to the cubic system. The carbon atoms in a carbon dioxide crystal form a face-centered cubic lattice, which is composed of four primitive cubic lattices. On each primitive cubic lattice, the axes of the CO_2 molecules are all parallel to a particular body diagonal; the direction of axes is different for each of the four primitive lattices (cf. Fig. 8). The distance d between nearest-neighbor carbon atoms is 4.16 Å; the cohesive energy of the crystal is -6.4 kcal/mol.

The contribution of the quadrupole-quadrupole interaction to the cohesive energy is

$$-\tfrac{1}{2} \times 21.4Q^2/d^5$$

per molecule, Q being the quadrupole moment.[4] For carbon dioxide this quantity is -1.51×10^{-13} erg per molecule or -2.2 kcal/mole, which is one third the cohesive energy mentioned above.

In the gaseous state, this type of interaction does not play an important role, because a multipole interaction almost vanishes when it is averaged with respect to the molecular orientations. In fact, the "core potential" of the intermolecular force with parameters determined from the second virial coefficient gives -4.7 kcal/mol as the cohesive energy, which is very close to the difference between the actual cohesive energy and the contribution of the quadrupole interaction.[5]

Thus we may note that the quadrupole interaction lowers the energy of the cubic crystal structure of carbon dioxide, which does not correspond to a close-packed structure for this prolate molecule. In fact, if carbon dioxide crystallized into a close-packed structure, keeping the molecular axes parallel, then the quadrupole interaction, which is repulsive in this case, would reduce its total cohesive energy.

Such multipole interactions can be simulated by molecular models with magnetic multipoles. The simplest model of this type is a quadrupolar sphere composed of two hemispheres of magnetized barium ferrite as shown in Figure 2. By assembling several such spheres, we can represent the crystal structure of carbon dioxide as shown in Figure 3.

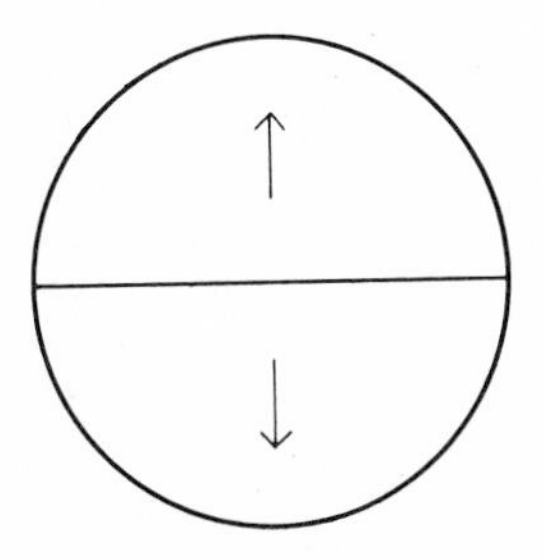

Fig. 2. Quadrupolar sphere.

Fig. 3. Cubic *Pa*3 structure composed of quadrupolar spheres. The left configuration is formed with 19 spheres and the right with 14 spheres.

A structure formed by the models is quite stable against gravitation and other disturbances. It is preferable, however, to assemble the models step by step as shown in Figures 4 and 5.

The spherical shape of the model is an idealization of the shape of the CO_2 molecule; a spherocylinder adopted in Section V will be closer to reality.

Types of molecular crystals that can be reproduced by barium ferrite models are limited in number. This limitation, however, is removed essentially by using manganese-zinc ferrite pieces in combination with barium ferrite magnets. This is because the large permeability of Mn-Zn ferrite plays the role of the electric polarizability of a molecule.

Examples selected in a systematic way will be given after an explanation of the notion of space groups.

Fig. 4. Steps in assembling 19 quadrupolar spheres. (i) On a desk, make two three-"molecule" systems that are congruent without reflection. (ii) Hold an additional molecule above the center of one system with the molecular axis kept in the vertical direction. Then the three underlying molecules make a slight inclination, as a result of which the vertically held molecule is attracted and a stable four-molecule system is formed. (iii) Put the four-molecule system on the three-molecule system. Then the composite system automatically takes the configuration in which the upper part is rotated by 60° with respect to the lower. (iv) Add three molecules. (v) Further add four molecules. (vi) Finally add five molecules in the top layer.

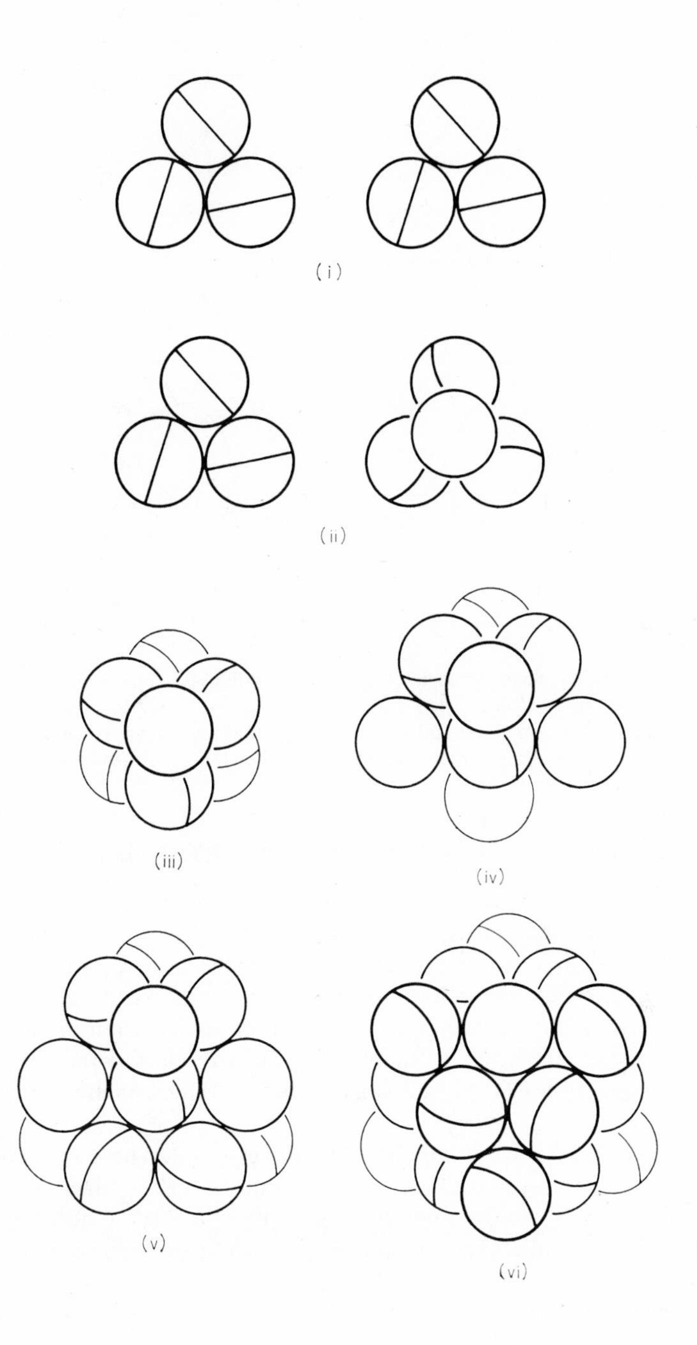
(i)
(ii)
(iii)
(iv)
(v)
(vi)

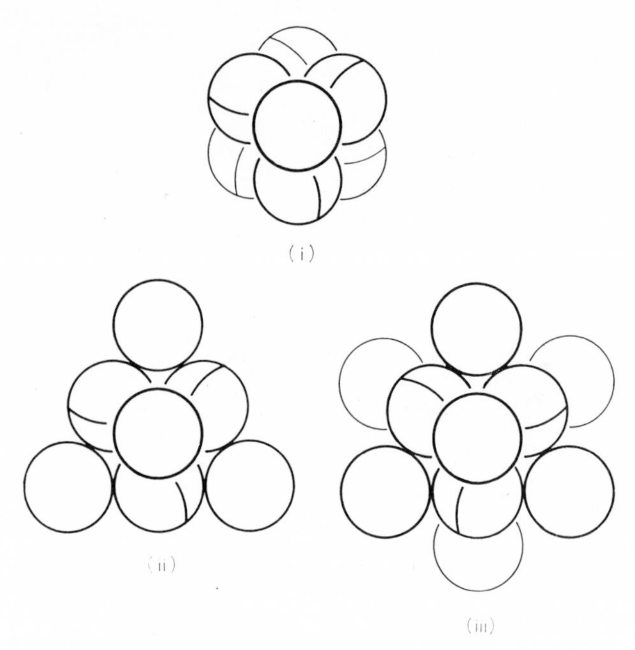

Fig. 5. Steps in assembling 14 quadrupolar spheres. (i) Make the seven-molecule system as in Figure 4. (ii) Add three molecules in a vertical orientation around the upper system. (iii) Invert the ten-molecule system on the palm of one hand, and put four more molecules on as in (ii).

IV. THE SYMMETRY OF CRYSTALS

We have seen that the crystal lattice of carbon dioxide is composed of four primitive cubic lattices, on each of which the molecules are in the same direction and can be brought into coincidence by a translation.

A lattice formed by such equivalent points that can be brought into coincidence by a translation (namely, without any rotation or reflection) is called a Bravais lattice of the crystal. Thus the lattice formed by the carbon atoms in a CO_2 crystal consists of four interpenetrating Bravais lattices.

There are fourteen types of Bravais lattices in all. These are shown in Figure 6. Here the symbol *P* stands for "primitive," *F* for "face-centered," *I* for the German word "Innenzentrum" (body center), *C* for "centered on the *C*-face," and *R* for "rhombohedral." These symbols are used in the notation of space groups.

The complete symmetry of a crystal structure can be expressed in terms

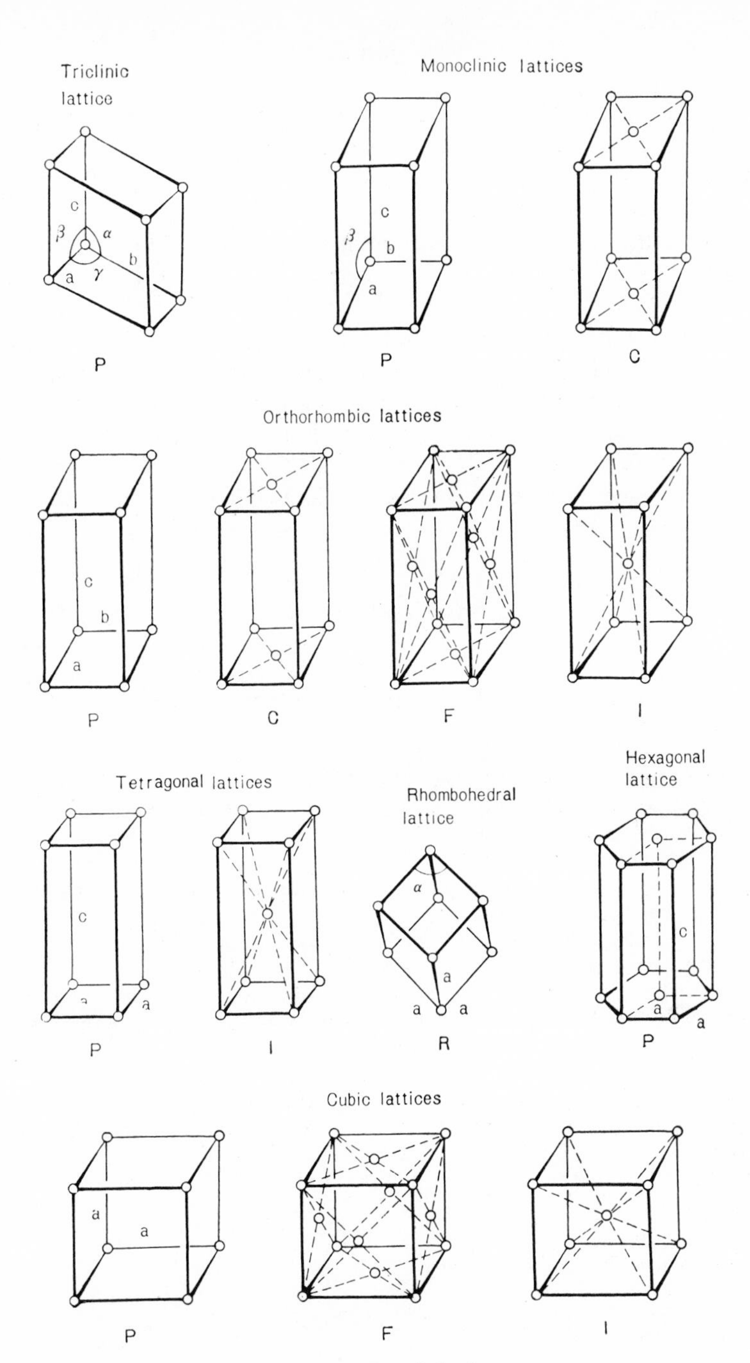

Fig. 6. The Bravais lattices.

of its space group. The space group for the carbon dioxide crystal is denoted by $Pa3$. Here the capital letter P means that the Bravais lattice is primitive; $a3$ indicates that the structure has the symmetry of a-glide plane (see below) and threefold axis.

The structure has a glide-reflection plane or glide plane if it is unchanged by a reflection in this plane, accompanied by a translation through a certain distance d in a particular direction lying in this plane. A glide plane is denoted by a if the translation is parallel to the a-axis and d is half the period; b- and c-glide planes are similarly defined. If the translation is parallel to a face diagonal and d is $\frac{1}{2}$ or $\frac{1}{4}$ of the diagonal period, the glide plane is called a net or a diamond glide plane, respectively.

The structure has an n-fold screw axis if it is unchanged by rotation through an angle $2\pi/n$ about this axis, accompanied by a translation of

TABLE III

Typical Space Groups

Monoclinic system. $P2_1/c$ (C_{2h}^5).

$P2_1/c$: The Bravais lattice is primitive. The b-axis is a twofold screw axis; a c-glide plane is perpendicular to this axis.

Orthorhombic system. $Pnnm$ (D_{2h}^{12}), $Pbca$ (D_{2h}^{15}), $Pnma$ (D_{2h}^{16}), $Cmca$ (D_{2h}^{18}).

Pnnm: A net is perpendicular to the a-axis, a net is perpendicular to the b-axis, and a mirror plane is perpendicular to the c-axis.

Cmca: The Bravais lattice is base-centered on the C-face. A mirror plane is perpendicular to the a-axis, a c-glide plane is perpendicular to the b-axis, and an a-glide plane is perpendicular to the c-axis.

Tetragonal system. $I\bar{4}2m$ (D_{2d}^{11}), $P4_2/nmc$ (D_{4h}^{15}).

$I\bar{4}2m$: The Bravais lattice has body-centers (Innenzentren). The c-axis is a fourfold rotary-inversion axis, the a-axis is a twofold axis, and a diagonal plane through the c-axis is a mirror.

Trigonal system: $R\bar{3}$ (C_{3i}^2), $P\bar{3}1c$ (D_{3d}^2), $P\bar{3}ml$ (D_{3d}^3), $R\bar{3}m$ (D_{3d}^5).

$P\bar{3}m1$: The Bravais lattice is hexagonal. The c-axis is a threefold rotary-inversion axis and lies in a mirror plane which is perpendicular to the a-axis.

$R\bar{3}m$: The Bravais lattice is rhombohedral. A threefold rotary-inversion axis lies in a mirror plane.

Hexagonal system. $P6_3/m$ (C_{6h}^2), $P6_3/mmc$ (D_{6h}^4).

Cubic system. $Im3$ (T_h^5), $Pa3$ (T_h^6), $I\bar{4}3m$ (T_d^3), $Fd3m$ (O_h^7).

$Fd3m$: The Bravais lattice is face-centered. The structure is called the diamond structure and is characterized by a diamond glide plane, a threefold axis, and a mirror.

distance d along the axis. An n-fold screw axis is denoted by n_p ($p = 1, 2, \ldots, n-1$) if d is equal to p/n of the period.

The structure has an n-fold rotary-inversion axis denoted by $\bar{n}$ if it is unchanged by rotation through an angle $2\pi/n$ about this axis followed by an inversion with respect to a point on the axis.

Space groups that will be used in this article are listed in Table III with some explanations.

V. MODELS 1–6 FOR RODLIKE MOLECULES

The molecular Models 1–6, which are shown in Figure 7, represent the crystal structures of typical rodlike molecules:

carbon dioxide	O=C=O
acetylene	H—C≡C—H
cyanogen	N≡C—C≡N
1, 2-dichloroethane	Cl—CH_2—CH_2—Cl
trans-1, 4-dibromocyclohexane	Br—CH(CH_2—CH_2)(CH_2—CH_2)CH—Br (ring: H_2C—CH_2 above, H_2C—CH_2 below)
dicyanoacetylene	N≡C—C≡C—C≡N
dimethyltriacetylene	CH_3—C≡C—C≡C—C≡C—CH_3
heavy halogens	Cl_2, Br_2, I_2

Each of these models is patterned after the electric multipole of a molecule as well as the shape of the molecule, which is governed by the van der Waals radii of atoms and the interatomic distances. When the electric multipole is sufficiently strong and the shape is not far from a sphere, a model composed only of barium ferrite magnets can be used. In general, however, the polarizability of the molecule plays an important part, and this effect can be taken into the model by mixing Mn-Zn ferrite. An appropriate pattern of this mixing is to put a magnet between two pieces of Mn-Zn ferrite.

The crystal structures are compared in Table IV with corresponding constructions formed by the models. These constructions are shown in Figures 8–16.

In Tables IV–VIII, *Crystal Structures*, written by Wyckoff,[6] and *Crystal Data*,[7] published by the American Crystallographic Association, are cited. The reader should compare the photographs in this article with the beautiful illustrations in the cited sections of Wyckoff's books.

A quadrupolar prolate spheroid can be made simply by combining two hemispheroidal magnets. If the spheroid is not far from a sphere, an assembly of such molecular models gives the cubic structure $Pa3$ (T_h^6), which corresponds to the crystal structure of acetylene, C_2H_2, above −140°C.[8] On the other hand, an assembly of more rodlike spheroids gives the orthorhombic $Pnnm$ (D_{2h}^{12}) structure. An assembly of critical spheroids with the axis ratio 3/4 gives both the cubic and orthorhombic structures mentioned above.

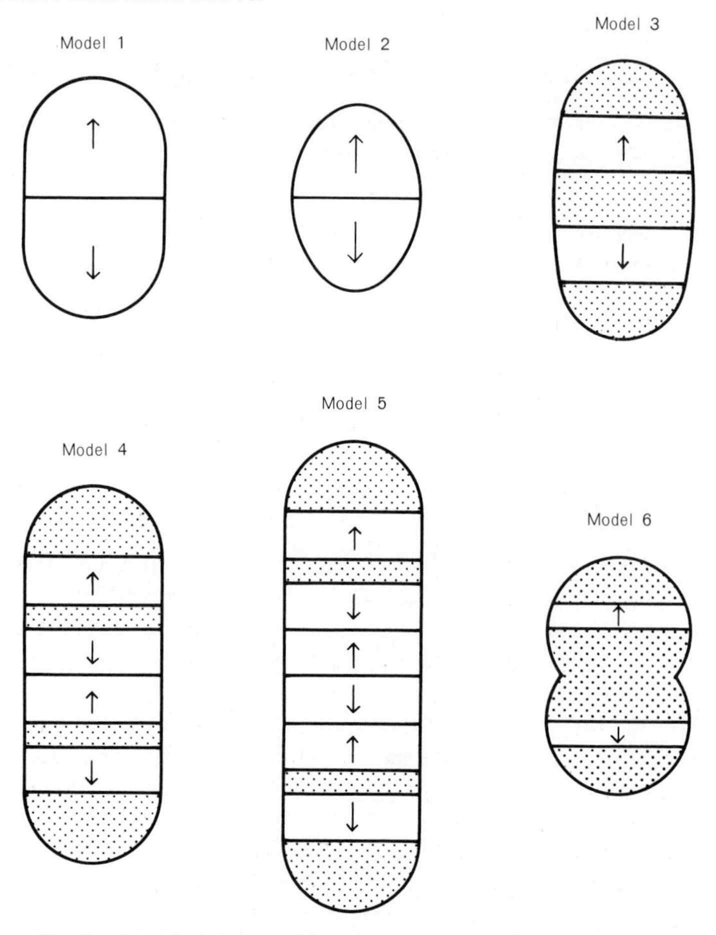

Fig. 7. Models 1–6 for rodlike molecules. Model 1 is a quadrupolar spherocylinder. Model 2 is a quadrupolar prolate spheroid. Model 3 is for N≡C—C≡N. Model 4 is for N≡C—C≡C—C≡N. Model 5 is for CH_3—C≡C—C≡C—C≡C—CH_3. Model 6 is for Cl_2, Br_2, and I_2. Manganese-zinc ferrite is shaded with dots.

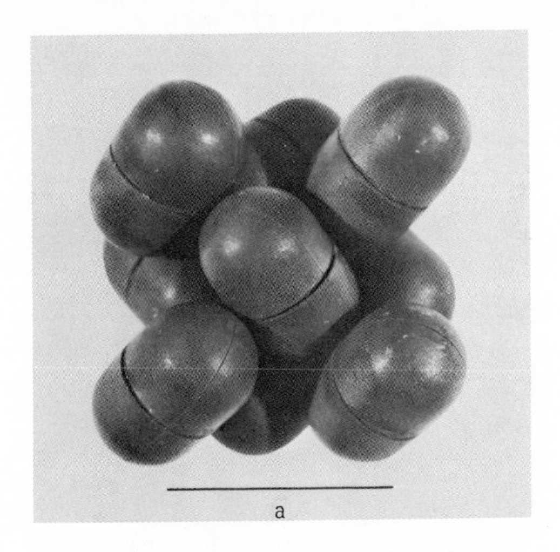

Fig. 8. Cubic *Pa*3 structure simulating the crystal structure of carbon dioxide (Model 1).

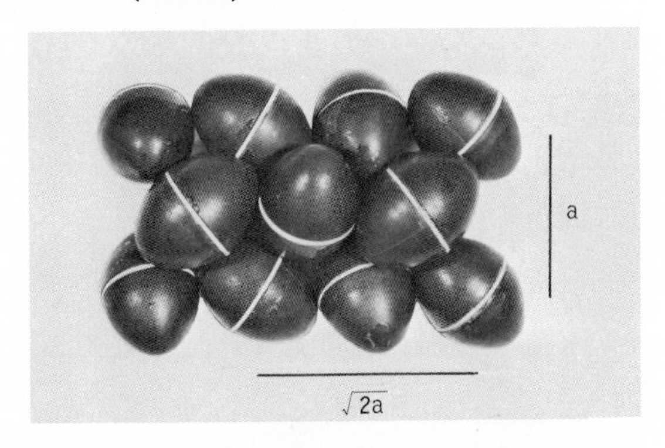

Fig. 9. Cubic *Pa*3 structure seen from (110) (Model 2).

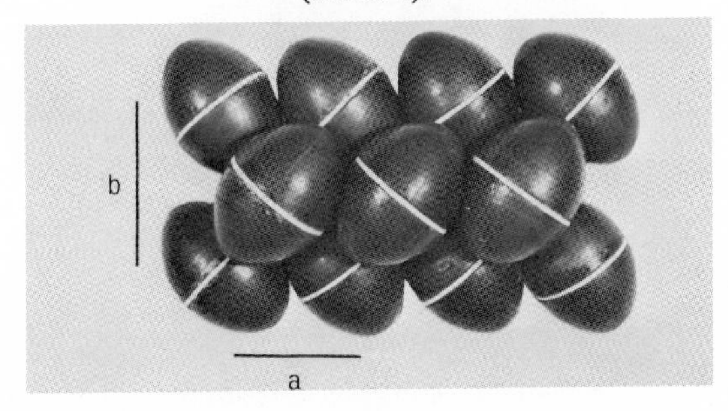

Fig. 10. Orthorhombic *Pnnm* structure (Model 2).

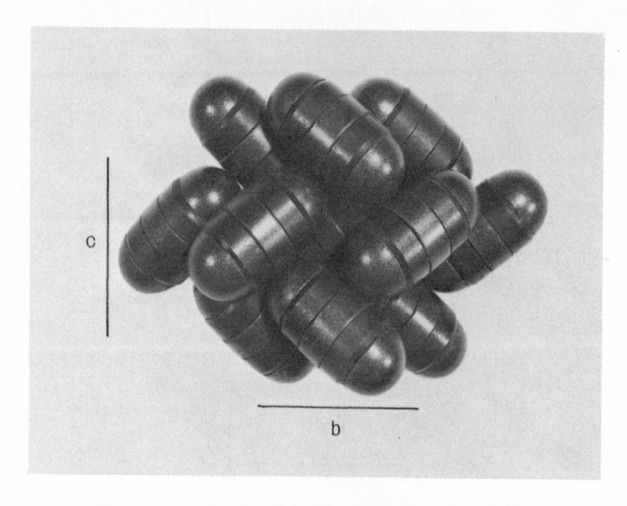

Fig. 11. Orthorhombic *Pbca* structure simulating the crystal structure of cyanogen (Model 3).

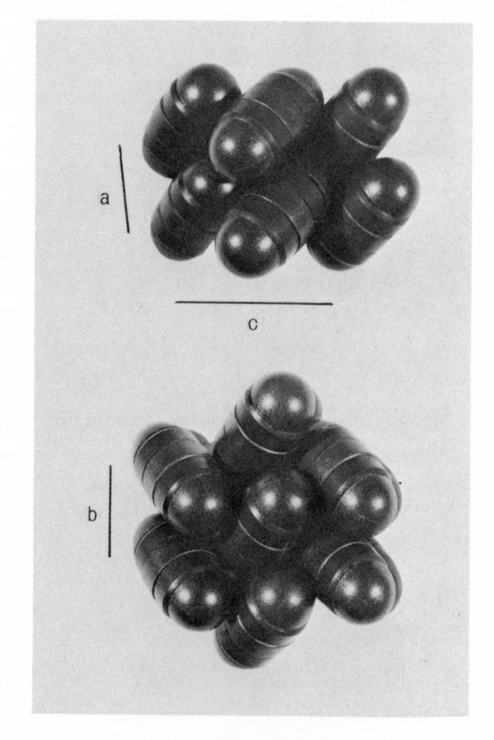

Fig. 12. Monoclinic $P2_1/c$ structure simulating the crystal structure of 1,2-dichloroethane (Model 3).

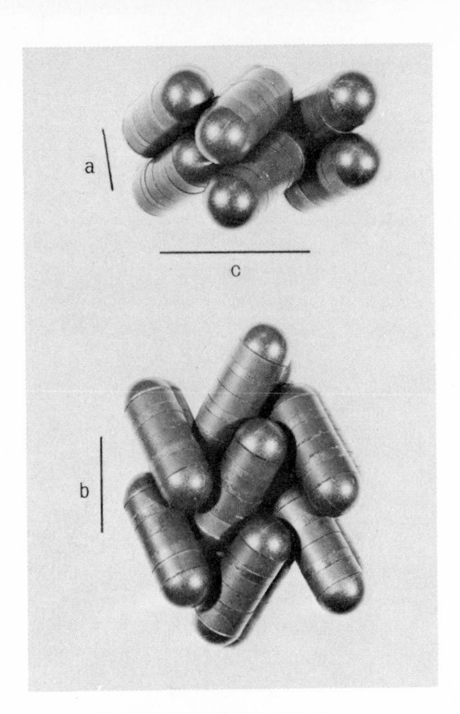

Fig. 13. Monoclinic $P2_1/c$ structure simulating the crystal structure of dicyanoacetylene (Model 4).

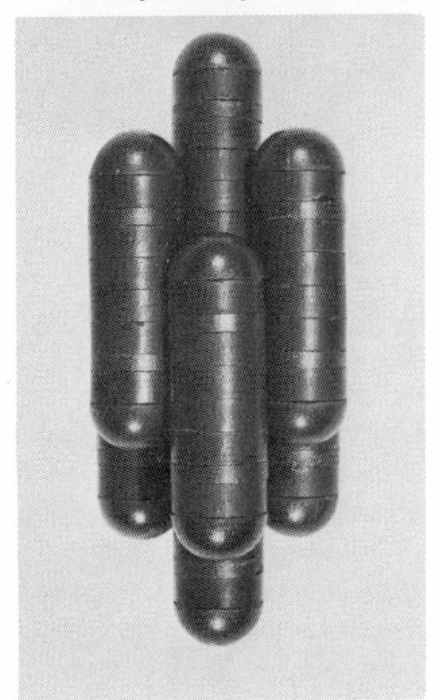

Fig. 14. Trigonal $R\bar{3}m$ structure simulating the crystal structure of dimethyltriacetylene (Model 5).

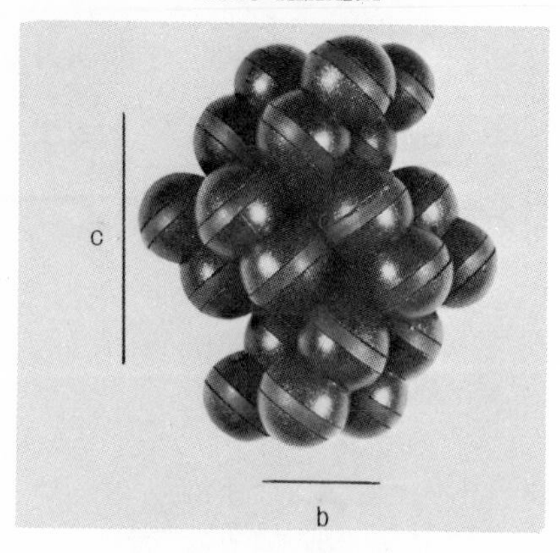

Fig. 15. Orthorhombic *Cmca* structure simulating the crystal structure of chlorine, bromine, and iodine (Model 6).

Fig. 16. Monoclinic $P2_1/c$ structure (Model 6).

The crystal structure of acetylene below a transition at $-140°C$ is orthorhombic and belongs to the holohedral class *mmm* (D_{2h}), the molecules being located at sites of $2/m$ (C_{2h}) symmetry.[8,9]

The molecule of acetylene may be approximated by a model of spheroidal shape with an axis ratio close to 3/4. Since the thermal motions result in the enlargement of the effective size of molecules, the molecular shape at high temperatures is effectively closer to a sphere than at low temperatures. Thus one may infer from our molecular model that the orthorhombic structure of acetylene at low temperatures may possibly be *Pnnm*.

Figures 9 and 10 show how the molecules change their orientations at the phase transition. In fact, one can demonstrate that the assembly of fourteen spheroids with the axis ratio 3/4 makes the transition from one structure to the other when a few spheroids are set into appropriate orientations.

It should be noted, however, that another possibility, an orthorhombic structure with the space group *Cmca* (D_{2h}^{18}) is not entirely excluded, as suggested by Ito, Yokoyama, and Suzuki.[10] In fact, the structure

$$Cmca,\ Z = 4,\ a/c = 0.85,\ b/c = 0.85$$

can be formed by use of somewhat more complicated models.

VI. MODELS 7–9 FOR DISKLIKE MOLECULES

The molecular Models 7–9 shown in Figure 17 represent the crystal structures of typical disklike molecules:

β-hexachlorocyclohexane	$C_6H_6Cl_6$
β-hexabromocyclohexane	$C_6H_6Br_6$
naphthalene	
tetracyanoethylene	$(CN)_2C{=}C(CN)_2$
benzene	C_6H_6

The results are given in Table V and Figures 18–20.

A quadrupolar oblate spheroid representing the molecule of β-hexachlorocyclohexane can be made by composing six radially magnetized divisions. If the spheroid is not far from a sphere, an assembly of such molecular models forms the cubic structure *Pa*3, which is the crystal structure of β-hexachlorocyclohexane or β-hexabromocyclohexane.

If the spheroid is very flat, a monoclinic structure with the space group $P2_1/c$ is formed, representing the crystal structures of naphthalene and tetracyanoethylene.

The crystal structure of benzene has the same symmetry elements as that of cyanogen, the space group being *Pbca*. Our model simulating the benzene molecule consists of two pieces of magnet and a plastic disk whose rim corresponds to the six hydrogen atoms in C_6H_6.

TABLE IV

Crystal Structures of Rodlike Molecules

Substance	Space group	Parameters	Reference
CO_2	$Pa3$ (T_h^6)	$Z = 4$[a]	(IV, il)[b]
Model 1	$Pa3$ (T_h^6)	$Z = 4$	Fig. 8
C_2H_2 above $-140°C$	$Pa3$ (T_h^6)	$Z = 4$	
C_2H_2 below $-140°C$	Crystal class	mmm (D_{2h})	
Model 2	$Pa3$ (T_h^6)	$Z = 4$	Fig. 9
Model 2	$Pnnm$ (D_{2h}^{12})	$Z = 2, a/c = 0.8, b/c = 1.0$	Fig. 10
$(CN)_2$	$Pbca$ (D_{2h}^{15})	$Z = 4, a/c = 0.874, b/c = 0.891$	(XIV, b2)
$ClCH_2CH_2Cl$	$P2_1/c$ (C_{2h}^5)	$Z = 2, a/b = 0.860, c/b = 1.454, \beta = 103° 30'$ ($-140°C$)	
		$a/b = 0.906, c/b = 1.439, \beta = 109° 30'$ ($-50°C$)	(XIV, b4)
	Above $-96°C$ the molecule is rotating about its Cl-Cl axis.		
trans-1,4-dibromocyclohexane	$P2_1/c$ (C_{2h}^5)	$Z = 2, a/b = 1.083, c/b = 2.144, \beta = 101° 49'$	(C.D.)[c]
trans-1,4-dichlorocyclohexane and trans 1,4-diiodocyclohexane are similar.			
Model 3	$Pbca$ (D_{2h}^{15})	$Z = 4, a/c = 0.85, b/c = 0.85$	Fig 11
	$Pnnm$ (D_{2h}^{12})	$Z = 2, a/c = 0.7, b/c = 0.9$	Similar to Fig. 10
	$P2_1/c$ (C_{2h}^5)	$Z = 2, a/b = 1.0, c/b = 1.8, \beta = 100°$	Fig. 12
$N{\equiv}C{-}C{\equiv}C{-}C{\equiv}N$	$P2_1/c$ (C_{2h}^5)	$Z = 2, a/b = 0.639, c/b = 1.478, \beta = 99° 20'$	(XIV, b32)
Model 4	$P2_1/c$ (C_{2h}^5)	$Z = 2, a/b = 0.65, c/b = 1.70, \beta = 100°$	Fig. 13
$CH_3{-}C{\equiv}C{-}C{\equiv}C{-}C{\equiv}C{-}CH_3$	$R\bar{3}m$ (D_{3d}^5)	$Z = 1, \alpha = 70° 58'$	(XIV, b33)
Model 5	$R\bar{3}m$ (D_{3d}^5)	$Z = 1, \alpha = 70°$	Fig. 14

Cl_2	$Cmca\ (D_{2h}^{18})$	$Z=4,\ a/c=0.755,\ b/c=0.524$	(II, u)
Br_2	$Cmca\ (D_{2h}^{18})$	$Z=4,\ a/c=0.765,\ b/c=0.514$	(II, u)
I_2	$Cmca\ (D_{2h}^{18})$	$Z=4,\ a/c=0.742,\ b/c=0.490$	(II, u)
Model 6	$Cmca\ (D_{2h}^{18})$	$Z=4,\ a/c=0.60,\ b/c=0.50$	Fig. 15
	$P2_1/c\ (C_{2h}^{5})$	$Z=2,\ a/b=0.53,\ c/b=0.78,\ \beta=120°$	Fig. 16

[a] Z denotes the number of molecules in a unit cell.
[b] IV, il stands for Chapter IV, il of *Crystal Structures* by Wyckoff.[6]
[c] C.D. stands for *Crystal Data* by Donnay et al.[7]

TABLE V

Crystal Structures of Disklike Molecules

β-$C_6H_6Cl_6$	$Pa3\ (T_h^6)$	$Z=4$	(C.D.)
β-$C_6H_6Br_6$	$Pa3\ (T_h^6)$	$Z=4$	(C.D.)
Model 7	$Pa3\ (T_h^6)$	$Z=4$	Fig. 18
naphthalene	$P2_1/c\ (C_{2h}^{5})$	$Z=2,\ a/b=1.35,\ c/b=1.37,\ \beta=116°$	(C.D.)
$(CN)_2C{=}C(CN)_2$	$P2_1/c\ (C_{2h}^{5})$	$Z=2,\ a/b=1.13,\ c/b=1.75,\ \beta=137°$	(XIV, b25)
Model 8	$P2_1/c\ (C_{2h}^{5})$	$Z=2,\ a/b=1.0,\ c/b=1.4,\ \beta=120°$	Fig. 19
C_6H_6	$Pbca\ (D_{2h}^{15})$	$Z=4,\ b/a=0.728,\ c/a=0.772\ (-3°C)$	(C.D.)
Model 9	$Pbca\ (D_{2h}^{15})$	$Z=4,\ b/a=0.8,\ c/a=0.9$	Fig. 20

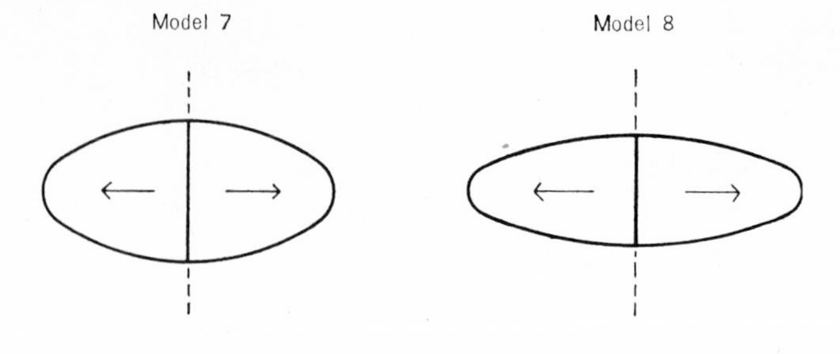

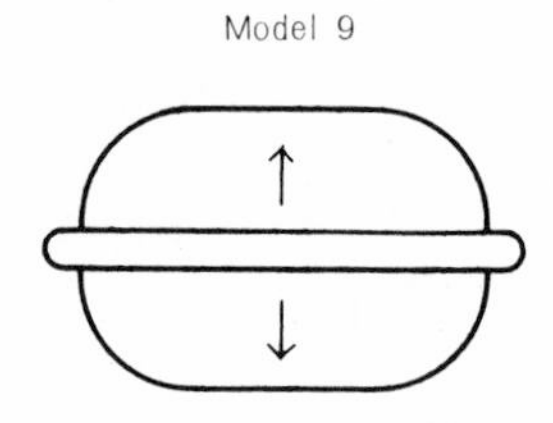

Fig. 17. Models 7–9 for disklike molecules. Model 7 is a quadrupolar oblate spheroid with the axis ratio 2/1. Model 8 is a quadrupolar oblate spheroid with the axis ratio 3/1. Model 9 is for benzene.

Fig. 18. Orthorhombic *Pa*3 structure simulating the crystal structure of β-hexachlorocyclohexane (Model 7).

Fig. 19. Monoclinic $P2_1/c$ structure simulating the crystal structure of tetracyanoethylene (Model 8).

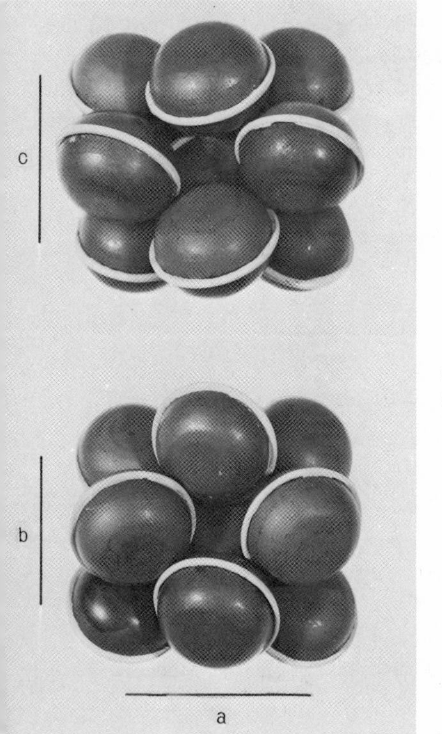

Fig. 20. Orthorhombic *Pbca* structure simulating the crystal structure of benzene (Model 9).

VII. MODELS 10–12 FOR OCTOPOLAR MOLECULES

Models 10–12 shown in Figure 21 are for octopolar molecules:

hexamethylenetetramine	$(CH_2)_6N_4$
silicon tetrafluoride	SiF_4
silicon tetraiodide	SiI_4, and so on
tetraboron tetrachloride	B_4Cl_4

An octopolar sphere simulating the hexamethylenetetramine molecule can be manufactured by combining together eight pieces of magnet, four pieces being magnetized outward and the others inward.

The structure obtained by assembling such models represents the real crystal structure of hexamethylenetetramine (Fig. 22). It should be noted that this body-centered cubic structure is not a closest-packing arrangement of spheres.

The macroscopic shape of this molecular crystal, which is a rhombic dodecahedron as shown in Figure 23, can also be reproduced as shown in Figure 24.

Representations of crystal structures by use of Models 11 and 12 are listed in Table VI and shown in Figures 25–28.

The *Pa*3 structure of SiI_4, GeI_4, and so forth, which contains eight molecules in a unit cell, is similar to the crystal structure of carbon dioxide if we let a two-molecule group correspond to one molecule of CO_2.

In solid methane (melting point 90°K), two second-order transitions have been observed at 20.4°K and 8°K. The structure between 20.4°K and 90°K and the structure between 8°K and 20.4°K are face-centered cubic.[11]

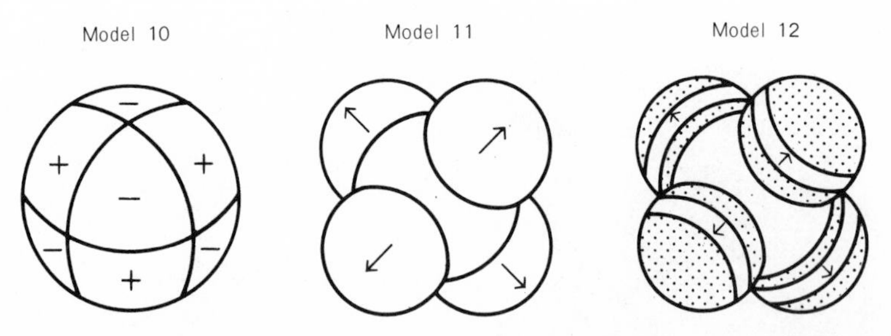

Fig. 21. Models 10–12 for octopolar molecules. Model 10 is an octopolar sphere. Model 11 is for SiF_4; the central part is made of plastic. Model 12 is for SnI_4.

TABLE VI

Crystal Structures of Octopolar Molecules

$(CH_2)_6N_4$	$I\bar{4}3m\ (T_d^3)$	$Z = 2$	(C.D.)
Model 10	$I\bar{4}3m\ (T_d^3)$	$Z = 2$	Fig. 22
SiF_4	$I\bar{4}3m\ (T_d^3)$	$Z = 2$	(V, cl)
Model 11	$I\bar{4}3m\ (T_d^3)$	$Z = 2$	Fig. 25
SiI_4, GeI_4, SnI_4, $TiBr_4$, TiI_4, $ZrCl_4$, $ZrBr_4$, and HfI_4	$Pa3(T_h^6)$	$Z = 8$	(V, c5)
B_4Cl_4	$P4_2/nmc\ (D_{4h}^{15})$	$Z = 2,\ a/c = 0.674$	(III, h7)
Model 12	$Pa3\ (T_h^6)$	$Z = 8$	Fig. 26
	$P4_2/nmc\ (D_{4h}^{15})$	$Z = 2,\ c/a = 0.71$	Fig. 27
	$1\bar{4}2m\ (D_{2d}^{11})$	$Z = 2,\ c/a = 1.8$	Fig. 28

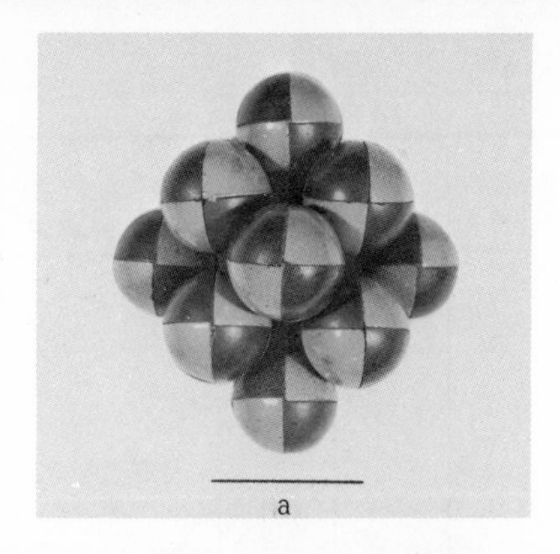

Fig. 22. Cubic $I\bar{4}3m$ structure simulating the crystal structure of hexamethylenetetramine (Model 10).

Fig. 23. A crystal of hexamethylenetetramine, $(CH_2)_6N_4$.

Fig. 24. Simulation of the crystal of hexamethylenetetramine (Model 10).

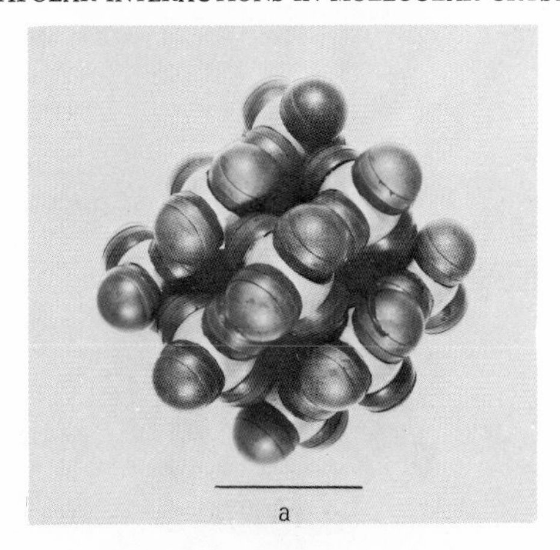

Fig. 25. Cubic $I\bar{4}3m$ structure simulating the crystal structure of silicon tetrafluoride (Model 11).

The structure below 8°K is still unestablished. A tetragonal structure with the space group $I\bar{4}2m$ was supposed by James and Keenan[12] and Yamamoto[13] in their theoretical analyses. In fact, this structure with $a/c = 1.8$ is formed by our model; when the temperature is raised it may become a face-centered cubic structure in a straightforward way, the ratio a/c decreasing from 1.8 to $\sqrt{2}$.

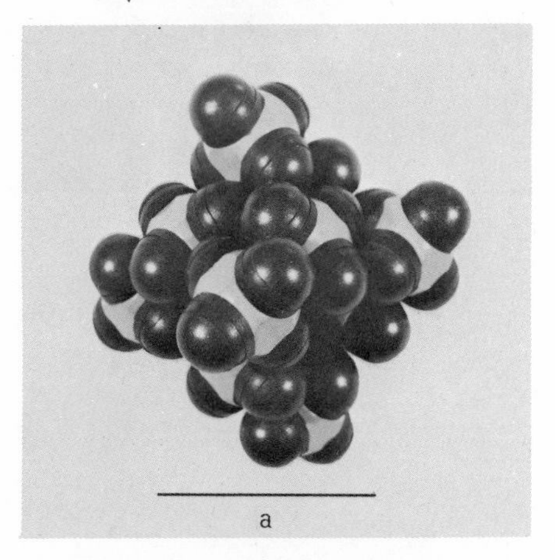

Fig. 26. Cubic $Pa3$ structure simulating the crystal structure of SnI_4 (Model 12).

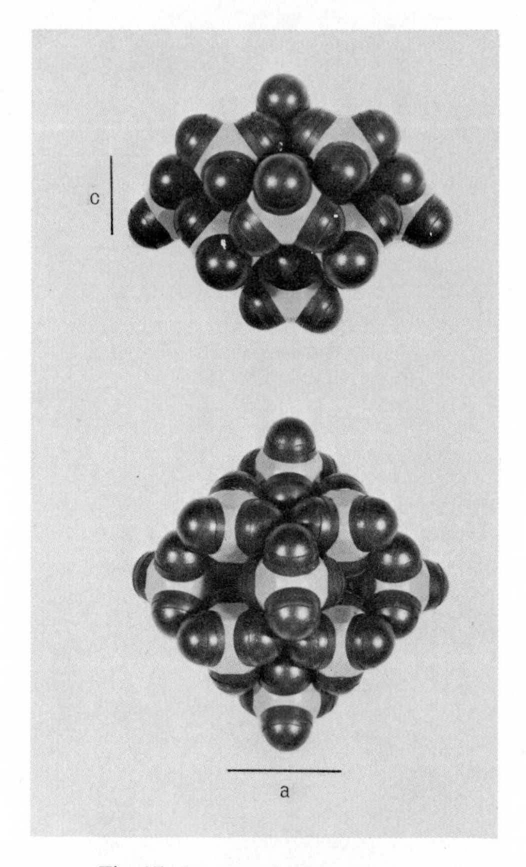

Fig. 27. Tetragonal *P*4_2/*nmc* structure simulating the crystal structure of B_4Cl_4 (Model 12).

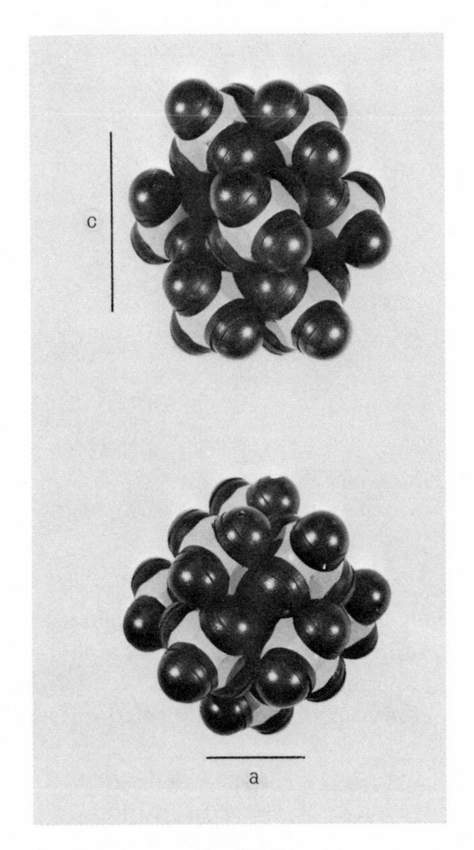

Fig. 28. Tetragonal $I\bar{4}2m$ structure (Model 12).

VIII. MODELS 13–17 FOR CUBELIKE AND OCTAHEDRONLIKE MOLECULES

Models 13–17 shown in Figure 29 simulate the molecules of

octa (silsesquioxane)	$(HSi)_8O_{12}$

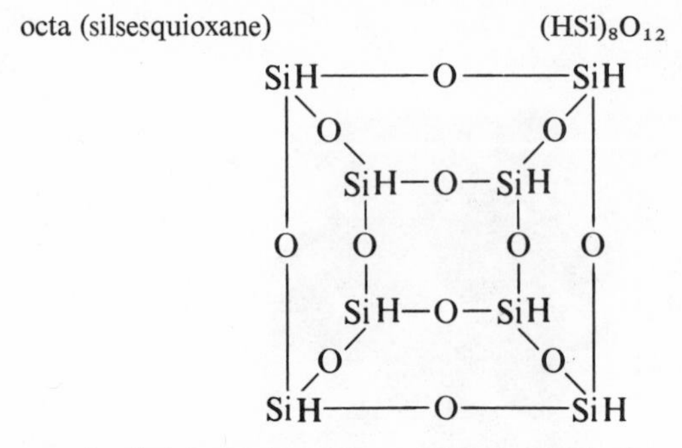

octa (methylsilsesquioxane)	$(CH_3Si)_8O_{12}$
tungsten hexachloride	WCl_6
uranium hexachloride	UCl_6
uranium hexafluoride	UF_6
hexabromoethane	C_2Br_6
tetraarsenic hexaoxide (arsenolite)	As_4O_6
tetraantimony hexaoxide (senarmontite)	Sb_4O_6
sulfur-6	S_6

The shape of the molecule of octa (silsesquioxane), as well as that of octa (methylsilsesquioxane), is like a cube; the eight SiH or $SiCH_3$ radicals are on the vertexes, forming eight poles of positive charge. These molecules are simulated by an appropriately magnetized cube, Model 13.

Representations of crystal structures are listed in Table VII and shown in Figures 30–36.

The crystal structure of osmium octafluoride, OsF_8, is probably rhombohedral $R\bar{3}$ structure, represented by Model 13.

The crystal structures of ethane, C_2H_6, and diborane, B_2H_6, are reported to be

$$P6_3/mmc\ (D_{6h}^4) \quad Z = 2,\ c/a = 1.84 \quad \text{and} \quad 1.91$$

with respect to the positions of the carbon and boron atoms, respectively. If the positions of the hydrogen atoms are taken into account, the space group at low temperatures is probably $P\bar{3}1c$; this structure corresponds to

$$P\bar{3}1c\ (D_{3d}^2) \quad Z = 2,\ c/a = 1.89$$

represented by Models 14 and 15 (Fig. 32).

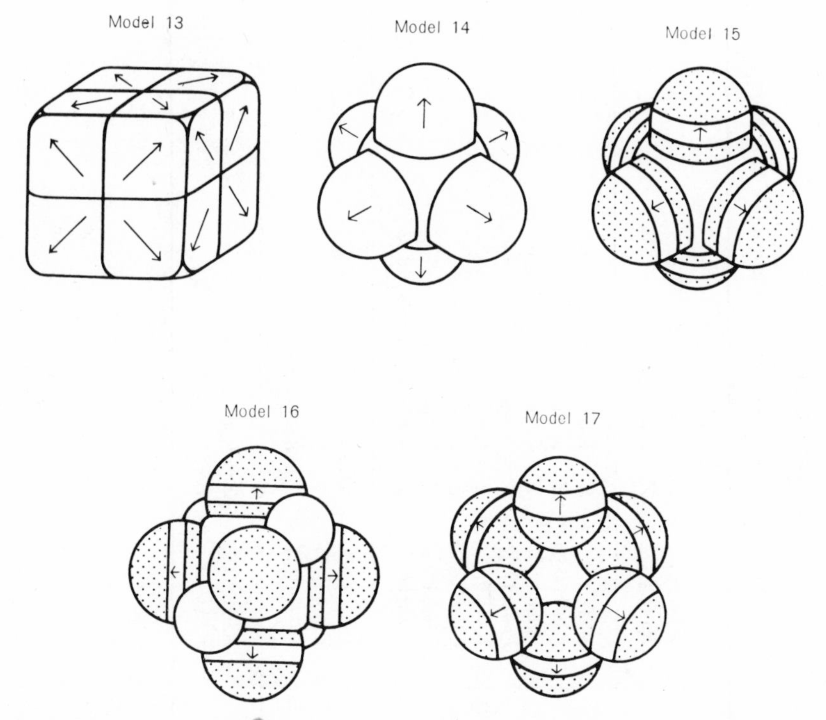

Fig. 29. Models 13–17 for cubelike and octahedronlike molecules. Model 13 is a hexadecapolar cube. Models 14 and 15 are for WCl_6, UCl_6, and UF_6. Model 16 is for As_4O_6 and Sb_4O_6; the small spheres representing the atoms of As or Sb are made of plastic. Model 17 is for S_6.

TABLE VII

Crystal Structures of Cubelike and Octahedronlike Molecules

$(HSi)_8O_{12}$	$R\bar{3}$ (C_{3i}^2)	$Z = 1$, $\alpha = 76° 50'$	(C.D.)
$(CH_3Si)_8O_{12}$	$R\bar{3}$ (C_{3i}^2)	$Z = 1$, $\alpha = 95° 39'$	(XIV, a62)
Model 13	$R\bar{3}$ (C_{3i}^2)	$Z = 1$, $\alpha = 100°$	Fig. 30
WCl_6	$R\bar{3}$ (C_{3i}^2)	$Z = 1$, $\alpha = 55° 0'$	(V, f2)
UCl_6	$P\bar{3}m1$ (D_{3d}^3)	$Z = 3$, $c/a = 0.551$	(V, f3)
UF_6	$Pnma$ (D_{2h}^{16})	$Z = 4$, $b/a = 0.905$, $c/a = 0.526$	(V, f4)
C_2Br_6	$Pnma$ (D_{2h}^{16})	$Z = 4$, $b/a = 0.887$, $c/a = 0.557$	(XIV, b7)
Models 14, 15	$R\bar{3}$ (C_{3i}^2)	$Z = 1$, $\alpha = 53°$	Fig. 31
Models 14, 15	$P\bar{3}1c$ (C_{3d}^2)	$Z = 2$, $c/a = 1.89$	Fig. 32
Models 14, 15	$P\bar{3}m1$ (D_{3d}^3)	$Z = 3$, $c/a = 0.55$	Fig. 33
Models 14, 15	$Pnma$ (D_{2h}^{16})	$Z = 4$, $b/a = 0.92$, $c/a = 0.53$	Fig. 34
As_4O_6	$Fd3m$ (O_h^7)	$Z = 8$	(V, all)
Sb_4O_6	$Fd3m$ (O_h^7)	$Z = 8$	(V, all
Model 16	$Fd3m$ (O_h^7)	$Z = 8$	Fig. 35
S_6	$R\bar{3}$ (C_{3i}^2)	$Z = 1$, $\alpha = 115° 18'$	(II, n2)
Model 17	$R\bar{3}$ (C_{3i}^2)	$Z = 1$, $\alpha = 113°$	Fig. 36

Fig. 30. Trigonal $R\bar{3}$ structure simulating the crystal structure of octa(methylsilsesquioxane) (Model 13).

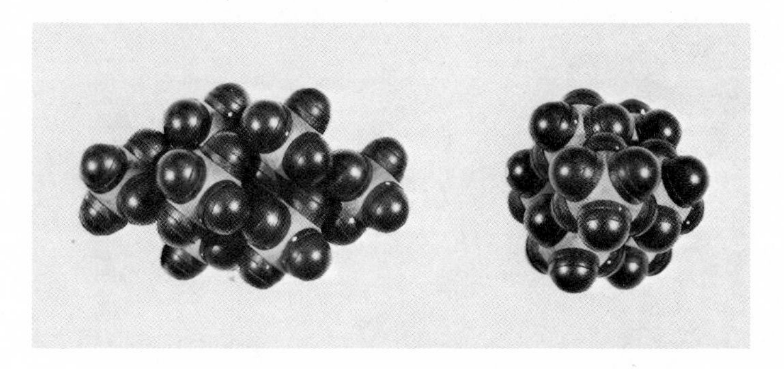

Fig. 31. Trigonal $R\bar{3}$ structure simulating the crystal structure of WCl_6 (Model 15).

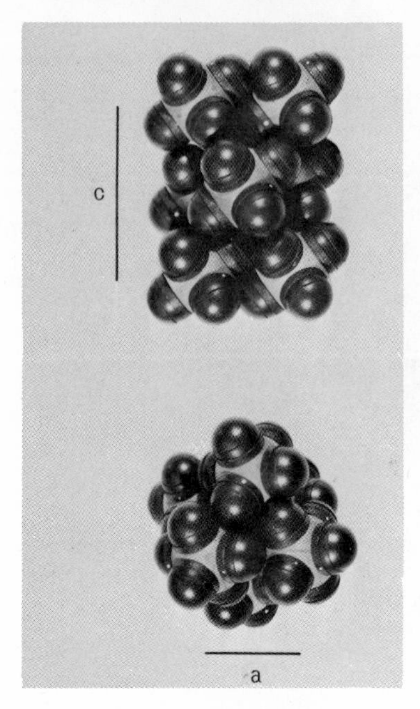

Fig. 32. Trigonal $P\bar{3}1c$ structure (Model 15).

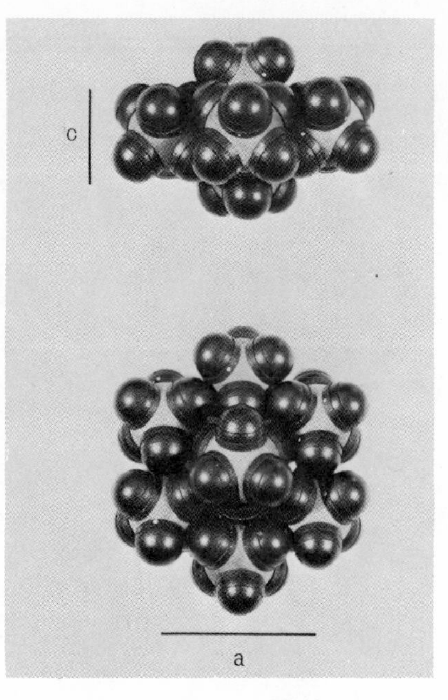

Fig. 33. Trigonal $P\bar{3}m1$ structure simulating the crystal structure of UCl_6 (Model 15).

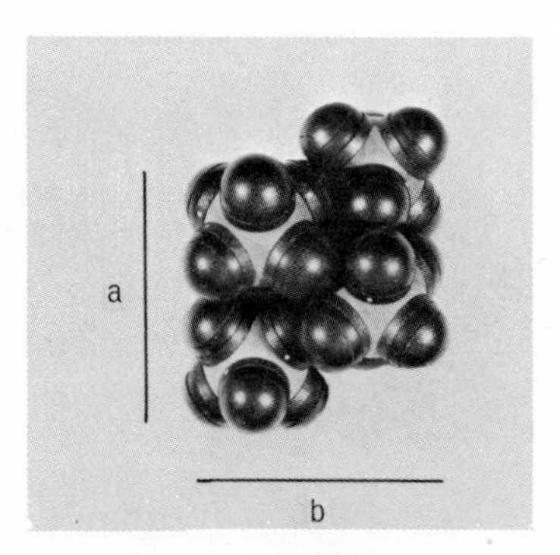

Fig. 34. Orthorhombic *Pnma* structure simulating the crystal structure of of UF_6 (Model 15).

Fig. 35. Cubic *Fd*3*m* structure simulating the crystal structure of As_4O_6 and Sb_4O_6 (Model 16).

Fig. 36. Trigonal $R\bar{3}$ structure simulating the crystal structure of S_6 (Model 17).

IX. MODELS 18–20 FOR OTHER SIMPLE MOLECULES

The molecular Models 18–20 shown in Figure 37 represent the crystal structures of

antimony pentachloride	$SbCl_5$
boron trichloride	BCl_3 (also BBr_3, BI_3)
dinitrogen tetroxide	N_2O_4

```
O       O
 \\    //
   N—N
 //    \\
O       O
```

The results are given in Table VIII and Figures 38–40.

TABLE VIII

Crystal Structures of Other Simple Molecules

$SbCl_5$	$P6_3/mmc$ (D_{6h}^4)	$Z=2$, $c/a=1.069$	(V, e4)
Model 18	$P6_3/mmc$ (D_{6h}^4)	$Z=2$, $c/a=1.3$	Fig. 38
BCl_3	$P6_3/m$ (C_{6h}^2) or $P6_3$ (C_6^6)	$Z=2$, $c/a=1.077$	(V, b16)
BBr_3	$P6_3/m$ (C_{6h}^2) or $P6_3$ (C_6^6)	$Z=2$, $c/a=1.071$	(V, b16)
BI_3	$P6_3/m$ (C_{6h}^2) or $P6_3$ (C_6^6)	$Z=2$, $c/a=1.066$	(V, b16)
Model 19	$P6_3/m$ (C_{6h}^2)	$Z=2$, $c/a=0.94$	Fig. 39
N_2O_4	$Im3$ (T_h^5)	$Z=6$	(IV, i4)
Model 20	$Im3$ (T_h^5)	$Z=6$	Fig. 40

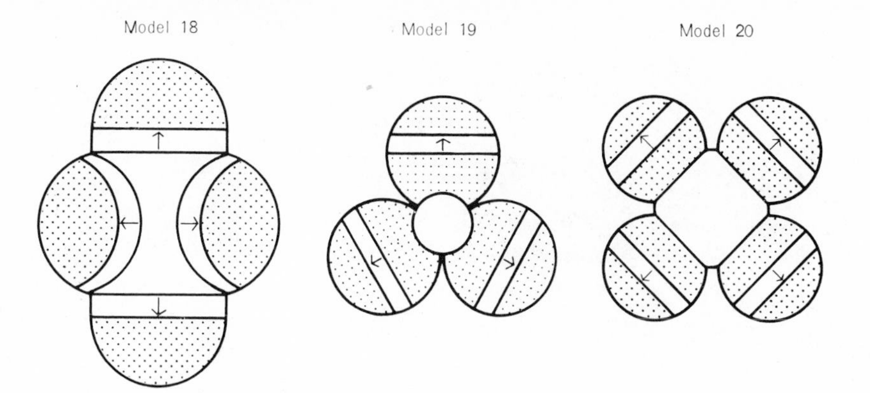

Fig. 37. Models 18–20 for other simple molecules. Model 18 is for $SbCl_5$. Model 19 is for BCl_3, BBr_3, and BI_3. Model 20 is for N_2O_4.

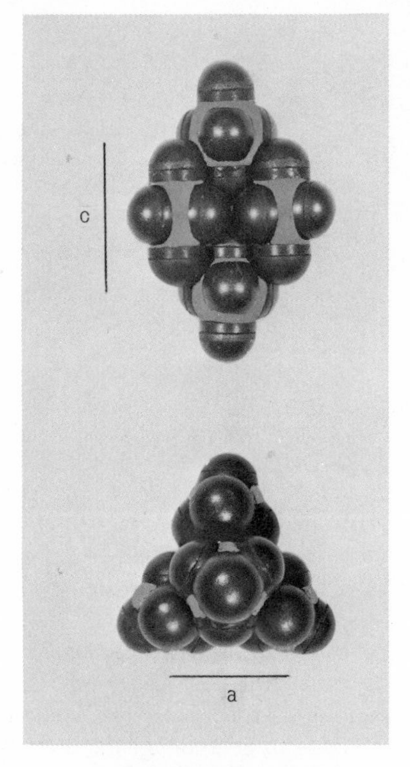

Fig. 38. Hexagonal *P*6_3/*mmc* structure simulating the crystal structure of $SbCl_5$ (Model 18).

Fig. 39. Hexagonal $P6_3/m$ structure simulating the crystal struc- of BCl_3, BBr_3, and BI_3 (Model 19).

Fig. 40. Cubic *Im*3 structure simulating the crystal structure of N_2O_4 (Model 20).

X. CONCLUSION

Molecular models made of barium ferrite magnets, in combination with manganese-zinc ferrite pieces if necessary, are useful for understanding the relation between intermolecular forces and crystal structures. We have seen that the structures of molecular crystals are governed greatly by the multipolar interactions between molecules.

The present collection of representations of crystal structures by means of molecular models may be regarded in a sense as a " theory " of structures of molecular crystals, and in another sense as "computer experiments," a set of models corresponding to an analog computer.

Acknowledgments

The author wishes to thank Professor J. H. Freed who has kindly improved the manuscript.

References

1. T. Kihara, *J. Phys. Soc. Japan*, **15,** 1920 (1960); *Acta Cryst.*, **16,** 1119 (1963); *Acta Cryst.*, **21,** 877 (1966); *Acta Cryst.*, **A26** 315 (1970).
2. A. D. Buckingham, and R. L. Disch, *Proc. Roy. Soc. London*, **A273,** 275 (1963). See Buckingham, *Advances in Chemical Physics*, Vol. 12, Wiley-Interscience, New York, 1967, p. 107.
3. D. E. Strogryn, and A. P. Strogryn, *Mol. Phys.*, **11,** 371 (1966).
4. L. Jansen, A. Michels, and J. M. Lupton, *Physica*, **20,** 1235 (1954).
5. T. Kihara, *Advances in Chemical Physics*, Vol. 5, Wiley-Interscience, New York, 1963, p. 147.
6. R. W. G. Wyckoff, *Crystal Structures*, 2nd Ed., Vols. 1, 2, and 5, Wiley-Interscience, New York, 1963–1966.
7. J. D. H. Donnay, G. Donnay, E. G. Cox, O. Kennard, and M. V. King, *Crystal Data*, 2nd Ed., American Crystallographic Association, 1963.
8. T. Sugawara, and E. Kanda, *Sci. Repts. Research Insts. Tohoku Univ. Ser. A*, **4,** 607 (1952).
9. G. L. Bottger, and J. Eggers, *J. Chem. Phys.*, **40,** 2010 (1964).
10. M. Ito, T. Yokoyama, and M. Suzuki, *Spectrochim. Acta*, **26** *A*, 695 (1970).
11. A. Schallamach, *Proc. Roy. Soc. London*, **A171,** 569 (1939).
12. H. M. James, and T. A. Keenan, *J. Chem. Phys.*, **31,** 12 (1959).
13. T. Yamamoto, *J. Chem. Phys.*, **48,** 3193 (1968).

THE COMPUTATION OF VIRIAL COEFFICIENTS

JOHN E. KILPATRICK

Department of Chemistry, Rice University, Houston, Texas

CONTENTS

I. GENERAL THEORY

Shortly before the beginning of this century van der Waals[1] deduced his celebrated equation of state for hard-sphere molecules with short-range attractive forces. The best-known version of his equation includes only two-body interactions, but he extended his arguments to include three- and then four-body effects. He used somewhat intuitive arguments involving the increment in entropy of a large system due to the addition of one molecule. His reasoning in regard to attractive forces does not extend beyond two-body interactions and is mainly of historical interest. His results for the hard-sphere contribution to the second and third virial coefficients is exact.

Trouble began with the fourth virial coefficient. van der Waals published his analysis but without actually evaluating a certain rather difficult integral. van Laar evaluated the integral and gave the "van der Waals" value for the fourth virial coefficient. Boltzmann[2] objected that certain details of van der Waals' expression were not handled properly and gave his result. The details of the argument are hard to follow: both men obtained their results by a series of approximations. Neither advanced arguments that seem completely convincing to us now; rather, they figuratively waved their arms and used the theorem "what I tell you three times is true." Boltzmann, however, was right. On another occasion we shall analyze the arguments for both sides in modern notation.

Later, Ursell[3] and Mayer[4] developed a rigorous theory of the virial equation, based on the canonical ensemble. Uhlenbeck and Beth[5] and Kahn[6] investigated the theory for a real quantum gas.

The easiest and most direct way to establish the virial expansion for the equation of state of a gas is by means of the grand canonical ensemble. The grand canonical partition function is defined by

$$\Xi(\alpha, \beta, V) = \sum_N \sum_E \Omega(N, E, V) e^{-\beta E} e^{-\alpha N}$$

where Ω is the number of quantum states for N molecules in volume V and at energy E. Since the canonical partition function is defined by

$$Z(N, \beta, V) = \sum_E \Omega(N, E, V) e^{-\beta E}$$

we may write

$$\Xi = \sum_N Z(N, \beta, V) e^{-\alpha N}$$

or

$$\Xi = 1 + Z_1 z + Z_2 z^2 + Z_3 z^3 + \cdots$$

where $z = e^{-\alpha}$, the absolute activity, $Z_N \equiv Z(N, \beta, V)$, and Z_0, the canonical partition function for zero molecules, is taken to be unity.

All the equilibrium thermodynamic properties of the system depend on $\ln \Xi$. Writing $\ln \Xi$ as a power series in z, we have

$$\ln \Xi = \sum_{j=1}^{\infty} V g_j z^j$$

The coefficients g_j are given by

$$\begin{aligned}
Vg_1 &= Z_1 \\
Vg_2 &= Z_2 - \tfrac{1}{2} Z_1^{\,2} \\
Vg_3 &= Z_3 - Z_2 Z_1 + \tfrac{1}{3} Z_1^{\,3} \\
Vg_4 &= Z_4 - Z_3 Z_1 - \tfrac{1}{2} Z_2^{\,2} + Z_2 Z_1^{\,2} - \tfrac{1}{4} Z_1^{\,4}
\end{aligned}$$

or in general,

$$V g_N = \sum_{i=1} (-)^{i-1} (i-1)! \sum_{(k)} \prod_s \frac{Z_s^{\,k_s}}{k_s!}$$

$$\begin{cases} \sum_{s=1} k_s = i \\ \sum_{s=1} s k_s = N \end{cases}$$

where the sum (k) is over all sets of k_s that satisfy the two restrictive conditions. Alternatively, the g_N are given by the Nth-order determinant:

$$Vg_N = \frac{(-1)^{N-1}}{N} \begin{vmatrix} Z_1 & 1 & 0 & 0 & 0 \\ 2Z_2 & Z_1 & 1 & 0 \cdots & 0 \\ 3Z_3 & Z_2 & Z_1 & 1 & 0 \\ 4Z_4 & Z_3 & Z_2 & Z_1 & 0 \\ & \cdots & & \cdots & \\ NZ_N & Z_{N-1} & Z_{N-2} & Z_{N-3} & Z_1 \end{vmatrix}_N$$

Two of the fundamental properties of the grand canonical partition function are that $pV = kT \ln \Xi$ and that the average number of molecules is the partial derivative of $\ln \Xi$ with respect to $\ln z$. That is,

$$p/kT = \sum_{j=1}^{\infty} g_j z^j$$

$$N/V = \sum_{j=1}^{\infty} j g_j z^j$$

The quotient of these two functions, pV/NkT, may be expanded as a series in p/kT or in N/V. The first is the pressure virial series; the second is the density expansion:

$$pV/NkT = 1 + \beta_2(p/kT) + \beta_3(p/kT)^2 + \cdots$$
$$= 1 + \alpha_2(N/V) + \alpha_3(N/V)^2 + \cdots$$

where α_n and β_n are the molecular density and pressure virial coefficients, respectively.

It can readily be shown that β_{n+1} is the coefficient of z^n in $(g_1 + g_2 z + g_3 z^2 + g_4 z^3 + \cdots)^{-n}$ and that α_{n+1} is the coefficient of z^n in $(g_1 + 2g_2 z + 3g_3 z^2 + \cdots)^{-n}/(n+1)$. That is, if

$$\beta_{n+1} = F(g_1, g_2, g_3, \ldots, g_n)$$

then

$$\alpha_{n+1} = F(g_1, 2g_2, 3g_3, \ldots, ng_n)/(n+1)$$

Both kinds of viral coefficients have equally simple relation to the g_j's and therefore to the first n canonical partition functions:[7,8]

$$\beta_1 = 1$$

$$\beta_2 = -g_2/g_1^2,$$

$$\beta_3 = (-2g_3g_1 + 3g_2^2)/g_1^4$$

$$\beta_4 = (-3g_4g_1^2 + 12g_3g_2g_1 - 10g_2^3)/g_1^6$$

$$\beta_5 = (-4g_5g_1^3 + 20g_4g_2g_1^2 + 10g_3^2g_1^2 - 60g_3g_2^2g_1 + 35g_2^4)/g_1^8$$

$$\alpha_1 = 1$$

$$\alpha_2 = -g_2/g_1^2$$

$$\alpha_3 = (-2g_3g_1 + 4g_2^2)/g_1^4$$

$$\alpha_4 = (-3g_4g_1^2 + 18g_3g_2g_1 - 20g_2^3)/g_1^6$$

$$\alpha_5 = (-4g_5g_1^3 + 32g_4g_2g_1^2 + 18g_3^2g_1^2 - 144g_3g_2^2g_1 + 56g_2^4)/g_1^8$$

The $n + 1$ pressure virial coefficient is given explicitly as a polynominal in g_1 through g_{n+1} as

$$\beta_{n+1} = \sum_{(s)} (-)^{n-s_0} \frac{(2n-1-s_0)!}{(n-1)!\, {g_1}^{2n-s_0}} \prod_{j=1} \frac{(g_{j+1})^{s_j}}{s_j!}$$

$$\begin{cases} \sum_{j=0} s_j = n \\ \sum_{j=0} j s_j = n \end{cases}$$

or as

$$\beta_{n+1} = \frac{(-1)^n}{n!\, {g_1}^{2n}} \begin{vmatrix} ng_2 & g_1 & 0 & 0 & \\ 2ng_3 & (n+1)g_2 & 2g_1 & 0 & \\ 3ng_4 & (2n+1)g_3 & (n+2)g_2 & 3g_1 & \cdots \\ 4ng_5 & (3n+1)g_4 & (2n+2)g_3 & (n+3)g_2 & \\ & & \cdots & & \\ n^2g_{n+1} & (n(n-1)+1)g_n & & \cdots & \end{vmatrix}_n$$

The corresponding expressions for the density virial coefficients follow from the relation given above.

II. CLASSICAL THEORY

The classical theory of virial coefficients results from evaluating the canonical partition functions Z_N classically:

$$Z_N = \frac{1}{N!\,h^{3N}} \int \cdots \int e^{-\beta H(p,q)}\, dp_1 \cdots dq_{3N}$$

where $H(p, q)$ is the classical Hamiltonian of the N particle system, $\beta = 1/kT$, and the integral is over the phase space of the system. If the system has internal degrees of freedom in each molecule, naturally they must be included in the integral Z_N. However, to the extent that there is no coupling between the internal degrees of freedom of one molecule and another, the effect of such internal degrees of freedom cancel out identically from the expressions for virial coefficients of all orders and may be ignored. That is, complex molecules with independent internal degrees of freedom have the same classical equation of state as a system of point mass molecules with the same potential of interaction.

The Hamiltonian will consist of $3N$ terms in momentum squared for the kinetic energy and some function of the relative coordinates for the the potential energy:

$$H = \sum_{j=1}^{3N} p_j^2/2m + U(r_1, r_2, \ldots, r_N)$$

The momentum integrations may be performed at once, yielding

$$Z_N = (N!)^{-1} \lambda_T^{-3N} Q_N$$

where $\lambda_T = h/(2\pi mkT)^{1/2}$, the thermal wave length, and

$$Q_N = \int \cdots \int e^{-\beta U}\, d\tau_1 \cdots d\tau_N$$

a $3N$-dimensional integral over the Boltzmann factor of the potential energy, which is called the configurational integral.

To make any further progress, something must be assumed about the form the interaction potential. From a hard, practical point of view, only a pure mathematician would bother to calculate the equation of state for a completely abstract Hamiltonian. Physical scientists are usually interested in the properties of real substances, the real material of this universe. A potential chosen primarily for simple mathematical properties and with no regard for its correspondence with the real universe provides merely a mathematical exercise.

On the other hand, we do not know the exact interaction potential (or even a fairly accurate approximation) for any real set of molecules. We must make some approximations or idealizations. The usual first idealization is to assume that the total interaction potential is the sum of all pairwise interactions with no triple- or higher-ordered interaction terms. We do not know yet just how important such three- or higher-bodied terms may be in their effect on the third and higher virial coefficients.

We assume

$$U = \sum_{i>j}\sum u(r_{ij})$$

where $u(r_{ij})$ is the true two-body potential, a function only of the distance between molecules i and j. In practice we do not know even u very well. Various empiric forms (Lennard-Jones, exp-six, Morse, etc.) or extreme idealizations (hard sphere, square-well hard sphere, etc.) have been used. Naturally little or nothing can be said about the form or importance of three-body interactions in even as simple a gas as helium until we know fairly well the simple two-body interaction. By the term "three-body interaction" we mean, of course, any excess three-body potential over and above the sum of the three pair interactions.

Introduction of the Mayer f function,

$$f_{ij} \equiv f(r_{ij}) = e^{-\beta u(r_{ij})} - 1$$

enables us to use some elementary graph theory and markedly simplify our virial relations. The Boltzmann factor for the N-body potential becomes

$$e^{-\beta U} = \prod e^{-\beta u_{ij}} = \prod (1 + f_{ij})$$

where the products are over all possible pairs of the N molecules. We have, then,

$$Q_N = \int \cdots \int d\tau_1 \cdots d\tau_N \{1 + \sum f_{ij} + \sum f_{ij} f_{kl} + \cdots\}$$

Each one of these integrals corresponds to a set of N labeled points (in ordinary three-dimensional space)—every pair of which either are or are not connected with a line. The presence of a factor f_{ij} in an integral corresponds to a line between points i and j. Even stronger, we shall say that an integral over N points in volume V with argument a particular product of f functions is represented by or is equal to the corresponding graph.

All of such graphs that have the same topological structure or connectivity have the same value. For example,

$Q_1 =$

$Q_2 =$ +

$Q_3 =$ + 3 ( ) + 3 ( ) +

$Q_4 =$ + 6 ( ) + 3 ( ) + 12 ( )

+ 4 ( ) + 4 ( ) + 12 ( ) + 3 ( )

+ 12 ( ) + 6 ( ) +

The weight or coefficient of a graph on N points of a given topological type is the number of distinct ways the points may be labeled. This weight, w, is $N!/\sigma$, where σ is the symmetry number of the graph. The symmetry number is the order of the group of permutations (of the vertex labels) that leaves the connectivity of the graph invariant. The total weight of all the graphs on N points with l lines is the binominal coefficient $\binom{N(N-1)/2}{l}$ so that Q_N, being the sum of all labeled graphs on N points, has a total sum of weights of $2^{(N(N-1)/2}$.

For example, the four-point graph

has the weight 6 since it can be labeled in 6 distinct ways:

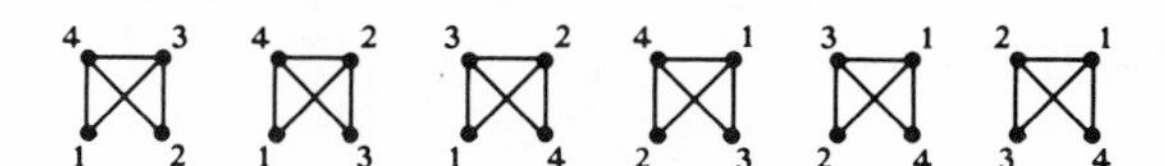

Alternatively, the weight is $6=4!/4$ since the symmetry number $\sigma = 4$. This fact can readily be deduced by the use of two rules: (1) the symmetry number of a graph and its complement, that is, a graph in which lines and no lines have been interchanged, are the same; and (2) the symmetry number of a disconnected graph (consisting of two or more disjoint pieces) is the product of the symmetry numbers of the separate pieces times a factor of $j!$ for each set of j identical pieces. Thus,

(by complementation)

whose $\sigma = (2)(1)\ 2!$ since σ for a point is 1, for a line is 2, and there are two points. As a further example, we use a graph on 5 points:

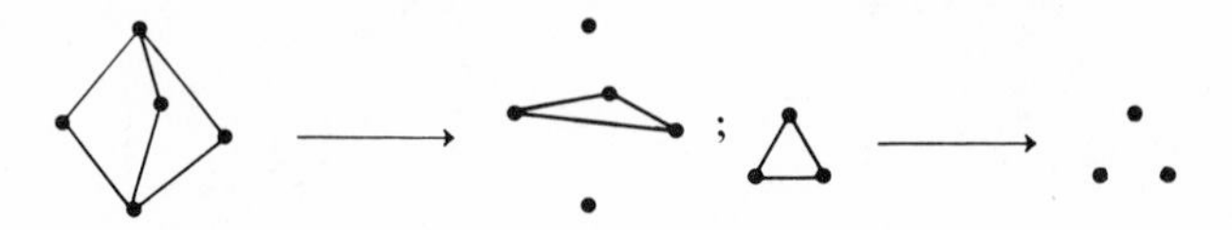

so $\sigma = (3!)(2!) = 12$, $w = 5!/12 = 10$.

A labeled graph is equal to an N point integral with the corresponding set of f functions as arguments. Following Rushbrooke,[9] we define an unlabeled graph as equal to $1/\sigma$ times the corresponding labeled graph. We can summarize many of our preceding remarks as follows:

1. Q_N equals the sum of all distinctly labeled graphs on N points.
2. Z_N equals λ^{-3N} times the sum of all unlabeled graphs on N points.
3. Ξ equals the sum of all unlabeled graphs of all orders, each weighted with the factor $(\lambda^{-3}z)^N$.

The configuration integral Q_N was defined as the space portion of the partition function Z_N:

$$Z_N = Q_N/\lambda^{3N}N!$$

It is convenient to define a set of coefficients b_N that bear a similar relation to the g_N:

$$g_N = b_N/\lambda^{3N}$$

Note that the permutational factor $N!$ has not been removed. The b_N are simply related to the Q_N:

$$\begin{aligned} Vb_1 &= Q_1 \\ 2!Vb_2 &= Q_2 - Q_1{}^2 \\ 3!Vb_3 &= Q_3 - 3Q_2Q_1 + 2Q_1{}^3 \end{aligned}$$

If the graph theory expansions for each of the Q_N are inserted, a truly remarkable simplification ensues:

$Vb_1 = \bullet$

$2!\,Vb_2 =$

$3!\,Vb_3 = 3 \ + $

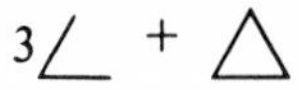

$4!\,Vb_4 = 4 \ + 12 \ + 12$

$+ 3 \ + 6 \ + $

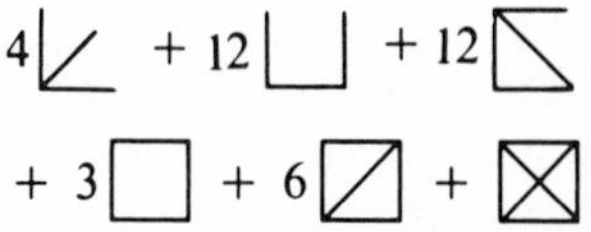

In general, $N!Vb_N$ is exactly the sum of all distinct, connected, labeled graphs. A further simplification results from dividing out the factors $N!$ and V. Division by V converts our free or unrooted graphs to singly rooted graphs, defined as an integral over the relative coordinates of the N points. The rooted nature of a graph is often shown graphically by making any one vertex an open circle. Division by σ converts our labeled graphs to unlabeled graphs. Note that the actual integration must be done for a labeled graph (the variables of integration must be labeled). The result is then divided by σ.

In summary, b_N is the sum of all singly rooted, labeled, connected graphs on N points, each weighted with the reciprocal of its symmetry number. For example,

$$b_4 = \frac{1}{6} \ + \frac{1}{2} \ + \frac{1}{2} \ + \frac{1}{8} \ + \frac{1}{4} \ + \frac{1}{24}$$

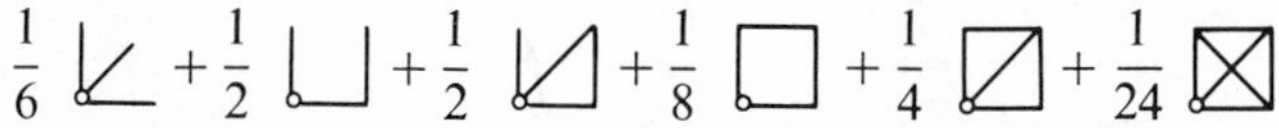

This is equivalent to the sum of all distinct, singly rooted, unlabeled, connected graphs on N points. We see then that only singly rooted, unlabeled connected graphs occur in the expansion of $\ln \Xi$.

The expressions for both the pressure and the density virial coefficients are homogeneous functions of degree zero in λ_T so α_N and β_N are the same functions of the b_N as they are of the g_N. When the density virial coefficients are written as functions of the component graphs of the b_N, a further striking reduction in diversity of types occurs:

$$\alpha_2 = -\frac{1}{2}$$

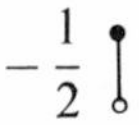

$$\alpha_3 = -2 \cdot \frac{1}{6}$$

$$\alpha_4 = -3 \left\{ \frac{1}{8} \ + \frac{1}{4} \ + \frac{1}{24} \right\}$$

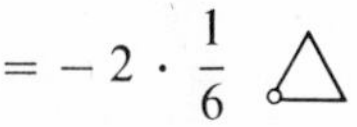

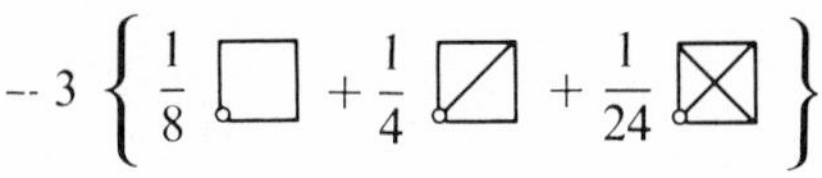

Every graph with an articulation point disappears from the expansions. The result is, in general: α_N equals $-(N-1)$ times the sum of all singly rooted, nonarticulated (i.e., at least doubly connected) unlabeled graphs on N points, each weighted with the reciprocal of its symmetry number. Rushbrooke[9] has given a very simple and picturesque graph theory argument that this well-known result holds in general.

The corresponding expansion for the β_N is not quite as simple. Not all of the articulated graphs cancel out. This is no great disadvantage, though, for the value of any articulated graph is merely the product of several lower-ordered nonarticulated graphs.

One further transformation—the Hoover-Ree[10] transformation—is sometimes advantageous. Each of the cluster integrals (star graphs) that appear in the expression for α_N has as its argument between N and $N(N-1)/2$ Mayer f functions. The vertexes of the graph, that is, the variables of integration, are, of course, labeled. Let us introduce the unit factor

$$1 = e^{-\beta u_{ij}} - f_{ij}$$

for every pair of points not connected by a line. Each Mayer integral then expands into the sum of a set of a new type of integral: every integral has $N(N-1)/2$ factors as its integrand, one for every possible line in an N-point graph. Some of the factors will be f functions; the remainder will be exponentials. One could use graphs with solid lines and wavy or dotted lines to represent these two kinds of factors. A less cluttered representation is obtained, however, (at the cost of some possible confusion) by using lines and no lines. The advantage of the transformation is that a large fraction of the remaining diagram types in the complete expression for a virial coefficient disappear. For example, expressed in Mayer graphs,

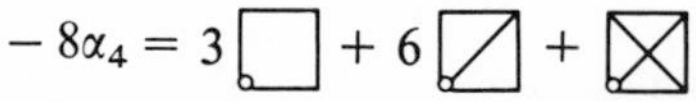

$$-8\alpha_4 = 3\,\square + 6\,\square + \square$$

Each of these three Mayer graphs expands into a set of Hoover graphs:

(Mayer) (Hoover)

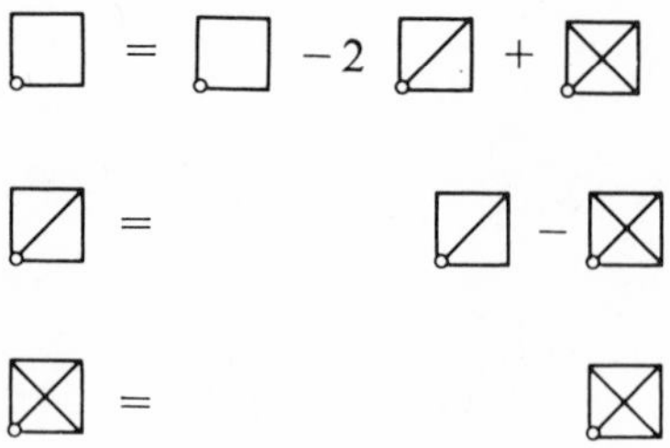

$$\square = \square - 2\,\square + \square$$

$$\square = \square - \square$$

$$\square = \square$$

We find, then, the fourth density virial coefficient involves only two Hoover graphs:

$$-8\alpha_4 = 3\,\square \; -2\,\boxtimes$$

The reduction is more striking in α_5. The ten Mayer graphs are reduced to only five. In α_6, 56 Mayer diagrams become 23.

In some unpublished work, Kilpatrick, Ford, and Yu have found a sufficient condition for a Hoover graph not to appear in the expression for a virial coefficient, namely, the presence of an articulation star of any order. An articulation star in a graph is a subset of points, all completely connected together pairwise, whose removal causes the graph to become disjoint. An articulation star of the first order is an ordinary articulation point. We have already seen that graphs with such points do not occur in the Mayer graph expression for a virial coefficient. Clearly a graph with one less than the maximum number of lines, $N(N-1)/2 - 1$, has such a star of order $N-2$. This rule accounts for all five Hoover graphs that do not appear in the fifth virial coefficient and 32 of the 33 that do not appear in the sixth. The one graph of zero weight that is not accounted for by this rule is

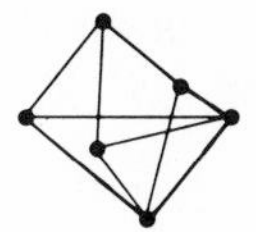

The number of such graphs (with zero Hoover weight but not excluded by the articulation star rule) slowly increases with N. It is not clear yet whether these exceptional cases have a simple geometric or topological explanation. It would be nicer to have one simple geometric rule that cover all cases.

Before closing this section we should like to emphasize that the clusters (cluster integrals or graphs) that we have been dealing with do not represent physical clusters. The nature of Mayer's original approach left this impression. We have used them here in the sense of convenient abbreviations for certain definite integrals that occur in the necessary expansions. That the structure of these graphs has a simple and intriguing topological form is of no particular significance. See, for example, the discussions in Hill[10a] and Kilpatrick.[10b]

III. CLASSICAL APPLICATION

The second virial coefficient is given by

$$\alpha_2 = -\frac{1}{2}\; \vert$$

that is,

$$\alpha_2 = -\frac{1}{2}\iiint f(r_{12})\, d\mathbf{r_2}$$

The potential being a function of the internuclear distance, the two angular integrations can be performed, yielding the well-known expression

$$\alpha_2 = -2\pi \int_0^\infty (\exp(-\beta u(r)) - 1) r^2\, dr$$

The evaluation of the second virial coefficient involves only a one-dimensional quadrature. For most potentials this cannot be accomplished in closed form; some variety of series expansion and term by term integration need be used. As a last resort, one can always use numerical integration.

For hard spheres of diameter σ the result is

$$\alpha_2 = \tfrac{2}{3}\pi\sigma^3 \equiv b$$

This unit is commonly used for the higher hard-sphere virials. For the hard-sphere square-well potential,

$$\begin{aligned} u &= \infty \quad &&\text{for} \quad r < \sigma_1 \\ u &= -\varepsilon &&\qquad \sigma_1 < r < \sigma_2 \\ u &= 0 &&\qquad r > \sigma_2 \end{aligned}$$

we obtain

$$\alpha_2 = \tfrac{2}{3}\pi[\sigma_1{}^3 e^{\beta\varepsilon} + \sigma_2{}^3(1 - e^{\beta\varepsilon})]$$

At high temperature the hard core σ_1 dominates and the expression approaches the hard-sphere result. At low temperature, the depth and width of the square well dominate and α_2 becomes negative. This behavior is fairly realistic. Qualitatively, any realistic potential will show the same behavior. The limiting high-temperature behavior will be essentially hard sphere. The same will be true of the higher virial coefficients. This, of course, is one of the reasons for the interest in the higher hard-sphere virial coefficients.

There is nothing in the derivation of the above working expression for α_2 that requires the two molecules to be point masses or that the potential even be independent of their relative orientation. A number of calculations (see Hirschfelder et al.[11]) have been made for orientation-dependent potentials. Essentially nothing is known about the real form of such potentials for nonspherical molecules. In general, after a lot of mathematical trouble, they turn out to give results very similar to a single average spherical potential.

The third virial coefficient offers more difficulty. As we have seen, the basic integral is

$$\alpha_3 = -\frac{1}{3} \triangle$$

The hard-sphere case was evaluated by van der Waals,

$$\alpha_3 = \tfrac{5}{8}b^2$$

The cluster integral can be found easily by elementary methods. The Mayer f function is -1 for $0 < r < \sigma$ and zero elsewhere. Were it not for the third line in the graph, the value would be $4b^2$, the product of the volumes of two spheres with radii σ. The effect of the requirement that the two moving points be at a distance of σ or less is to reduce the result by the factor 15/32.

Kihara[12] obtained the third virial for the hard-core square-well potential in closed analytic form. Numerous authors[13–19] have used various continuous potentials and expressed their results as numerical tables or expansions, both high and low temperature.

Beginning with the fourth virial, difficulties increase rapidly. For the fourth virial, one must evaluate three Mayer clusters or graphs; for the fifth, 10 clusters; and for the sixth, 56 clusters. The clusters with the maximum number of lines are always the most difficult. Even for the hard-sphere gas, which would seem to be the simplest, moderately realistic potential, such integrals are difficult. The maximum number of lines, $N(N-1)/2$, is quadratic in N but the mathematical difficulties increase far more rapidly.

Boltzmann was the first to correctly evaluate the fourth virial coefficient for hard spheres. One can deduce all three cluster integrals from his work even though he did not possess this formulation. An indication of the difficulty of further progress is that over fifty years elapsed before even a first approximation to the hard-sphere fifth virial coefficient was obtained or before the remarkably complicated algebra of the van Laar-van der Waals integration was confirmed.

The values of the three fourth-order cluster integrals are

$$\square = \left(\frac{272}{105}\right) b^3 = 2.5904761905 b^3$$

$$\boxslash = -\left(\frac{6347}{3360}\right) b^3 = -1.8889880952 b^3$$

$$\boxtimes = \left(\frac{2707}{1120} - \frac{219\sqrt{2} + 4131(\tan^{-1}\sqrt{2} - \pi/4)}{280\pi}\right) b^3$$
$$= 1.266903952 b^3$$

and the fourth virial coefficient is

$$\alpha_4 = 0.2869495059 b^3$$

Besides the Mayer and Hoover expansions for the virial equation, there are two expansions based upon the properties of the radial distribution function, the so-called pressure and compressibility expansions (see Hill[20] and Yu[21]). Both of these expansions are in terms of two-center graphs. A two-center graph has two fixed vertexes and represents an integral over all relative positions of the remaining $N-2$ vertexes, expressed as a function of the distance between the two fixed centers.

One of these expansions assumes a particularly simple form for the special case of hard spheres. This is due to the fact that the derivative of the potential with respect to r is then a Dirac delta function:

$$p/\rho k\mathrm{T} = 1 + b\rho + b\left(\wedge\right)\rho^2 + \frac{b}{2!}\left([\text{two-center graph 1}] + [\text{two-center graph 2}] + 4\,[\text{two-center graph 3}] + 2\,[\text{two-center graph 4}]\right)\rho^3 + \cdots$$

where $\rho = N/V$, $b = \frac{2}{3}\pi\sigma^3$ as usual, and the two-center clusters are evaluated at the distance $r = \sigma$. Both the Mayer one-center cluster expansion and this two-center expansion are power series in ρ; the coefficients must be identical. Yu[21] has shown that there is also a relation between individual clusters. Suppose we define the two-center expansion of a given Mayer cluster as the sum of all two-center clusters that can be formed by removing a line (in every possible way) followed by the insertion of open circles at the removed line terminals. The original one-center circle is of course dropped. The value of a Mayer cluster is $-2b/(N-1)$ times the sum of the expansion. For example, the fourth virial in terms of Mayer clusters is given by

$$\alpha_4 = -\frac{3}{4!}\left\{\boxtimes + 6\,\boxslash + 3\,\square\right\}$$

The three clusters expand as follows:

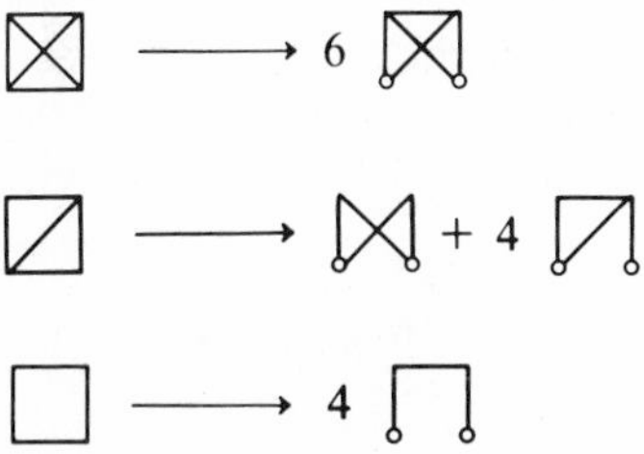

The above expression for α_4 becomes

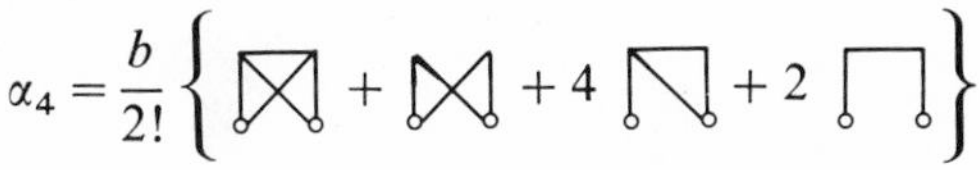

$$\alpha_4 = \frac{b}{2!}\left\{ \quad + \quad + 4 \quad + 2 \quad \right\}$$

or

$$\alpha_4 = 2b\left\{ \frac{1}{4} \quad + \frac{1}{4} \quad + \frac{1}{1} \quad + \frac{1}{2} \quad \right\}$$

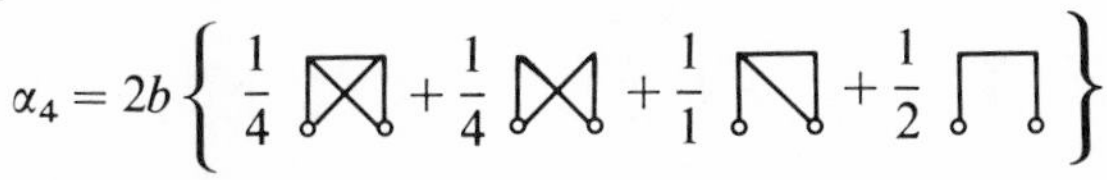

where the denominators are the symmetry numbers of the graphs. This result is true in general and the multiplicative factor on the front is always $2b$.

For a complete star (maximum number of lines) we get an even simpler result. The Mayer one-center hard-sphere graph equals $-Nb$ times the two-center complete star graph. For example,

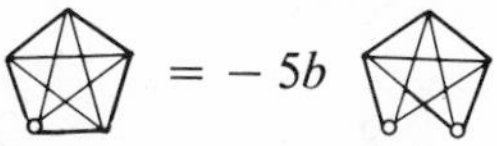

$$= -5b$$

where the two-center five-point star is evaluated at the two-center distance σ. The obvious advantage is that a $3(N-1)$-fold integration is converted to a $3(N-2)$-fold integration.

In 1952, more than fifty years after van der Waals, van Laar, and Boltzmann had evaluated the fourth virial coefficient for hard spheres, Nijboer and Van Hove[22] obtained the four two-center star cluster integrals on four points. As we have seen, this at once gives us the three Mayer integrals of the fourth virial coefficient. van Laar's integration was finally checked.

The fourth virial for the hard-sphere square-well potential has been treated by a number of authors[23–28] using a variety of methods. Boys and Shavitt,[29] using the Lennard-Jones 12-6 potential, and by means of a gaussian expansion, were able to obtain the fourth virial for the first realistic continuous potential. Barker and Monaghan[30] have devised a method which converges much faster and is applicable, in addition, to any potential. After some well-chosen analytic transformations, the problem is reduced to the numerical evaluation of some fairly low dimensionality integrals.

The fifth virial coefficient involves ten Mayer clusters or five Hoover clusters. Of the Mayer clusters, there is one with five bonds, two with six bonds, three with seven, two with eight, one with nine, and one with ten:[31]

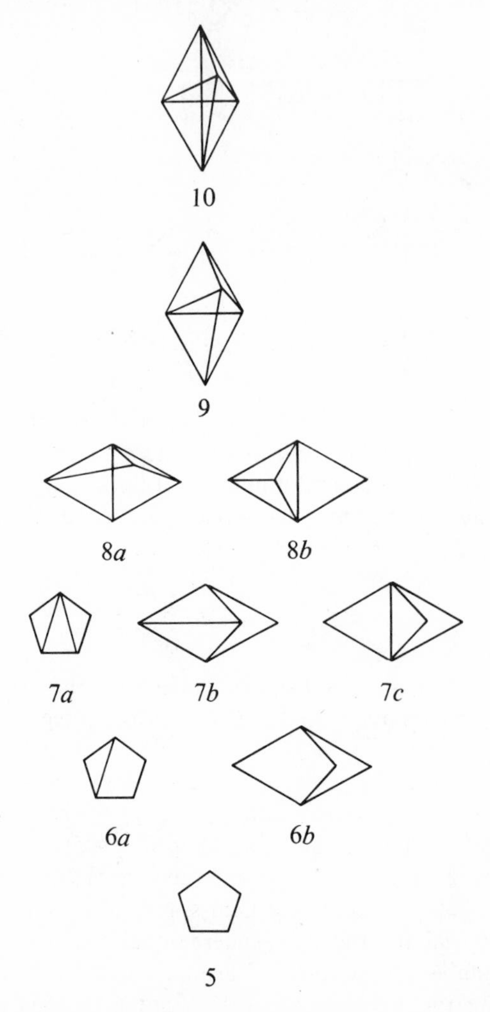

There are some advantages in representing these clusters by their complementary graphs. In graphs with as many as five points and with close to the maximum number of lines, the difference in connectivity between one graph and another is a little hard to see. In the complementary graph only the missing lines of the direct graph are shown. Furthermore, one need not show explicitly any points that do not terminate lines. The requisite number of additional points is assumed:

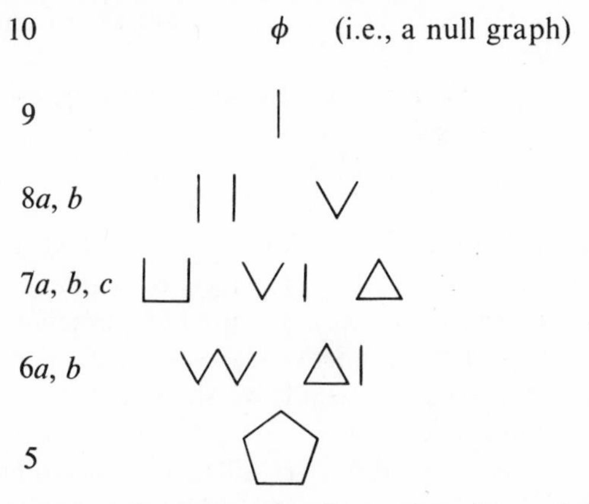

In this representation the connecting lines of the "correlation diagram" are easy to see. Notice that the correlation diagram for the complementary graphs on N points is a subgraph of the diagram for $N + 1$ points.

So far these ten clusters have not all been evaluated analytically for the hard sphere potential nor for any other reasonably realistic potential. In 1954 Rosenbluth and Rosenbluth[32] succeeded in getting moderately accurate values ($\sim 5\%$ error) for the hard-sphere potential using a Monte Carlo integration system. Very few details of their calculation are given in their paper but their final result was

$$\alpha_5/b^4 = 0.115 \pm 0.005$$

Rushbrooke and Hutchinson[33] have found the exact, rational values of clusters 5, 6*a*, 6*b*, 7*a*, and 7*c* for hard spheres by the use of convolution integrals:

$$\begin{aligned}
(5)/b^4 &= -40949/10752 = -3.808501 \\
(6a)/b^4 &= 68419/26880 = 2.545350 \\
(6b)/b^4 &= 82/35 = 2.342857 \\
(7a)/b^4 &= -34133/17920 = -1.904743 \\
(7c)/b^4 &= -73491/35840 = -2.050530
\end{aligned}$$

Recently Kim and Henderson[34] have found exact closed expressions for clusters 7*b* and 8*b*. Using all the known exact values and best values of the others from various authors, they find

$$\alpha_5/b^4 = 0.11027 \pm 0.00005$$

Katsura and Abe[35] have used Monte Carlo integration to determine all ten clusters. There is reason to suspect that all ten of these values are a little low in magnitude, in that the first five are systematically lower than the known exact values and the others are lower than some other results we shall presently cite.

Ree and Hoover[36] have used Monte Carlo to evaluate the five necessary Hoover clusters. Their result is

$$\alpha_5/b^4 = 0.1103 \pm 0.0003$$

Katsura[37] and Kilpatrick[38] have developed a method (often called the multiple summation method) that in principle will handle some of the more complicated clusters for any potential. In practice it works well only for hard spheres or hard sphere-square well. It would take too much space to explain the system in detail, we shall content ourselves with a partial outline and examples.

The original cluster integral is over all relative positions of N points with the integrand being some product of Mayer f functions, one for each line in the graph. The variables are vectors describing the vertex points. One can think of the graph as being a polyhedron in 3-space with the lines being the edges. A transformation is made to a new set of vectors which correspond to the faces (any closed cycle of lines) of the polyhedron. If there are N points and B lines, $B - N + 1$ of the face vectors are needed. The new integral has a continuous integrand, even for hard spheres. It consists of B factors, each one the spherical Fourier transform of the Mayer f function. For hard spheres of unit diameter this transform is

$$F(|\mathbf{t}|) = -|\mathbf{t}|^{-3/2} J_{3/2}(|\mathbf{t}|)$$

For example, for the five-point graph 5, the pentagon, we obtain

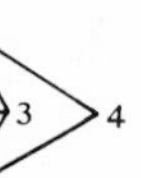

$$\begin{aligned} &= (2\pi)^{1/2} \int F^5(|\mathbf{t}|)\, d\mathbf{t} \\ &= 2^{5/2}\pi^{3/2} \int_0^\infty J_{3/2}^5(t) t^{-11/2}\, dt \end{aligned}$$

where the needed face is the pentagon and

5

2 3 4

1

$$\begin{aligned} &= (2\pi)^{-5/2} \iiint F(t_1)F(t_2)F^2(t_3) \\ &\quad F(|t_1 - t_2|)F(|t_2 - t_3|)F(|t_3 - t_1|)\, dt_1\, dt_2\, dt_3 \\ &= \frac{1}{2} \sum_l \sum_m \sum_n (\tfrac{3}{2} + l)(\tfrac{3}{2} + m)(\tfrac{3}{2} + n) M^* W_{0lm} W_{0ln} W_{00mn} \end{aligned}$$

and l is even, $m + n$ is even, $M^* = \min(m, n)$,

$$W_{abc} = 4\sqrt{2\pi} \int_0^\infty J_{3/2+a}(t) J_{3/2+b}(t) J_{3/2+c}(t) t^{-5/2}\, dt$$

$$W_{abcd} = 2\pi \int_0^\infty J_{3/2+a}(t) J_{3/2+b}(t) J_{3/2+c}(t) J_{3/2+d}(t) t^{-7}\, dt$$

The three faces used are (1, 2, 5), (1, 3, 5), and (1, 3, 5, 4). Notice that face vectors appear as arguments in each F function (corresponding to an edge) that is part of that face. The W integrals can readily be evaluated by complex variable and the series converge very rapidly if there are not too many summations. Best results are obtained by choosing the faces so that no edge appears in more than two faces.

To use this method one must have (1) the spherical Fourier transform of the f function, and (2) an addition formula for the transform. We have both for hard spheres.

Kilpatrick and Katsura[38] evaluated the first nine hard-sphere clusters by the multiple summation method. Yu[21] checked this work, using the same expansions (i.e., the same choice in faces) and in addition, the next to the most efficient. Some small errors were found in the original work. The two different expansions checked each other very well.

Yu[21] obtained by Monte Carlo what is probably the most accurate value of cluster 10 using the two center representation. The decrease from 12 to 9 dimensions considerably speeds up convergence. His value is

$$(10) = 0.7103 \pm 0.00016$$

This may be compared with Katsura's[35] value, 0.7078 ± 0.0018; Hoover's[36] value, 0.711 ± 0.001; Ree's[39] 0.7115 ± 0.0015; and Barker's,[40] 0.7118. All these results are by Monte Carlo except that of Barker.

Some of the results for the values of the five-point clusters are compared in Table I.

The best value we can give now for the fifth hard-sphere virial coefficient is found using the seven exactly known clusters together, with Yu's values for 8a, 9, and 10:

$$\alpha_5/b^4 = 0.110277 \pm 0.000014$$

For the first time, the dominating error is not from cluster 10; it now is from cluster 9.

Barker, Leonard, and Pompe have extended the method of Reference 30 to get the fifth virial for the Lennard-Jones 12-6 potential. Their results in this case show a very curious contrast to the monotomic behavior of the

TABLE I

Hard-Sphere Fifth Virial Clusters (unit, b^4)

Type	Weight	Exact value	Katsura[34] (M.C.)	Yu and Kilpatrick[21] (M.S.)
5	12	−3.808501	−3.80564 ± 0.0036	−3.808501
6*a*	60	2.545350	2.54277 ± 0.0033	2.545350
6*b*	10	2.342857	2.34098 ± 0.0033	2.342858 ± 0.000001
7*a*	60	−1.904743	−1.89942 ± 0.0029	−1.904744 ± 0.000003
7*b*	30	−1.583772	−1.58107 ± 0.0027	−1.583773 ± 0.000003
7*c*	10	−2.050530	−2.04866 ± 0.0029	−2.050528 ± 0.000003
8*a*	15		1.13579 ± 0.0024	1.139317 ± 0.000002
8*b*	30	1.329807	1.32741 ± 0.0024	1.32980 ± 0.00001
9	10		−0.91180 ± 0.0021	−0.91471 ± 0.00004
10	1		0.70782 ± 0.0018	

second, third, and fourth LJ virials: there is a negative minimum above the critical temperature.

Ree and Hoover,[36,41] extending the method used so successfully for the fifth hard-sphere virial, have found the sixth and the seventh

$$\alpha_6/b^5 = 0.0386 \pm 0.0004$$
$$\alpha_7/b^6 = 0.0138 \pm 0.0004$$

There are several lines of argument indicating that eventually some negative hard-sphere virial coefficients must occur. This may be as high as the eighteenth.[21]

IV. QUANTUM THEORY

There are two principal differences between a quantum and a classical calculation of a virial coefficient: (1) purely quantum effects, that is, those that stem directly from the uncertainty principle, primarily the discrete nature of energy states; and (2) symmetry effects, those arising from the fact that all real particles (whether elementary or large, complex molecules) are either even (Bose-Einstein) or odd (Fermi-Dirac). For a light molecule such as helium the purely quantum effects are distinctly noticeable at temperatures as high as 2000°K. We cite data from Boyd, Larsen, and Kilpatrick,[42] the second virial coefficient of ^{4}He in ml/mol calculated for the De Boer Lennard-Jones potential:

Temp, °K	$B_{\text{classical}}$	B_{quantum}	Difference
20	−13.982	−0.843	13.139
30	−2.874	3.573	6.447
50	4.897	8.161	3.264
100	9.628	10.994	1.366
300	11.109	11.480	0.371
500	10.7344	10.9402	0.2058
1000	9.9353	9.9048	−0.0305
2000	8.7617	8.7476	−0.0141

The quantum-classical difference is even larger for the lighter ^{3}He and is by no means negligible even at room temperature for the common permanent gases. On the other hand, purely symmetry or exchange effects are quite unimportant for ^{4}He above 5°K:

Temp, °K	B	B_{exch}
1	−429.030	−52.400
2	−177.419	−3.427
3	−110.134	−0.396
4	−78.105	−0.060
5	−59.214	−0.011
10	−21.321	−0.00000

We can sum up these possibly surprising results, again using ^{4}He as an example: the uncertainty principle, the direct difference between classical and quantum mechanics, makes an important contribution to the second virial coefficient at temperatures as high as 2000°K. Purely exchange effects, the difference between Bose and Fermi statistics, contribute literally nothing except below 5°K. With increasing molecular weight, both temperature limits drop but the exchange limit drops much the faster. Qualitatively, similar remarks are justified about the third and higher virial coefficients.

We shall now discuss the source of these two types of difference between classical and quantum theory for the second virial coefficient. Not much has yet been done quantum-mechanically with the third virial. The theory is a straightforward extension of the two-body theory but the computational difficulties are severe.

The second virial coefficient (per molecule) is given by

$$\alpha_2 = -g_2/g_1^2 = V(Z_2 - \tfrac{1}{2}Z_1^2)/Z_1^2$$

The source of a nonzero value of α_2 is the difference between Z_2 and $\frac{1}{2}Z_1^2$. Both Z_1 and Z_2 must be taken as the sum over states of one and two molecules in a container of volume V, respectively. Clearly the pair potential is irrelevant for the one body system. If we assume the single molecule is a point mass with nuclear spin s, application of the Euler-Maclaurin summation formula yields the classical result for Z_1 except for the spin degeneracy factor $2s + 1$:

$$Z_1 = (2s + 1)V/\lambda_T^3$$

This result is wholly from the Euler-Maclaurin integral. The first correction term is negligible if, as is normally the case, the dimensions of the container are large compared with λ_T.

Evaluation of Z_2 requires a knowledge of the quantum states of two interacting particles in a volume V. The two particles are either Bose or Fermi. Perhaps the matter will be made clearer if one observes that $\frac{1}{2}Z_1^2$ is exactly the Z_2 of a pair of noninteracting particles that obey corrected Maxwell-Boltzmann statistics. In a CMB system of independent particles, states that involve n_1 particles in one quantum state, n_2 in another, and so on, receive a weight of $1/n_1!n_2!\cdots$ in contrast with independent BE or FD particles, where the weight is $+1$ or 0, respectively.

We can define the purely quantum contribution to the second virial coefficient as that due to the difference between the two cases, two CMB particles, interacting and then noninteracting. The exchange contribution is due to the difference between the two cases, two interacting particles, first either BE or FD and then CMB:

$$B_{\text{direct}} = -NV(Z_{2(\text{CMB})} - \tfrac{1}{2}Z_1^2)/Z_1^2$$

$$B_{\text{exch}} = -NV(Z_{2(\text{BE or FD})} - Z_{2(\text{CMB})})/Z_1^2$$

$$B = B_{\text{direct}} + B_{\text{exch}}$$

Another way to view this division is that the transformation properties of the terms of Z_2 are studied under the two-element permutation group on two things. Since there is a 1 : 1 correspondence between the irreducible representations and the classes, we can associate some of Z_2 with the identity class and the remainder with the twofold permutation. This point of view holds for three or more body systems also. The direct virial coefficient is associated with the identity class; the exchange virial results from all other classes.

In the older literature[43,44] another division was used which has not proved as useful. B was broken into two terms, (1) the second virial for a noninteracting but BE or FD gas and (2) a remainder:

$$B = -NV(Z^*_{2(\text{BE or FD})} - \tfrac{1}{2}Z_1^2)/Z_1^2 - NV(Z_2 - Z_2^*)/Z_1^2$$

where the asterisk means no interaction. The disadvantage of this division is that the rapid disappearance of exchange effects for real, hard-core potentials is concealed.

A two-body partition function may be separated into two factors: a center of gravity factor which is evaluated just as in the case of a single particle, and a relative motion factor. For any type of Z_2, we obtain the result:

$$\frac{VZ_2}{Z_1^{\,2}} = \frac{2^{3/2}\lambda_T^{\,3} Z_2^{\,(\mathrm{rel})}}{(2s+1)}$$

The factor $2^{3/2}$ comes from the reduced mass for the center of gravity (we keep the meaning of λ_T as originally defined, based on the mass of one molecule). $Z_2^{(\mathrm{rel})}$ still contains the nuclear spin degrees of the two particle system.

$Z_{2(\mathrm{CMB})}^{(\mathrm{rel})}$ is exactly half the sum of Boltzmann factors for the energy states of relative motion multiplied by the nuclear spin degeneracy factor of $(2s+1)^2$. The factor of two is the symmetry number for the exchange group. The direct second virial becomes:

$$B_{\mathrm{direct}} = -2^{1/2} N \lambda_T^{\,3} (Z_{2(\mathrm{CMB})}^{(\mathrm{rel})} - Z_{2(\mathrm{CMB})}^{(\mathrm{rel})}{}^*)$$

All the energy states of our two-body system are discrete since the two particles are enclosed in a volume V. There are two qualitatively different kinds of states, however. Negative, discrete, bound states can occur only for two particles interacting with a potential which is somewhere attractive. This contribution must always be summed into Z_2. Such states of course never appear in $Z_{2(\mathrm{CMB})}^{(\mathrm{rel})}{}^*$. No bound states actually occur in ^{3}He. ^{4}He is almost exactly on the borderline for the appearance of a first bound state. The number of bound states rapidly increases as one goes up the list of heavier molecules.

The second class of relative states is the whole positive "continuum." These levels are very densely packed (again, at least if λ_T is very small compared with the size of the container) and may be summed with an Euler-Maclaurin integral. We need, then, the difference in density of states between the two cases, interacting and noninteracting particles.

Groper[45] showed by a very simple argument (recapitulated in Ref. 44) that this information is contained in the variation of the phase shift for the relative, interacting, two-body system with the energy (or wave number) and discrete angular momentum of collision. The radial Schrödinger equation for relative motion must be solved (analytically if possible but usually numerically), the solution carried out beyond the effective range of the potential, and the phase shift determined. In principle, this must be done for all integral values of the angular quantum number l and for all positive energies. In practice, the maximum value of the temperature at which B is desired sets an upper bound on both parameters. That is, relative energies or angular momenta too high to have an appreciable probability of occurrence at a given temperature cannot affect the second virial coefficient. Furthermore, the energies may be taken at discrete, small intervals adequate for the subsequent numerical integration.

The results of these considerations is the following working equation:[42]

$$B_{\text{direct}} = -2^{1/2} N \lambda_T^3 \left\{ \frac{\lambda_T^2}{\pi^2} \int_0^\infty k\, dk G_+(k) \exp\left(-\frac{\lambda_T^2 k^2}{2\pi} \right) + \sum_l (2l+1) \sum_n [\exp(-\beta E_{n,l}) - 1] \right\}$$

where

$$G_+(k) = \sum_l (2l+1)\delta_l(k)$$

and E_n, l is the energy of the nth bound state with angular momentum l, k is the wave number, and $\delta_l(k)$ the phase shift as a function of l and k. The -1 that occurs in the discrete level sum comes from the behavior of the phase shifts at zero energy with the number of discrete levels and from an integration by parts. This expression is the quantum-mechanical equivalent of the classical configuration space integral and indeed passes into that form at sufficiently high temperature. Note that a large part of the quantum-classical deviation is due to the discrete nature of angular momentum. No integral can replace this sum except at very high temperature. Furthermore, the quantum f function in the discrete level sum is of just the right form to prevent any sudden change in the second virial if the depth of the potential or the mass of the molecules is changed so as to make a discrete level appear or disappear. We shall never be able to tell whether or not a pair of ^{4}He atoms really have a single bound state from experimental measurements of the second virial coefficient. The effect is completely averaged or smeared out.

The exchange second virial coefficient has very much the same form:

$$B_{\text{exch}} = \mp \frac{2^{1/2} N \lambda_T^3}{2s+1} \left\{ \frac{\lambda_T^2}{\pi^2} \int_0^\infty k\, dk \left(G_-(k) + \frac{\pi}{8} \right) \exp\left(-\frac{\lambda_T^2 k^2}{2\pi} \right) + \sum_l (-)^l (2l+1) \sum_n [\exp(-\beta E_{n,l}) - 1] \right.$$

where

$$G_{-}(k) = \sum_{l} (-)^l (2l + 1)\delta_l(k)$$

As usual, $\mp$ is associated with BE or FD statistics. The alternation in sign G_{-} comes from the reversal of symmetry with respect to exchange of the angular factors of the relative wave function with the parity of l. The term $\pi/8$ comes from the ideal (BE or FD) gas contribution to $Z_{2(\mathrm{BEorFD})}$.

If our quantum gas had a zero potential, the entire value for B would come from the $\pi/8$ (both G_{+} and G_{-} would be zero):

$$B = \mp 2^{-5/2} N \lambda_T{}^3/(2s + 1)$$

On the other hand, if the potential has a hard core (as all realistic potentials do), G_{-} becomes very nearly $-\pi/8$ except at the lowest energies. The net result[46] is that B_{exch} approaches zero very rapidly as the temperature increases, in fact, in a nonanalytic fashion, faster than any finite negative power of T. In general, at a temperature high enough that λ_T is small compared to the radius of the potential core, B_{exch} is negligible and all spin-statistics effects have disappeared from the second virial coefficient.

There is a very close analogy to the exchange partition function and a classical problem in diffraction. Consider a point source of radiation of wave length λ_T just outside an opaque sphere of radius a. The exchange partition function for two hard spheres is closely related to the intensity of the radiation at a point equidistant from the center of the opaque sphere but diametrically opposite. Obviously the intensity will begin to fall off very rapidly just when λ_T becomes smaller than the diameter of the sphere. This picture comes from the Feynman interpretation of wave mechanics. An equally valid optical model can be set up for a continuous potential. If there is a hard core, diffraction effects force Z_{exch} to vanish very rapidly with increasing temperature.

The third virial coefficient is a much more complicated problem. Three particles require three vector coordinates or nine scalar coordinates. The wave functions are nine dimensional. The coordinates can be chosen in a variety of ways to represent three center of gravity coordinates, one overall scale coordinate R and five angular coordinates (giving the three orientation angles of the triangle and two other angles giving the proportions or shape of the triangle). Center of gravity degrees of freedom separate off just as in the two-body case. The asymptotic solution of the six internal degrees of freedom wave equation is

$$\psi \sim (c_1 J_{l+2}(kR) + c_2 N_{l+2}(kR)) Y_{lmabc}$$

where J and N are Bessel functions and the Y are spherical harmonics on the five-dimensional surface of a six-dimensional hypersphere. The same type of argument that led in the two-body case to a connection between phase shifts and density of states works in this case. Unfortunately, however, there are five angular quantum numbers and the phase shifts depend on three of them. Only two can be used as degeneracy factors. Furthermore, as the potential energy depends on the three internuclear distances, it is not a function of R alone and the wave equation is not separable. The difficulties in actually producing a few phase shifts appear to be formidable.

There is no particular problem in handling exchange effects. It is certain that above about 5°K (for ^4He) Boltzmann statistics may be used. There will be two exchange third virials which will vanish very rapidly with temperature, corresponding to the permutations 3^1 and 1, 2. Of these, the first will be the more important since it involves the smaller angle. The situation is exactly the same as in Ford's[47] treatment of the partition function of the rotating symmetric top.

V. QUANTUM APPLICATION

No real gas except for the two stable helium isotopes and their mixtures has had its second virial coefficient calculated quantum-mechanically. The phase-shift method has not yet been applied to the third virial.

De Boer, van Kranendonk, and Compaan[43] performed the first quantum calculation on the helium isotopes, using the De Boer Lennard-Jones potential. This potential was adjusted by De Boer to agree with the available high temperature second virial data, using the classical expression for B plus the first several high-temperature quantum corrections. Their calculation was apparently carried out on a desk calculator. We have the greatest of admiration for them in that this tedious and difficult task was performed so successfully. Their calculated values agree with the very-low-temperature experimental data very well.

A few years later Kilpatrick, Keller, Hammel, and Metropolis,[44] using one of the earliest internally programed electronic computers, MANIAC I, considerably improved the technique for finding phase shifts and greatly extended the range of the calculation. Both the Lennard-Jones and the exp-six potential were used. The most disturbing result of these calculations is that the particular potentials of the two types used that best fit all the experimental data for B differ from each other by at least 30%. We still don't know the true two-body He potential at all well.

Larsen, Witte, and Kilpatrick[48] have calculated the pair-correlation function for ^4He, using the Lennard-Jones potential. This function is closely related to the second virial coefficient and shows the same quantum-statistical properties. The calculation is much more difficult, however, because the radial wave functions for all values of R, k, and l must be stored until the end of the calculation.

Boyd, Larsen, and Kilpatrick[49] explored the second virial coefficient for hard spheres both analytically and numerically. The wave functions and phase shifts can be obtained in exact analytic form in this case, in contrast to the numerical solutions necessary for realistic, continuous potentials. A great deal of difficult and tedious numerical work was still necessary to evaluate the necessary summations, to obtain the direct and exchange second virials at temperatures high enough for the quantum and statistical effects on the classical results to be small.

At low temperature they obtained the results, in units of $B_{cl} = \frac{2}{3}\pi N\sigma^3$,

$$B_{\text{direct}} = \frac{3}{2\pi}\left(\frac{\lambda_T}{\sigma}\right)^2\left[1 + 3\pi\left(\frac{\sigma}{\lambda_T}\right)^2 - \frac{22\pi^2}{3}\left(\frac{\sigma}{\lambda_T}\right)^4 + \frac{1921\pi^3}{45}\left(\frac{\sigma}{\lambda_T}\right)^6 + \cdots\right]$$

$$B_{\text{exch}} = \mp\frac{1}{2s+1}\frac{3}{2^{7/2}\pi}\left(\frac{\lambda_T}{\sigma}\right)^3\left[1 - 2^{5/2}\left(\frac{\sigma}{\lambda_T}\right) + 2^{5/2}3\pi\left(\frac{\sigma}{\lambda_T}\right)^3 + \frac{2^{15/2}\pi}{3}\left(\frac{\sigma}{\lambda_T}\right)^5 + \ldots\right]$$

At high-temperature, least-square fitting of their results to a series in λ_T/σ gave

$$B_{\text{direct}} = 1 + \frac{3}{2\sqrt{2}}\left(\frac{\lambda_T}{\sigma}\right) + \frac{1}{\pi}\left(\frac{\lambda_T}{\sigma}\right)^2 + \frac{1}{16\sqrt{2\pi}}\left(\frac{\lambda_T}{\sigma}\right)^3 - \frac{1}{105\pi^2}\left(\frac{\lambda_T}{\sigma}\right)^4 + \cdots$$

This expansion is unusual, being in all powers of $\hbar$. The Wigner-Kirkwood[50] treatment of high-temperature quantum effects leads one to expect a series in even powers of $\hbar$. This result holds, however, only for continuous potentials. Uhlenbeck and Beth[5] had predicted correctly that the first coefficient would be $3/2\sqrt{2}$. Mohling[51] had erroneously calculated the second coefficient to be $1/2\pi$. Handelsman and Keller[52] and Hill[53] have deduced (analytically) the values of the first six coefficients. Very recently Arruda and Hill have sent us a preprint in which the first seven coefficients of B_{direct} are calculated by a method much simpler than previously used. The next three terms are

$$+\frac{1}{640\pi^2\sqrt{2}}\left(\frac{\lambda_T}{\sigma}\right)^5-\frac{2}{3003\pi^3}\left(\frac{\lambda_T}{\sigma}\right)^6+\frac{47}{215040\pi^3\sqrt{2}}\left(\frac{\lambda_T}{\sigma}\right)^7+\cdots$$

Although at first sight there seems to be a number of clues to the general form of these coefficients, no one has been able to find it yet.

The form of B_{exch} for hard spheres at high temperature is a more difficult problem. None of the classical solutions for the corresponding diffraction problem are valid at the diametrically opposite point where we need them. Boyd, Larsen, and Kilpatrick[49] had expected that the dominant term would be $\exp(-(\pi^3/2)(\sigma/\lambda_T)^2$ from the correspondence of the exchange partition function for two hard spheres (mass concentrated at their centers) and that of the rigid rotor. Numerical fitting of their data gave results like

$$\ln B_{\text{exch}} = C_1 + C_2 \ln(\lambda_T/\sigma) + C_3(\sigma/\lambda_T) + C_4(\sigma/\lambda_T)^2$$

$$C_1 \sim -\ \ 3.25 \pm 0.18$$

$$C_2 \sim \quad 3.55 \pm 0.12$$

$$C_3 \sim -\ \ 5.02 \pm 0.25$$

$$C_4 \sim -15.55 \pm 0.06$$

The value of $\pi^3/2$ is 15.50. Lieb[54] has shown that this is indeed the form of the dominant term.

Hill[53] has deduced an asymptotic expansion for B_{exch} in the form

$$B_{\text{exch}} = \mp\frac{N}{2s+1}\sum_{n=1} 4\pi^3\sigma^3 h_n \exp f_n$$

$$h_n = 1 + \frac{14}{9}\beta_n\left(\frac{\lambda_T}{\pi\sigma}\right)^{4/3} + \frac{130}{405}\beta_n{}^2\left(\frac{\lambda_T}{\pi\sigma}\right)^{8/3} + \cdots$$

$$f_n = -\pi\left\{\frac{1}{2}\left(\frac{\pi\sigma}{\lambda_T}\right)^2 + \beta_n\left(\frac{\pi\sigma}{\lambda_T}\right)^{2/3} + \frac{4}{45}\beta_n{}^2\left(\frac{\pi\sigma}{\lambda_T}\right)^{-2/3} + \cdots\right.$$

and the β_n are zeros of Airy functions. Lieb[54] had predicted that the second term in the exponent would be 2/3 power in σ/λ_T as is indeed the case. The most important part of this series for high temperature (large σ/λ_T) is $n = 1$. It gives a very good fit to the directly calculated values of Reference 49.

Quite recently Boyd, Larsen, and Kilpatrick[55] have published a very complete calculation of the second virial coefficient of the helium isotopes, using the De Boer Lennard-Jones potential and covering the range 0.1°K to 2000°K. The phase-shift calculations were carried well above room temperature, generously overlapping the region of validity of the classical equation with Wigner-Kirkwood quantum corrections. The results agree very well with the older calculations. The program used in this work is available from the Los Alamos Scientific Laboratory T-Division in FORTRAN language. It will accept any realistic (i.e., hard core, moderate range) potential, and it calculates the phase shifts and G functions, determines the presence and location of discrete levels, evaluates the second virial coefficients as high as the phase shifts will permit, then finishes off with a classical calculation with quantum corrections.

Larsen has supplied us with a preprint of a paper in which he analyzes the direct third virial coefficient using the phase-shift technique and has applied his equations to the state of zero angular momentum to obtain the limiting low-temperature behavior of this function for the hard-sphere square-well potential.

Fosdick and Jordan[56] have used the Feynman-Weiner path integral technique to get the second and the third virials for a Lennard-Jones 12-6 potential. The results are inferior to those from phase shifts (due of course to the tremendous numerical difficulty of this method) but are all that we have for the third virial. There is a considerable difference with the classical third virial and neither agree at all well with the limited experimental data. This probably indicates, as has been long suspected, that the Lennard-Jones potential is not particularly close to the real helium potential.

References

1. J. D. van der Waals, *Koninkl. Ned. Akad. Wetenschap. Proc.* **1,** 138 (1899); J. J. van Laar, *Koninkl. Ned. Akad. Wetenschap. Proc.*, **1,** 273 (1899).
2. L. von Boltzmann, *Koninkl. Ned. Akad. Wetenschap. Proc.*, **1,** 398 (1899).
3. H. D. Ursell, *Proc. Cambridge Phil. Soc.*, **23,** 685 (1927).
4. J. E. Mayer and M. G. Mayer, *Statistical Mechanics*, Wiley, New York, 1940.
5. G. E. Uhlenbeck and E. Beth, *Physica*, **3,** 729 (1936); E. Beth and G. E. Uhlenbeck, *Physica*, **4,** 915 (1937).
6. B. Kahn, Thesis, Amsterdam, 1938; B. Kahn and G. E. Uhlenbeck, *Physica*, **5,** 399 (1938).
7. J. E. Kilpatrick, *J. Chem. Phys.*, **21,** 274 (1953).
8. J. E. Kilpatrick and D. I. Ford, *Am. J. Phys.*, **37,** 881 (1969).
9. G. S. Rushbrooke, *Lectures on Diagram Techniques for Classical Fluids and Lattice Models*, Rice University, Houston, Texas, 1967.
10. F. H. Ree and W. G. Hoover, *J. Chem. Phys.*, **41,** 1635 (1964).

10a. T. L. Hill, *J. Chem. Phys.*, **23,** 617 (1955).

10b. J. E. Kilpatrick, *Ann. Rev. Phys. Chem.*, **7,** 67 (1956).

11. J. D. Hirschfelder, C. F. Curtiss, and R. B. Bird, *Molecular Theory of Gases and Liquids*, Wiley, New York, 1954.

12. T. Kihara, *Proc. Phys.-Math. Soc. Japan*, **17,** 11 (1943); also Ref. 11.
13. R. B. Bird, E. L. Spotz, and J. O. Hirschfelder, *J. Chem. Phys.*, **18,** 1395 (1950).
14. T. Kihara, *J. Phys. Soc. Japan*, **3,** 265 (1948); *ibid.*, **6,** 184 (1951).
15. R. Bergeon, *Compt. Rend.*, **234,** 1039 (1952).
16. J. S. Rowlinson, F. H. Sumner, and J. R. Sutton, *Trans. Faraday Soc.*, **50,** 1 (1954).
17. R. Bergeon, *J. Rech. Centre Natl. Rech. Sci. Lab. Bellevue* (*Paris*), **9,** 171 (1958).
18. J. S. Rowlinson, *Mol. Phys.*, **6,** 75 (1963).
19. A. E. Sherwood and J. M. Prausnitz, *J. Chem. Phys.*, **41,** 413 (1964).
20. T. L. Hill, *Statistical Mechanics*, McGraw-Hill, New York, 1956.
21. A. P. Yu, Thesis, Rice University, Houston, Texas, 1966.
22. B. R. A. Nijboer and L. Van Hove, *Phys. Rev.*, **85,** 777 (1952).
23. S. Katsura, *Phys. Rev.*, **115,** 1417 (1959); *ibid.*, **118,** 1667 (1960).
24. J. A. Barker and J. J. Monaghan, *J. Chem. Phys.*, **36,** 2558 (1962).
25. E. H. Hauge, *J. Chem. Phys.*, **39,** 389 (1963).
26. D. A. McQuarrie, *J. Chem. Phys.*, **40,** 3455 (1964).
27. S. Katsura, *J. Chem. Phys.*, **45,** 3480 (1966).
28. J. A. Barker and J. J. Monaghan, *J. Chem. Phys.*, **45,** 3482 (1966).
29. S. F. Boys and I. Shavitt, *Proc. Roy. Soc.* (*London*), **A254,** 487 (1960).
30. J. A. Barker and J. J. Monaghan, *J. Chem. Phys.*, **36,** 2564 (1962).
31. In my laboratory I have two models suspended in the air, one of the 10 fifth-virial clusters and one of the 56 sixth-virial clusters. They were constructed by D. Ford, J. Hubbard, and myself from stainless steel bar silver soldered together. The cluster lines are about two inches in length, correlation lines (i.e., connecting clusters that may be converted into each other by removal or addition of one line) hold the models together, and clusters of zero weight in the Hoover interpretation are painted a special color.
32. M. Rosenbluth and A. Rosenbluth, *J. Chem. Phys.*, **22,** 881 (1954).
33. G. S. Rushbrooke and P. Hutchinson, *Physica*, **27,** 647 (1961).
34. S. Kim and D. Henderson, *Phys. Lett.* **27A,** 378 (1968).
35. S. Katsura and Y. Abe, *J. Chem. Phys.*, **39,** 2068 (1963).
36. F. H. Ree and W. G. Hoover, *J. Chem. Phys.*, **40,** 939 (1964).
37. S. Katsura, *Phys. Rev.*, **115,** 1417 (1959); *ibid.*, **118,** 1667 (1960).
38. J. E. Kilpatrick and S. Katsura, *J. Chem. Phys.*, **45,** 1866 (1966); J. E. Kilpatrick, S. Katsura, and Y. Inove, *Math. Comp.*, **21,** 407 (1967).
39. F. H. Ree, R. N. Keeker, and S. L. McCarthy, *J. Chem. Phys.*, **44,** 3407 (1966).
40. J. A. Barker and D. Henderson, *Can. J. Phys.*, **45,** 3959 (1967).
41. F. H. Ree and W. G. Hoover, *J. Chem. Phys.*, **46,** 4181 (1967).
42. M. E. Boyd, S. Y. Larsen, and J. E. Kilpatrick, *J. Chem. Phys.*, **50,** 4034 (1969).
43. J. De Boer, J. van Kranendonk, and K. Compaan, *Phys. Rev.*, **76,** 998 (1949); *ibid.*, **79,** 1728 (1949); *Physica*, **16,** 545 (1950).
44. J. E. Kilpatrick, W. E. Keller, E. F. Hammel, and N. Metropolis, *Phys. Rev.*, **94,** 1103 (1954); *ibid.*, **97,** 9 (1955).
45. L. Groper, *Phys. Rev.*, **50,** 963 (1936); *ibid.*, **51,** 1108 (1937).
46. S. Y. Larsen, J. E. Kilpatrick, E. H. Lieb, and H. Jordan, *Phys. Rev.*, **140** (1965).
47. D. I. Ford, Thesis, Rice University, Houston, Texas, 1969.
48. S. Y. Larsen, K. Witte, and J. E. Kilpatrick, *J. Chem. Phys.*, **44,** 213 (1966).
49. M. E. Boyd, S. Y. Larsen, and J. E. Kilpatrick, *J. Chem. Phys.*, **45,** 499 (1966).
50. E. Wigner, *Phys. Rev.*, **40,** 747 (1932); J. G. Kirkwood, *Phys. Rev.*, **44,** 31 (1933).

51. F. Mohling, *Phys. Fluids*, **6,** 1097 (1963).
52. R. A. Handelsman and J. B. Keller, *Phys. Rev.*, **148,** 94 (1966); P. C. Hammer and K. J. Monk, *Phys. Rev.*, **158,** 114 (1967).
53. R. N. Hill, *J. Math. Phys.*, **9,** 1534 (1968).
54. E. Lieb, *J. Math. Phys.*, **7,** 1016 (1966).
55. M. E. Boyd, S. Y. Larsen, and J. E. Kilpatrick, *J. Chem. Phys.*, **50,** 4034 (1969).
56. L. D. Fosdick and H. F. Jordan, *Phys. Rev.*, **143,** 58 (1966); *ibid.*, **171,** 128 (1968).

THE ORIGIN OF HYSTERESIS IN SIMPLE MAGNETIC SYSTEMS

T. ERBER

Department of Physics, Illinois Institute of Technology, Chicago, Illinois

H. G. LATAL*

Department of Physics, Illinois Institute of Technology, Chicago, Illinois

and

B. N. HARMON

Department of Physics, Northwestern University, Evanston, Illinois

CONTENTS

* Permanent address: Institut für Theoretische Physik, Universität Graz, Austria.

Abstract

It has been found that a system of four magnets interacting solely through dipole forces exhibits a configurational hysteresis. This irreversible behavior can be traced to the existence of complementary instability points which are intercepted by the system during cyclic lattice deformations. The properties of this hysteresis cycle have been worked out analytically and verified with observations on carefully shielded model arrays mounted on deformable linkages. The variation of spontaneous magnetization during the cycle can be related to the appearance of symmetry-breaking interactions. Previous results on the population of metastable levels (state-area principle) have been extended to the occupation probabilities of the branches of the hysteresis cycle. Experiment confirms the existence of "anti-Boltzmann" population inversions.

Analogous results have been established for more complex systems. As the arrays are enlarged from 2×2 to 6×6 magnets, the density of instability points and the intricacy of the hysteresis loops increase rapidly. These trends can be extrapolated to still larger systems with the help of topological arguments (Morse-Lusternik-Schnirelmann). The results suggest that irreversibility in many macroscopic systems is a manifestation of dense sets of instability points of the Hessian. As a test of these ideas we derive a general law for energy losses in hysteresis cycles; the Rayleigh law (ferromagnetic hysteresis of iron, nickel, cobalt, etc.) and Dorey's rule (stress-strain hysteresis of iron, steel, bakelite, plywood, lucite, etc.) follow as special cases.

I. INTRODUCTION

An investigation of the stable configurations of a four-magnet system arranged on a square lattice[1,2,3] has revealed a mechanism for dynamical symmetry breaking due to the variation of the relative magnitudes of dipole and octopole forces as a function of the lattice spacing. In elaborating the ideas outlined in Reference 1, we have studied the basic four-magnet system corresponding to another regular tessellation of the plane, that is, the rhombus consisting of two equilateral triangles.[4] This showed that in addition to the dynamical symmetry breaking, there was another type which may aptly be called a geometric symmetry breaking since even for pure dipole interactions two patterns with differing symmetry properties coexist. This is in contrast to the square arrangement where only a single configuration is observed under similar circumstances. Since it was not at all obvious how the geometric symmetry breaking would be quenched as one approached the limit of a square tessellation we decided to study the variations of the stable patterns of four magnets located at the vertices of a continuously deformable rhomboidal lattice. The most surprising result of this investigation was the discovery that the system exhibited a configurational hysteresis in response to *continuous* lattice deformations. This was manifested in *discontinuous* and *irreversible* transitions in the patterns displayed by the four-magnet array. Fortunately this problem lends itself to rigorous analytical treatment in terms of pure dipole interactions, since the hysteresis effects persist even in this limit. Experimentally this can be realized with "wide-space" linkages in which the dipole forces dominate the octopole interactions. Conceptually clear-cut and consistent numerical agreement between experiment and theory has thereby been established.

Naturally one wonders whether this type of hysteresis is an affliction peculiar to four-magnet linkages or whether it is symptomatic of a more fundamental property of complex magnetic systems. Studies in this direction have already been carried out experimentally for arrays extending up to 6×6 magnets, and hysteresis has indeed been found to be a prominent effect. In fact as the systems are enlarged, the intricacy of the hysteresis networks increases rapidly.

In this paper, after a brief description of the experimental arrangements (Section II), we summarize the observations and compare with detailed calculations (Section III). The underlying Hessian instability theory is discussed in Section IV. Various statistical features of the hysteresis cycle are studied in Section V. The experimental results are interpreted in terms of the state-area principle. In Section VI, a number of useful topological concepts and lemmas are assembled. Finally in Section VII, we propose an abstract hysteresis theory for systems of arbitrary complexity. This provides a unified basis for Rayleigh's law of ferromagnetic hysteresis, and Dorey's rule for stress-strain hysteresis.

II. EXPERIMENTAL ARRANGEMENTS

The essential components of the experimental setup are shown schematically in Figure 1. The magnets are mounted on support pins which are positioned at the vertices of a continuously deformable linkage. The entire array slides on a plexiglass board, which in turn is held by a wooden support system (not shown in Fig. 1) inside a double-walled magnetic shield of molybdenum permalloy. There actually exist two different linkages: one capable of accommodating up to 6×6 magnet arrays with a lattice constant given by $a = 5.2 \pm 0.05$ cm ("wide" spacing); the other with a capability extending to 9×9 systems and a corresponding lattice constant

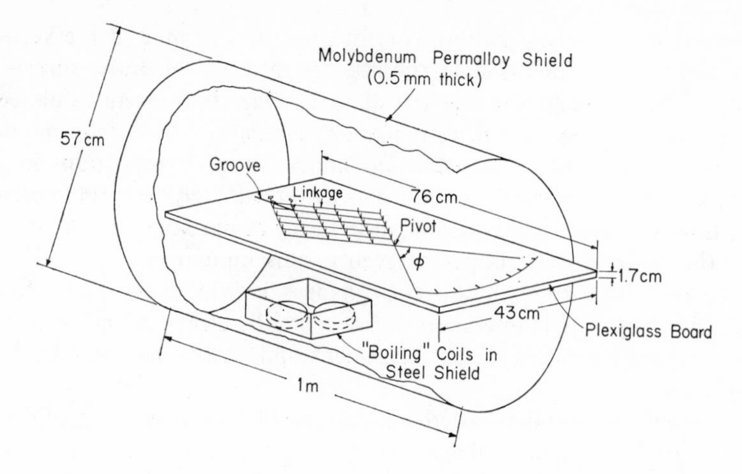

Fig. 1. Schematic view of the experimental arrangements. The permalloy shield indicated actually consists of two concentric cylinders separated by $\sim$3 mm. The auxiliary randomizing array ("boiling coils") is installed in a separate steel shield. Not shown is a front silvered mirror which is used for photographic pattern recording.

of $a = 3.2 \pm 0.05$ cm ("close" spacing). The choice of these particular lattice spacings was dictated by the desire to minimize the influence of octopole forces.[1] Both linkages are made out of (nonmagnetic) brass bars, each 26 cm long, 0.9 cm wide, and 0.25 cm thick—except for the two extended handles, which are about 50 cm long and serve as pointers in conjunction with a protractor scale etched on the plexiglass board. The linkage joints consist of tubular rivets which also hold the brass pins on which the individual magnets are pivoted. The corner of the linkage next to the protractor is maintained in a stationary position while the opposite corner is free to slide in a groove. The mechanical accuracy of this setup permits measurement of the opening angle ϕ of the linkage to within two degrees. This represents a conservative estimate of the cumulative uncertainties due to the play in the joints of the linkage. The corresponding error in the determination of the angular orientation of the individual magnets is not larger than three degrees.

The magnets are precisely machined cylinders of Cunife I (length 0.889 ± 0.002 cm, diameter 0.318 ± 0.004 cm, mass 0.587 ± 0.015 g) with bearing holes accurately matched to the brass support pins to allow rotation with minimal friction. The magnetic conditioning of these magnets and the associated instrumentation is described in detail in References 5, 6, and 7. For the present discussion it is sufficient to note that the dipole moments of the individual magnets are matched to better than 0.4%, and stabilized against fluctuating external fields with intensities up to 30 G. This stability permits a linear superposition of the individual magnetic fields to an accuracy exceeding 0.8%, as verified by extensive experimental checks.[6] In the present series of experiments two levels of magnetization were employed: one corresponding to a uniform dipole moment of 14.5 G-cm^3 for each magnet, the other to 18.4 G-cm^3.

In preliminary tests it was found that the earth's magnetic field and other stray sources produced significant disturbances in the system. Consistent results were obtained only when the experimental setup was placed inside a magnetic shield (see Fig. 1). Although one end of this double-cylinder arrangement was left open to provide access to the linkage, the residual magnetic field at the location of the array was less than 0.02 G, as measured with an RFL Model 1890. Even this very weak field is responsible for detectable variations in certain ranges. These are discussed in detail in Section IV C.

For studies of the relative populations of metastable states it was necessary to provide randomized agitation for the magnets. This was derived from an auxiliary "boiling" array of the type described in Reference 6. For the present application, special precautions were taken to isolate the molybdenum shield from these noise fields.

III. THE FOUR-MAGNET HYSTERESIS CYCLE

A. Observations

With the experimental setup as described in the preceding section, the four-magnet hysteresis cycle was studied in detail. The essential observations are presented in abbreviated form in Figure 2. In going through a complete cycle it is convenient to suppose that the linkage is initially fixed at an opening angle of $\phi = 90°$. For this square lattice we know from previous work[1] that the "neutral" pattern shown on the diagram is the only stable dipole configuration. If we now decrease the angle ϕ—the "hysteresis coordinate"—by physically deforming the linkage, the neutral pattern persists in *unchanged* form down to $\phi \cong 45°$. This sequence is represented by the heavy line indicated on the left-hand side of Figure 2. Obviously this process is completely reversible. Just below this angle ($\phi \lesssim 45°$) the system begins to exhibit the first of a number of remarkable features: As ϕ decreases, the two magnets across the short diagonal of the rhombus deviate from the "neutral" position and swing around, creating a net magnetic moment. A representative intermediate stage is sketched on the figure. This moment arises literally from a breaking of the symmetry of

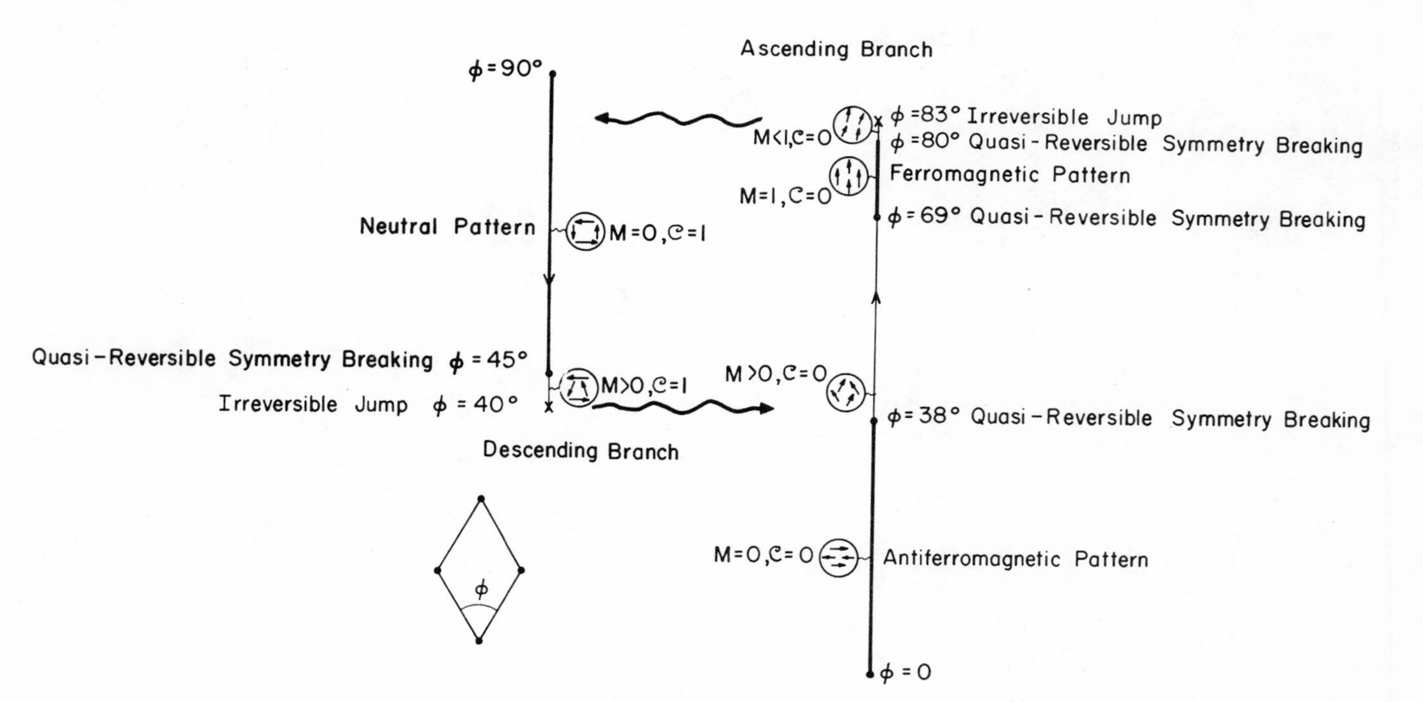

Fig. 2. The four-magnet hysteresis cycle (abbreviated). The heavy lines represent intervals of ϕ, where the patterns remain unchanged. Representative configurations are shown in the circular inserts. Values are also indicated for the magnetization, M (4.5), and circulation, $\mathscr{C}$ (6.1). The wavy arrows indicate the discontinuous jumps which join the branches of the hysteresis loop.

the neutral magnetic configuration. A little reflection will make it clear that there must be a corresponding pattern whose moment points in the opposite direction. The symmetry-breaking transition in the vicinity of $\phi \sim 45°$ must therefore be associated with the appearance of two degenerate patterns. Indeed, a glance ahead to Figure 4 will show the full complexity of the situation. Note that linkage deformations corresponding to upward transitions across $\phi = 45.6° \uparrow$ are associated with continuous and unique variations of the magnet patterns. Downward transitions across $\phi = 45.6° \downarrow$ will be continuous and dichotomic. This type of symmetry-breaking transformation will therefore be called "quasi-reversible." The analytical basis for these branching processes is discussed in Section III B.

If we now decrease the opening angle of the linkage further we observe a sudden drastic change in the system at $\phi \cong 40°$: the magnets twitch out of equilibrium and the pattern breaks into oscillation. This accompanies even the gentlest increments of linkage deformation in the vicinity of $\phi \simeq 40°$. Finally the magnets come to rest in a pattern of a distinctly different type. Under the present experimental conditions, the quiescence time is of the order of two seconds. This transition is indicated on Figure 2 by the wavy arrow directed to the right. In contrast to the symmetry breaking encountered at $\phi \cong 45°$, this transition is completely *irreversible*, since small increases in ϕ, after the twitch, do *not* lead back along the "descending branch" shown on Figure 2. Rather, as indicated on the diagram, the system has undergone a discontinuous jump into another sequence of equilibrium configurations.

The linkage can be deformed still further in the direction of decreasing ϕ. In particular, at $\phi \cong 38°$ we find that the pattern goes over into a perfect antiferromagnetic configuration which persists in *unchanged* form as the opening angle is decreased to the smallest experimentally realizable value of 20°. From analytical extrapolations (Section III B) we know that this is the only stable configuration down to $\phi = 0°$; and therefore no additional complications are expected between $\phi = 20°$ and 0°.

If we now reverse the process and increase the hysteresis coordinate, then in the range $20° \gtrsim \phi \lesssim 38°$, we retrace our steps identically as shown on the "ascending branch" of Figure 2. However, at $\phi \cong 38°$ we encounter the same type of quasi-reversible symmetry breaking as we observed at $\phi \cong 45°$ on the descending branch. At this point the antiferromagnetic configuration begins to deform, and as ϕ is increased further the individual magnets realign themselves into a sequence of patterns possessing a non-vanishing magnetic moment. Just as before, this corresponds to a splitting into two degenerate states. The complete sequence is indicated on Figure 4. It is natural to consider an upward transition across the symmetry-breaking point at 37.9° as a "fork" in the trajectories, and a downward crossing as a "confluence". These terms apply in an obvious sense to all similar situations.

As we continue to increase ϕ the magnets progressively become more aligned, and eventually at $\phi \cong 69°$, we encounter a third confluence which leads to a perfect ferromagnetic configuration. This is shown near the top of the ascending branch on Figure 2. The heavy line ($69° \gtrsim \phi \gtrsim 80°$) emphasizes that during further increases of ϕ the ferromagnetic pattern persists in *unchanged* form. At $\phi \cong 80°$ another point of quasi-reversible symmetry breaking is reached. In the range $80° \gtrsim \phi \gtrsim 83°$ the patterns exhibit a slight decrease from unit magnetization. Finally, at $\phi \cong 83°$ the magnets break into oscillation and the system jumps across to the neutral pattern on the descending branch. This is represented by the wavy arrow directed to the left on Figure 2. As before, one can experimentally demonstrate that this transition is *irreversible*. It should be noted that we have now closed the hysteresis loop, since retracing the path in the direction of increasing ϕ along the remaining portion of the descending branch ($83° \gtrsim \phi \leqslant 90°$) yields nothing new.

In Table I we summarize the experimental information on the location of the characteristic points in the four-magnet hysteresis cycle. For purposes of comparison, we also list the computed values which are derived from the ideal dipole model of Section III B. The agreement is remarkably good in case of the wide-spaced lattice ($a = 5.2$ cm) for both levels of magnetization. This shows that the ideal-dipole approximation is sufficiently accurate to describe all of the essential features of the observed hysteresis. As expected, we find the best agreement for the higher levels of magnetization since the magnetic dipole moments scale the potential energy of the system quadratically, and this diminishes the influence of external disturbances. (Section IV C). The deviations between theoretical predictions and the experimental findings that appear for the close-spaced lattice ($a = 3.2$ cm) can be attributed to the increased influence of octopole forces. This point is also discussed in detail in Section IV C.

Another comparison between theory and experiment that demonstrates the adequacy of the theoretical description is presented in Table II. This summarizes the angular orientations of all the individual magnets according to the conventions given in Figure 3. Representative patterns along the hysteresis cycle are listed together with the theoretically predicted configurations. It is obvious that the agreement is also remarkably good. The deviations nowhere exceed 5°, except for certain patterns poised near the irreversible jumps ($\phi \cong 40°$ and $\phi \cong 83°$). We shall show that in these cases external disturbances have an enhanced significance (Section IV C).

TABLE I

Transition Points Associated with the Four-Magnet Hysteresis Cycle. Comparison of Theory and Experiment. Table Entries are Values of ϕ in Degrees

Transition point in hysteresis cycle[b]	Theory	Experiment[a]		
		$\mu = 14.5$ G-cm^3 $a = 5.2$ cm	$\mu = 18.4$ G-cm^3 $a = 5.2$ cm	$\mu = 14.5$ G-cm^3 $a = 3.2$ cm
Quasi-reversible symmetry breaking (Terminates neutral path)	45.6	45.5	45.5	45.5
Irreversible jump (right arrow)	39.8	40.0	40.5	43.5
Quasi-reversible symmetry breaking (terminates antiferromagnetic path)	37.9	37.5	38.0	41.5
Reversible symmetry breaking (ferromagnetic confluence)	68.7	68.0[c]	68.0[c]	62.5[d]
Reversible symmetry breaking (terminates ferromagnetic path)	80.4	79.0	80.5	68.0
Irreversible jump (left arrow)	83.4	82.0	83.5	71.5

[a] The angle ϕ was measured to within 0.5° in all experiments. This error represents the observational scatter for runs taken in one direction only to minimize backlash of the linkage.

[b] Compare Figures 2 and 4.

[c] The ferromagnetic pattern was set up at this angle; see Section IV C.

[d] At this angle the two magnets across the longer diagonal lined up; see Section IV C.

TABLE II

Representative Patterns of the Four-Magnet Hysteresis Cycle. Comparison of Theory and Experiment[a]

Descending branch

Pattern	ϕ	θ_1 Th	θ_1 Ex	θ_2 Th	θ_2 Ex	θ_3 Th	θ_3 Ex	θ_4 Th	θ_4 Ex
Neutral	90	135	134	45	45	−45	−44	−135	−135
	80	130	129	40	41	−50	−50	−140	−141
	70	125	123	35	35	−55	−56	−145	−144
	60	120	120	30	30	−60	−59	−150	−149
	50	115	113	25	26	−65	−66	−155	−153
	46	113	113	23	24	−67	−68	−157	−155
Symmetry breaking	44	141	132	22	22	−97	−90	−158	−156
	42	153	162	21	15	−111	−116	−159	−142

Ascending branch

Pattern	ϕ	θ_1 Th	θ_1 Ex	θ_2 Th	θ_2 Ex	θ_3 Th	θ_3 Ex	θ_4 Th	θ_4 Ex
Symmetry breaking	82	−65	−78	−53	−60	−65	−74	−53	−56
Ferro-magnetic	80	−50	−54	−50	−49	−50	−51	−50	−51
	70	−55	−58	−55	−56	−55	−57	−55	−56
Symmetry breaking	60	−97	−99	−53	−53	−97	−95	−53	−55
	50	−125	−128	−43	−42	−125	−126	−43	−45
	40	−151	−154	−11	−5	−151	−154	−11	−8
Anti-ferro-magnetic	30	−165	−165	15	15	−165	−165	15	10
	20	−170	−170	10	10	−170	−169	10	6

[a] Experimental values derived from photographs of the $\mu = 18.4$ G-cm^3, $a = 5.2$ cm array. The angles are defined in Figure 3, and measured in degrees. The experimental errors are $\pm 1°$.

B. Theory

In this section we shall work out the theory of the four-magnet hysteresis cycle. The calculations will be based on the assumption that the magnets behave as ideal dipoles; or in a more precise sense that the influence of octopole contributions may be neglected. The comparisons summarized in Tables I and II, together with the results derived in References 1 and 5, give ample justification for this line of approach. The essential quantity of interest is the general expression for the potential energy of the four-magnet system as a function of the angular orientation of the individual magnets and the opening angle ϕ of the underlying lattice. The local minima of this energy function then represent the possible stable configurations corresponding to a particular deformation ϕ of the linkage. We shall see that the points where quasi-reversible symmetry breaking or irreversible jumps occur are explicitly determined by general stability criteria for the relative minima as functions of ϕ.

1. *Energy of a Four-Magnet System* (*Ideal Dipole Case*)

We begin with the general expression for the potential energy of two magnets with moments $\boldsymbol{\mu}_i$ and $\boldsymbol{\mu}_j$, whose centers are joined by a vector $\mathbf{r}_{ij}$,

$$U_{ij} = \frac{\boldsymbol{\mu}_i \cdot \boldsymbol{\mu}_j}{r_{ij}^{\ 3}} - 3\,\frac{(\boldsymbol{\mu}_i \cdot \mathbf{r}_{ij})(\boldsymbol{\mu}_j \cdot \mathbf{r}_{ij})}{r_{ij}^{\ 5}} \tag{3.1}$$

In a four-magnet system there are six distinct combinations of this type which must be summed to give the total energy. In Figure 3 we show an arbitrary configuration of the system which defines our conventions for

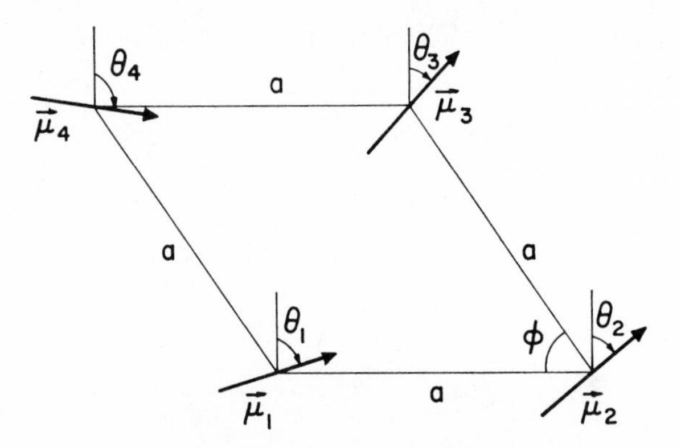

Fig. 3. Arbitrary configuration of the four-magnet system. Note that the angles θ_i are defined with respect to the movable arms of the linkage.

orienting and measuring the four magnet angles θ_i ($i = 1, 2, 3, 4$) and the opening angle ϕ. The lattice constant a corresponds to the separation between the vertices of the linkage. With these definitions, and noting that all the magnetic moments are equal in magnitude, the six contributions to the total potential energy are as follows:

$$U_{12} = -\frac{\mu^2}{a^3}[3\sin\theta_1 \sin\theta_2 - \cos(\theta_1 - \theta_2)] \tag{3.2a}$$

$$U_{13} = -\frac{\mu^2}{8a^3\sin^3\dfrac{\phi}{2}}\left[3\cos\left(\theta_1 - \frac{\phi}{2}\right)\cos\left(\theta_3 - \frac{\phi}{2}\right) - \cos(\theta_1 - \theta_3)\right] \tag{3.2b}$$

$$U_{14} = -\frac{\mu^2}{a^3}[3\sin(\theta_1 - \phi)\sin(\theta_4 - \phi) - \cos(\theta_1 - \theta_4)] \tag{3.2c}$$

$$U_{23} = -\frac{\mu^2}{a^3}[3\sin(\theta_2 - \phi)\sin(\theta_3 - \phi) - \cos(\theta_2 - \theta_3)] \tag{3.2d}$$

$$U_{24} = -\frac{\mu^2}{8a^3\cos^3\dfrac{\phi}{2}}\left[3\sin\left(\theta_2 - \frac{\phi}{2}\right)\sin\left(\theta_4 - \frac{\phi}{2}\right) - \cos(\theta_2 - \theta_4)\right] \tag{3.2e}$$

$$U_{34} = -\frac{\mu^2}{a^3}[3\sin\theta_3 \sin\theta_4 - \cos(\theta_3 - \theta_4)] \tag{3.2f}$$

By summing over these six terms and grouping the results appropriately we obtain a general expression for the potential energy of the four-magnet system:

$$\begin{aligned} U(\theta_1, \theta_2, \theta_3, \theta_4; \phi) = \frac{\mu^2}{2a^3}\Bigg\{ & 3[\cos(\theta_1 + \theta_2) + \cos(\theta_3 + \theta_4) \\ & + \cos(\theta_1 + \theta_4 - 2\phi) + \cos(\theta_2 + \theta_3 - 2\phi)] \\ & - [\cos(\theta_1 - \theta_2) + \cos(\theta_3 - \theta_4) \\ & + \cos(\theta_1 - \theta_4) + \cos(\theta_2 - \theta_3)] \\ & - \frac{1}{8\sin^3\dfrac{\phi}{2}}[3\cos(\theta_1 + \theta_3 - \phi) \\ & + \cos(\theta_1 - \theta_3)] + \frac{1}{8\cos^3\dfrac{\phi}{2}} \\ & \times [3\cos(\theta_2 + \theta_4 - \phi) - \cos(\theta_2 - \theta_4)]\Bigg\} \end{aligned} \tag{3.3}$$

At this point it is convenient to introduce a new set of variables which reflects the natural symmetries of the problem, namely

$$\begin{aligned} \alpha_1 &= \tfrac{1}{2}(\theta_1 - \theta_2 + \theta_3 - \theta_4) & \beta_1 &= \tfrac{1}{2}(\theta_1 + \theta_2 - \theta_3 - \theta_4) \\ \alpha_2 &= \tfrac{1}{2}(\theta_1 - \theta_2 - \theta_3 + \theta_4) & \beta_2 &= \tfrac{1}{2}(\theta_1 + \theta_2 + \theta_3 + \theta_4) \end{aligned} \tag{3.4}$$

In terms of these, the basic expression for the potential energy may be rearranged in the form

$$\begin{aligned} U(\alpha_1, \alpha_2, \beta_1, \beta_2; \phi) = \frac{\mu^2}{a^3}\Bigg\{ & 3 \cos \beta_1 \cos \beta_2 + 3 \cos \alpha_2 \cos (\beta_2 - 2\phi) \\ & - \cos \alpha_1 \cos \alpha_2 - \cos \alpha_1 \cos \beta_1 \\ & - \frac{\cos^3 \dfrac{\phi}{2} - \sin^3 \dfrac{\phi}{2}}{2 \sin^3 \phi} \\ & \times [3 \cos \alpha_1 \cos (\beta_2 - \phi) - \sin \alpha_2 \sin \beta_1] \\ & + \frac{\cos^3 \dfrac{\phi}{2} + \sin^3 \dfrac{\phi}{2}}{2 \sin^3 \phi} \\ & \times [3 \sin \alpha_1 \sin (\beta_2 - \phi) - \cos \alpha_2 \cos \beta_1]\Bigg\} \end{aligned} \tag{3.5}$$

We now have to find all the local minima of this function of four variables —regarding ϕ as a parameter—and trace their evolution as ϕ is varied in the range $0° < \phi \leqslant 90°$. For this purpose we need all the first- and second-order partial derivatives of (3.5). Using the notation $U_{\alpha_1} = \partial U/\partial \alpha_1$, and so forth, these may be written as

$$\begin{aligned} U_{\alpha_1} = \frac{\mu^2}{a^3}\Bigg[& \sin \alpha_1 \cos \alpha_2 + \sin \alpha_1 \cos \beta_1 + 3 \, \frac{\cos^3 \dfrac{\phi}{2} - \sin^3 \dfrac{\phi}{2}}{2 \sin^3 \phi} \\ & \times \sin \alpha_1 \cos (\beta_2 - \phi) + 3 \, \frac{\cos^3 \dfrac{\phi}{2} + \sin^3 \dfrac{\phi}{2}}{2 \sin^3 \phi} \cos \alpha_1 \sin (\beta_2 - \phi) \Bigg] \end{aligned} \tag{3.6a}$$

$$\begin{aligned} U_{\alpha_2} = \frac{\mu^2}{a^3}\Bigg[& -3 \sin \alpha_2 \cos (\beta_2 - 2\phi) + \cos \alpha_1 \sin \alpha_2 + \frac{\cos^3 \dfrac{\phi}{2} - \sin^3 \dfrac{\phi}{2}}{2 \sin^3 \phi} \\ & \times \cos \alpha_2 \sin \beta_1 + \frac{\cos^3 \dfrac{\phi}{2} + \sin^3 \dfrac{\phi}{2}}{2 \sin^3 \phi} \sin \alpha_2 \cos \beta_1 \Bigg] \end{aligned} \tag{3.6b}$$

$$U_{\beta_1} = \frac{\mu^2}{a^3}\left[-3\sin\beta_1\cos\beta_2 + \cos\alpha_1\sin\beta_1 + \frac{\cos^3\frac{\phi}{2} - \sin^3\frac{\phi}{2}}{2\sin^3\phi} \times \sin\alpha_2\cos\beta_1 + \frac{\cos^3\frac{\phi}{2} + \sin^3\frac{\phi}{2}}{2\sin^3\phi}\cos\alpha_2\sin\beta_1\right] \tag{3.6c}$$

$$U_{\beta_2} = \frac{\mu^2}{a^3}\left[-3\cos\beta_1\sin\beta_2 - 3\cos\alpha_2\sin(\beta_2 - 2\phi) + 3\,\frac{\cos^3\frac{\phi}{2} - \sin^3\frac{\phi}{2}}{2\sin^3\phi} \times \cos\alpha_1\sin(\beta_2 - \phi) + 3\,\frac{\cos^3\frac{\phi}{2} - \sin^3\frac{\phi}{2}}{2\sin^3\phi}\sin\alpha_1\cos(\beta_2 - \phi)\right] \tag{3.6d}$$

$$U_{\alpha_1\alpha_1} = \frac{\mu^2}{a^3}\left[\cos\alpha_1\cos\alpha_2 + \cos\alpha_1\cos\beta_1 + 3\,\frac{\cos^3\frac{\phi}{2} - \sin^3\frac{\phi}{2}}{2\sin^3\phi} \times \cos\alpha_1\cos(\beta_2 - \phi) - 3\,\frac{\cos^3\frac{\phi}{2} + \sin^3\frac{\phi}{2}}{2\sin^3\phi}\sin\alpha_1\sin(\beta_2 - \phi)\right] \tag{3.7a}$$

$$U_{\alpha_2\alpha_2} = \frac{\mu^2}{a^3}\left[-3\cos\alpha_2\cos(\beta_2 - 2\phi) + \cos\alpha_1\cos\alpha_2 - \frac{\cos^3\frac{\phi}{2} - \sin^3\frac{\phi}{2}}{2\sin^3\phi} \times \sin\alpha_2\sin\beta_1 + \frac{\cos^3\frac{\phi}{2} + \sin^3\frac{\phi}{2}}{2\sin^3\phi}\cos\alpha_2\cos\beta_1\right] \tag{3.7b}$$

$$U_{\beta_1\beta_1} = \frac{\mu^2}{a^3}\left[-3\cos\beta_1\cos\beta_2 + \cos\alpha_1\cos\beta_1 - \frac{\cos^3\frac{\phi}{2} - \sin^3\frac{\phi}{2}}{2\sin^3\phi} \times \sin\alpha_2\sin\beta_1 + \frac{\cos^3\frac{\phi}{2} + \sin^3\frac{\phi}{2}}{2\sin^3\phi}\cos\alpha_2\cos\beta_1\right] \tag{3.7c}$$

$$U_{\beta_2\beta_2} = \frac{\mu^2}{a^3}\left[-3\cos\beta_1\cos\beta_2 - 3\cos\alpha_2\cos(\beta_2 - 2\phi) + 3\,\frac{\cos^3\frac{\phi}{2} - \sin^3\frac{\phi}{2}}{2\sin^3\phi}\right.$$

$$\left.\times\cos\alpha_1\cos(\beta_2 - \phi) - 3\,\frac{\cos^3\frac{\phi}{2} + \sin^3\frac{\phi}{2}}{2\sin^3\phi}\sin\alpha_1\sin(\beta_2 - \phi)\right] \tag{3.7d}$$

$$U_{\alpha_1\alpha_2} = \frac{\mu^2}{a^3}[-\sin\alpha_1\sin\alpha_2] \tag{3.7e}$$

$$U_{\alpha_1\beta_1} = \frac{\mu^2}{a^3}[-\sin\alpha_1\sin\beta_1] \tag{3.7f}$$

$$U_{\alpha_1\beta_2} = \frac{\mu^2}{a^3}\left[-3\,\frac{\cos^3\frac{\phi}{2} - \sin^3\frac{\phi}{2}}{2\sin^3\phi}\sin\alpha_1\sin(\beta_2 - \phi)\right.$$

$$\left.+ 3\,\frac{\cos^3\frac{\phi}{2} + \sin^3\frac{\phi}{2}}{2\sin^3\phi}\cos\alpha_1\cos(\beta_2 - \phi)\right] \tag{3.7g}$$

$$U_{\alpha_2\beta_1} = \frac{\mu^2}{a^3}\left[\frac{\cos^3\frac{\phi}{2} - \sin^3\frac{\phi}{2}}{2\sin^3\phi}\cos\alpha_2\cos\beta_1\right.$$

$$\left.- \frac{\cos^3\frac{\phi}{2} + \sin^3\frac{\phi}{2}}{2\sin^3\phi}\sin\alpha_2\sin\beta_1\right] \tag{3.7h}$$

$$U_{\alpha_2\beta_2} = \frac{\mu^2}{a^3}[3\sin\alpha_2\sin(\beta_2 - 2\phi)] \tag{3.7i}$$

$$U_{\beta_1\beta_2} = \frac{\mu^2}{a^3}[3\sin\beta_1\sin\beta_2] \tag{3.7j}$$

A necessary condition for the potential energy (3.5) to have an extremum is the simultaneous vanishing of its first-order partial derivatives (3.6a–d). In practice, this is equivalent to locating the (possibly empty) set of intersections of the zeros of the individual equations. The stability criteria which we shall introduce later then may be used to select the actual local minima. In anticipation of this selection procedure we shall discuss only those zeros of (3.6a–d) that finally turn out to correspond to the stable states.

From (3.6b) and (3.6c) we find that one possible simultaneous zero of these two expressions is given by

$$\sin \alpha_2 = \sin \beta_1 = 0 \tag{3.8}$$

Without loss of generality we may choose as the elementary periodicity interval for our variables α_1, α_2, β_1, β_2 the range from $-180°$ to $+180°$. The four possible solutions for α_2 and β_1 resulting from (3.8) can then be grouped into two classes. The two combinations in each class represent the pole-reversed images of each other. It will therefore be sufficient to consider only one of these equivalent sets. For the further work we choose

$$C_1: \quad \alpha_2 = 0°, \quad \beta_1 = 180° \tag{3.9a}$$

$$C_0: \quad \alpha_2 = 0°, \quad \beta_1 = 0° \tag{3.9b}$$

Let us now investigate class C_1: by inserting (3.9a) into the remaining two expressions (3.6a) and (3.6d), and setting the results equal to zero, we get the equations

$$\frac{\cos^3 \dfrac{\phi}{2} - \sin^3 \dfrac{\phi}{2}}{\sin^3 \phi} \sin \alpha_1 \cos (\beta_2 - \phi) + \frac{\cos^3 \dfrac{\phi}{2} + \sin^3 \dfrac{\phi}{2}}{\sin^3 \phi} \cos \alpha_1 \sin (\beta_2 - \phi) = 0 \tag{3.10a}$$

and

$$\frac{\cos^3 \dfrac{\phi}{2} + \sin^3 \dfrac{\phi}{2}}{\sin^3 \phi} \sin \alpha_1 \cos (\beta_2 - \phi) + \frac{\cos^3 \dfrac{\phi}{2} - \sin^3 \dfrac{\phi}{2}}{\sin^3 \phi} \cos \alpha_1 \sin (\beta_2 - \phi) + 4 \sin \phi \cos (\beta_2 - \phi) = 0 \tag{3.10b}$$

These lead to the unique (physical) solution

$$\cos \alpha_1 = \cos (\beta_2 - \phi) = 0 \tag{3.11}$$

Actually (3.11) represents four distinct combinations of α_1 and β_2.

Anticipating the stability criteria of (3.25a–d), we only need to concern ourselves with the set

$$\alpha_1 = 90°, \quad \alpha_2 = 0°, \quad \beta_1 = 180°, \quad \beta_2 = \phi - 90° \tag{3.12}$$

This corresponds to the "neutral" pattern shown on Figure 2. It should be noted that this solution has only a trivial dependence on ϕ and thus will exist as an extremum throughout the entire range $0° < \phi \leqslant 90°$.

Another solution which we shall incorporate into class C_1 can be obtained by starting with (3.11), and inserting it into (3.6a) and (3.6d). This leads to the two equations

$$\cos \alpha_2 + \cos \beta_1 = 0 \tag{3.13a}$$

and

$$3 \cos \phi[\cos \alpha_2 + \cos \beta_1] = 0 \tag{3.13b}$$

These have the obvious solution

$$\cos \beta_1 = -\cos \alpha_2 \tag{3.14}$$

Again anticipating the constraints provided by the stability criteria, we find that (3.14) together with (3.11) fixes two of our four variables as

$$\alpha_1 = 90°, \quad \beta_2 = \phi - 90° \tag{3.15a}$$

and also yields

$$\beta_1 = \alpha_2 - 180° \tag{3.15b}$$

Inserting these solutions into (3.6b) and (3.6c), and setting the resulting expressions equal to zero, leads to the single equation

$$\sin \alpha_2 \left[-3 \sin \phi + \frac{\cos^3 \dfrac{\phi}{2}}{\sin^3 \phi} \cos \alpha_2 \right] = 0 \tag{3.16}$$

Obviously, the solution given by $\sin \alpha_2 = 0$ leads us back to (3.12). The new solution for α_2 obtained by setting the bracket in (3.16) equal to zero is the symmetry-breaking solution

$$\cos \alpha_2 = 24 \sin \phi \sin^3 \frac{\phi}{2} \tag{3.17}$$

This is limited to the range $0° \leqslant \phi \leqslant 45.63°$ in virtue of the condition $|\cos \alpha_2| \leqslant 1$. Exactly at the point $\phi = 45.63°$, the symmetry-breaking solution coincides with the neutral state. From the structure of (3.17) it is apparent that it represents two geometrically distinct patterns ($\alpha_2 \lessgtr 0$) for each value of ϕ. On the other hand, the neutral solution gives rise to a unique pattern throughout the entire (physical) range of ϕ.

In order to locate all the minima associated with the class C_0 we now insert (3.9b) into (3.6a) and (3.6d). Setting the results equal to zero leads to the following conditions:

$$2 \sin \alpha_1 + 3 \frac{\cos^3 \frac{\phi}{2} - \sin^3 \frac{\phi}{2}}{2 \sin^3 \phi} \sin \alpha_1 \cos (\beta_2 - \phi) + 3 \frac{\cos^3 \frac{\phi}{2} + \sin^3 \frac{\phi}{2}}{2 \sin^3 \phi} \cos \alpha_1 \sin (\beta_2 - \phi) = 0 \quad (3.18a)$$

and

$$-6 \cos \phi \sin (\beta_2 - \phi) + 3 \frac{\cos^3 \frac{\phi}{2} + \sin^3 \frac{\phi}{2}}{2 \sin^3 \phi} \sin \alpha_1 \cos (\beta_2 - \phi) + 3 \frac{\cos^3 \frac{\phi}{2} - \sin^3 \frac{\phi}{2}}{2 \sin^3 \phi} \cos \alpha_1 \sin (\beta_2 - \phi) = 0 \quad (3.18b)$$

By taking appropriate linear combinations of these equations we get

$$\cos \alpha_1 \sin (\beta_2 - \phi) = -\frac{8}{3} \left(\cos^3 \frac{\phi}{2} + \sin^3 \frac{\phi}{2} \right) \sin \alpha_1 - 8 \left(\cos^3 \frac{\phi}{2} - \sin^3 \frac{\phi}{2} \right) \cos \phi \sin (\beta_2 - \phi) \quad (3.19a)$$

and

$$\sin \alpha_1 \cos (\beta_2 - \phi) = \frac{8}{3} \left(\cos^3 \frac{\phi}{2} - \sin^3 \frac{\phi}{2} \right) \sin \alpha_1 + 8 \left(\cos^3 \frac{\phi}{2} + \sin^3 \frac{\phi}{2} \right) \cos \phi \sin (\beta_2 - \phi) \quad (3.19b)$$

It is then convenient to square these expressions. Subtracting the results and simplifying we get the relation

$$\sin (\beta_2 - \phi) = -c \sin \alpha_1 \quad (3.20a)$$

where

$$c = \left[\frac{1 + \frac{32}{9} \sin^3 \phi}{1 + 32 \sin^3 \phi \cos^2 \phi} \right]^{1/2} \quad (3.20b)$$

The negative sign in (3.20a) has been selected in the light of the stability criteria. Inserting (3.20) into (3.19) we obtain as our final equations

$$\sin \alpha_1 \left[c \cos \alpha_1 - \frac{8}{3}\left(\cos^3 \frac{\phi}{2} + \sin^3 \frac{\phi}{2}\right) + 8c\left(\cos^3 \frac{\phi}{2} - \sin^3 \frac{\phi}{2}\right) \cos \phi \right] = 0 \tag{3.21a}$$

and

$$\sin \alpha_1 \left[\cos (\beta_2 - \phi) - \frac{8}{3}\left(\cos^3 \frac{\phi}{2} - \sin^3 \frac{\phi}{2}\right) + 8c\left(\cos^3 \frac{\phi}{2} + \sin^3 \frac{\phi}{2}\right) \cos \phi \right] = 0 \tag{3.21b}$$

The solutions are either

$$\sin \alpha_1 = 0 \tag{3.22a}$$

or

$$\cos \alpha_1 = \frac{8}{3c}\left[\cos^3 \frac{\phi}{2} + \sin^3 \frac{\phi}{2} - 3c\left(\cos^3 \frac{\phi}{2} - \sin^3 \frac{\phi}{2}\right) \cos \phi \right] \tag{3.22b}$$

$$\cos (\beta_2 - \phi) = \frac{8}{3}\left[\cos^3 \frac{\phi}{2} - \sin^3 \frac{\phi}{2} - 3c\left(\cos^3 \frac{\phi}{2} + \sin^3 \frac{\phi}{2}\right) \cos \phi \right] \tag{3.22c}$$

From the condition that the absolute values of the cosines cannot exceed unity, we find that the solutions of (3.22b) and (3.22c) can exist only in the intervals $37.94° \leqslant \phi \leqslant 68.72°$ and $80.37° \leqslant \phi \leqslant 90°$. It is obvious that the other solution (3.22a) is independent of ϕ. The entire class C_0 will then comprise three types of minima: the first two are obtained by combining (3.22a), (3.20a), and (3.9b). Specifically, the antiferromagnetic configuration

$$\alpha_1 = 180°, \quad \alpha_2 = 0°, \quad \beta_1 = 0°, \quad \beta_2 = \phi - 180° \tag{3.23a}$$

which is indicated for the range $0° < \phi \gtrsim 38°$ on Figure 2; and the ferromagnetic configuration

$$\alpha_1 = 0°, \quad \alpha_2 = 0°, \quad \beta_1 = 0°, \quad \beta_2 = \phi - 180° \tag{3.23b}$$

which is shown for $69° \gtrsim \phi \gtrsim 80°$. The third combination (3.22b), (3.22c), (3.20), and (3.9b) completely specifies the symmetry-breaking solution which leads from the antiferromagnetic to the ferromagnetic state for $38° \gtrsim \phi \gtrsim 69°$, and extends beyond the ferromagnetic state in the range $80° \gtrsim \phi \gtrsim 83°$. This entire sequence constitutes the ascending branch shown on Figure 2. As before, a close examination of the symmetry-breaking solution shows that there are actually two geometrically distinct patterns for each value of ϕ. The complete sequence of solutions is shown in detail on Figure 4. See also Table VII.

2. *Stability Criteria*

We shall now introduce the stability criterion which can be used to select the local minima out of the extremal set determined by the zeros of (3.6a–d). A central role in this is played by the Hessian which is a symmetric, square matrix whose elements are the second-order partial derivatives of the potential energy (3.5), that is,

$$\mathscr{H} = (\partial_{ij}{}^2 U) \tag{3.24}$$

A sufficient condition for an extremal point of U to be a local minimum is given by the following general theorem.

Theorem.[8] Suppose that the n-place function U has continuous second-order partial derivatives on an open set in E_n, and let $\vec{\alpha}_0$ be a point of this set for which $\partial_1\ U(\vec{\alpha}_0) = \cdots = \partial_n\ U(\vec{\alpha}_0) = 0$. Assume that the associated Hessian $\mathscr{H}(\vec{\alpha}_0)$ is a nonsingular matrix. Let $D_n \equiv \det \mathscr{H}(\vec{\alpha}_0)$, and D_{n-k} be the determinant obtained from D_n by deleting the last k rows and columns. Then a necessary and sufficient condition that U have a local minimum at $\vec{\alpha}_0$ is that each of the sequence of numbers $\{D_{n-k}\}_{k=0}^{n-1}$ is positive.

In virtue of this theorem the stability conditions for the four-magnet system can then be expressed in terms of the following set of inequalities:

$$D_1 = U_{\alpha_1\alpha_1} > 0 \tag{3.25a}$$

$$D_2 = U_{\alpha_1\alpha_1}U_{\beta_2\beta_2} - U^2_{\alpha_1\beta_2} > 0 \tag{3.25b}$$

$$\begin{aligned} D_3 = {} & U_{\alpha_1\alpha_1}U_{\beta_2\beta_2}U_{\alpha_2\alpha_2} + 2U_{\alpha_1\alpha_2}U_{\alpha_1\beta_2}U_{\alpha_2\beta_2} - U_{\alpha_1\alpha_1}U^2_{\alpha_2\beta_2} \\ & - U_{\beta_2\beta_2}U^2_{\alpha_1\alpha_2} - U_{\alpha_2\alpha_2}U^2_{\alpha_1\beta_2} > 0 \end{aligned} \tag{3.25c}$$

$$\begin{aligned} D_4 = {} & U_{\alpha_1\alpha_1}U_{\beta_2\beta_2}U_{\alpha_2\alpha_2}U_{\beta_1\beta_1} - U_{\alpha_1\alpha_1}U_{\beta_2\beta_2}U^2_{\alpha_2\beta_1} - U_{\alpha_1\alpha_1}U_{\alpha_2\alpha_2}U^2_{\beta_1\beta_2} \\ & - U_{\alpha_1\alpha_1}U_{\beta_1\beta_1}U^2_{\alpha_2\beta_2} - U_{\beta_2\beta_2}U_{\alpha_2\alpha_2}U^2_{\alpha_1\beta_1} - U_{\beta_2\beta_2}U_{\beta_1\beta_1}U^2_{\alpha_1\alpha_2} \\ & - U_{\alpha_2\alpha_2}U_{\beta_1\beta_1}U^2_{\alpha_1\beta_2} + U^2_{\alpha_1\beta_2}U^2_{\alpha_2\beta_1} + U^2_{\alpha_1\alpha_2}U^2_{\beta_1\beta_2} + U^2_{\alpha_1\beta_1}U^2_{\alpha_2\beta_2} \\ & + 2U_{\alpha_1\alpha_1}U_{\alpha_2\beta_2}U_{\alpha_2\beta_1}U_{\beta_1\beta_2} + 2U_{\beta_2\beta_2}U_{\alpha_1\alpha_2}U_{\alpha_1\beta_1}U_{\alpha_2\beta_1} \\ & + 2U_{\alpha_2\alpha_2}U_{\alpha_1\beta_2}U_{\alpha_1\beta_1}U_{\beta_1\beta_2} + 2U_{\beta_1\beta_1}U_{\alpha_1\alpha_2}U_{\alpha_1\beta_2}U_{\alpha_2\beta_2} \\ & - 2U_{\alpha_1\alpha_2}U_{\alpha_1\beta_2}U_{\alpha_2\beta_1}U_{\beta_1\beta_2} - 2U_{\alpha_1\alpha_2}U_{\alpha_1\beta_1}U_{\alpha_2\beta_2}U_{\beta_1\beta_2} \\ & - 2U_{\alpha_1\beta_2}U_{\alpha_1\beta_1}U_{\alpha_2\beta_2}U_{\alpha_2\beta_1} > 0 \end{aligned} \tag{3.25d}$$

The second-order partial derivatives are given explicitly in (3.7a–j). There is some freedom of choice in these expressions since D_4 is invariant under interchanges of two rows and columns in the Hessian. The conventions shown in (3.25a–c) have the virtue of manipulative simplicity. For our purposes, we need not bother to write out these general expressions since we shall only be interested in the evolution of the local minima. Therefore we study $D_1 - D_4$ for the special values of α_1, α_2, β_1, β_2 which are given by (3.12), (3.15), and (3.17) for class C_1; and (3.9b), (3.20), (3.22b and c), and (3.23a and b) for class C_0. Since these are generally functions of ϕ, the set of inequalities (3.25a–d) will then result in additional constraints on the range of stability of the solutions.

3. *Results*

The detailed investigation of the ranges of ϕ where the various extremal solutions represent stable local minima is described in Appendix A. The essential results are as follows: minima of the class C_1 are restricted to the interval $90° \geqslant \phi > 39.84°$; this in turn is subdivided into a region $90° \geqslant \phi > 45.63°$ where we obtain the neutral pattern (3.12), and an adjoining region $45.63° > \phi > 39.84°$ where the symmetry-breaking solution (3.15), (3.17) is the minimum configuration. It is interesting to note that for $\phi > 45.63°$ the symmetry-breaking state not only ceases to be a stable minimum, but the extremal solution itself is extinguished.

On the other hand, minima of class C_0 are restricted to the interval $0° < \phi < 83.34°$ which is also subdivided into several regions: $0° < \phi < 37.94°$—antiferromagnetic configuration (3.23a); $37.94° < \phi < 68.72°$—symmetry-breaking state (3.9b), (3.20), (3.22b and c); $68.72° < \phi < 80.37°$—ferromagnetic configuration (3.23b); and finally, $80.37° < \phi < 83.34°$—symmetry-breaking state (3.9b), (3.20), (3.22b and c). This sequence corresponds to the ascending branch indicated on Figure 2. It should be noted that the symmetry-breaking extremum exists in the entire range $80.37° \leqslant \phi \leqslant 90°$, but the stability criterion limits the stable minimum to the interval $80.37° < \phi < 83.34°$. On the other hand, in the range $37.94° \leqslant \phi \leqslant 68.72°$ the existence of the extremum in the closed interval is accompanied by the persistence of stability in the open interval. See also Table VII.

A complete inventory of the various stable states of the four-magnet system, including also the pole-reversed patterns, is given in Figure 4. The heavy lines represent regions of ϕ where certain configurations, such as the neutral and ferro- and antiferromagnetic, preserve an unchanged aspect as the linkage is deformed. With each of the symmetry-breaking solutions of both classes (C_0, C_1) there are associated four geometrically distinct patterns. In contrast there are only two different configurations

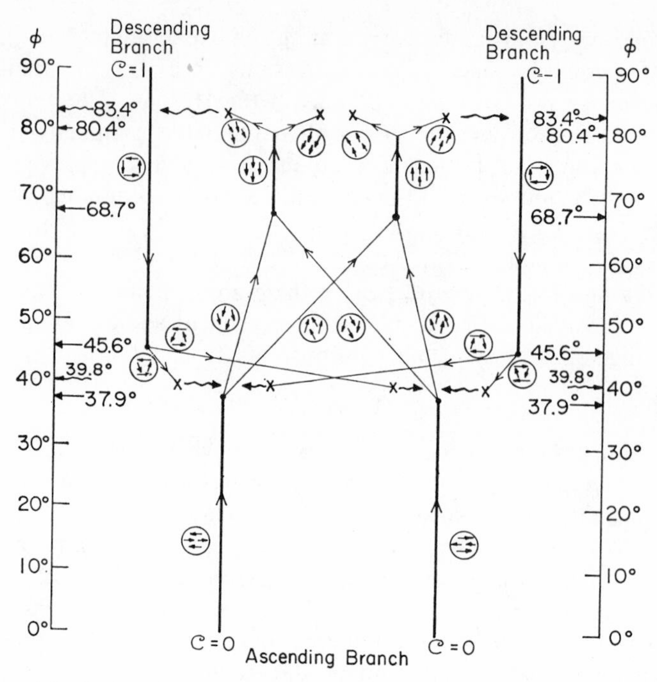

Fig. 4. Detailed map of the four-magnet hysteresis cycle. The heavy lines indicate regions of invariant pattern aspect. The wavy arrows denote the irreversible jump discontinuities. Representative patterns are shown in the circular inserts. Note that there is a continuous transition between the two pole-reversed antiferromagnetic patterns via the forks at 68.7°.

corresponding to each of the neutral ferro- and antiferromagnetic solutions. Since the symmetry-breaking patterns simultaneously emerge from the states of higher symmetry, as shown on Figure 4, the junction points have the character of forks, or confluences, depending on the direction of variation of ϕ. These continuous transitions are in marked contrast to the irreversible jumps which will concern us in the subsequent discussion. The patterns sketched on Figure 4 are supplemented in Table II by a list of representative patterns encountered during a hysteresis cycle.

IV. DISCUSSION

The results of the preceding section show that the four-magnet linkage is characterized by two types of instability points corresponding to singularities of the Hessian (3.24). These are irreversible jumps at $\phi = 39.84°$ and 83.34°; and quasi-reversible transitions at $\phi = 45.63°$, 37.94°, 68.72°, and 80.37°. Both of these are associated with changes in the symmetry aspects of the magnet patterns. These symmetries, however, do not have a simple correspondence to the transformation properties of the Hamiltonian, that is, the potential energy U of (3.3) or (3.5). This is plausible if we recall that the minimum solutions are obtained from the intersection of zeros of partial derivatives of U, which in general do not share the symmetry properties of U. The stability inequalities (3.25a–d) include expressions which are quadrilinear in the second partials of U. The connection of their symmetries with the transformation properties of the Hamiltonian is even more remote. This has the practical consequence that the stability conditions are able to redirect the system from one type of symmetry to another. There is a deep analogy here with well-known arguments concerning the relation of phase transitions and symmetry-breaking interactions in macroscopic systems.[9] In the four-magnet system this connection can be elucidated completely since we have an explicit analytical description of all states. In particular, we shall now explain the origin of the irreversible jumps which give rise to the hysteresis cycle.

A. Irreversible Instabilities

For the purpose of the discussion it will be convenient to introduce the notion of an "extended" energy surface: from the basic expression (3.3) for the potential energy we may in principle compute the general energy surface of the system which spans the four-dimensional configuration ("phase") space (θ_1, θ_2, θ_3, θ_4) for *each* value of the parameter ϕ. By joining a sequence of these surfaces corresponding to neighboring values of ϕ we obtain a higher-dimensional manifold which obviously has the character of an energy surface "extended" in the ϕ "direction." In order to visualize this process let us suppose that the four coordinates θ_i are collapsed into one "configuration" coordinate, as shown in Figure 5. The extended energy surface then lies in a three-dimensional space where the axes correspond to the "configuration" coordinate (θ_i), the "hysteresis" coordinate (ϕ), and the energy (U). The result is shown in Figure 5. The solid and dashed lines indicated on the drawing represent the entire set of stable states of class C_1 and C_0 respectively. The projection of these on the $\phi - \theta_i$ plane is a replica of Figure 2. For the sake of simplicity we have omitted the complications arising from the existence of the symmetry-breaking forks. In order to help visualize the situation, we have also shown the projection of the hysteresis trajectory in the $U - \phi$ plane. It is immediately clear why the relation $U = U(\phi)$ is not single valued. The precise relations can easily be calculated from the general expression (3.5). For the minima of class C_1 we obtain

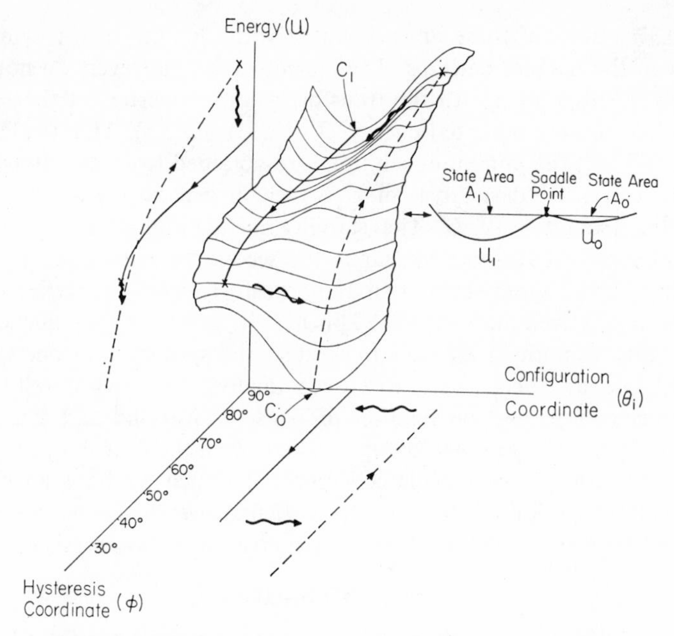

Fig. 5. Extended-energy surface. The configuration coordinate represents the angular orientation of the individual magnets. The hysteresis coordinate measures the linkage deformation (compare Fig. 3). The sections parallel to the $U - \theta_i$ plane are energy surfaces. The projection in the $\phi - \theta_i$ plane corresponds to the hysteresis cycle shown on Figure 2. The solid and dashed lines indicate two branches of the hysteresis loop; the wavy arrows show the irreversible jumps. One of the energy surfaces is excerpted and shown in detail. (Compare the state-area discussion, Section V.)

$$U_1 = -\frac{\mu^2}{a^3 \sin^3 \phi}\left[6 \sin^4 \phi + \cos^3 \frac{\phi}{2} + \sin^3 \frac{\phi}{2}\right]$$
$$90° \geqslant \phi \geqslant 45.63 \qquad (4.1a)$$

and

$$U_1 = -\frac{\mu^2}{a^3 \sin^3 \phi}\left[72 \sin^5 \phi \sin^3 \frac{\phi}{2} + 2\cos^3 \frac{\phi}{2} + \sin^3 \frac{\phi}{2}\right],$$
$$45.63° \geqslant \phi \geqslant 39.84° \qquad (4.1b)$$

The corresponding expressions for the minima of class C_0 are

$$U_0 = -\frac{\mu^2}{a^3 \sin^3 \phi}\left[2 \sin^3 \phi(3 \cos \phi - 1) + 2\cos^3 \frac{\phi}{2} - \sin^3 \frac{\phi}{2}\right],$$
$$0° < \phi \leqslant 37.94° \qquad (4.2a)$$

$$U_0 = -\frac{\mu^2}{a^3 \sin^3 \phi}\left\{\frac{1}{2}\left(\cos^3 \frac{\phi}{2} + \sin^3 \frac{\phi}{2}\right)\right.$$
$$\times [1 + \sqrt{(1 + 32 \sin^3 \phi \cos^2 \phi)(9 + 32 \sin^3 \phi)}]$$
$$\left. - 16 \sin^3 \phi \cos \phi \left(\cos^3 \frac{\phi}{2} - \sin^3 \frac{\phi}{2}\right)\right\},$$
$$\begin{cases} 37.94° \leqslant \phi \leqslant 68.72° \\ \text{and} \\ 80.37° \leqslant \phi \leqslant 83.34° \end{cases} \quad (4.2b)$$

$$U_0 = -\frac{\mu^2}{a^3 \sin^3 \phi}\left[2 \sin^3 \phi(3 \cos \phi + 1) + 2 \sin^3 \frac{\phi}{2} - \cos^3 \frac{\phi}{2}\right],$$
$$68.72° \leqslant \phi \leqslant 80.37° \quad (4.2c)$$

Obviously two local minima in the region of coexistence must be separated by an obstruction, for example, a saddle point, on the general extended energy surface. In the simplified version shown on Figure 5 this obstruction must obviously have the form of a relative maximum. It will appear that this saddle point progressively shifts its position in θ_i from coincidence with the class C_1 minimum at $\phi \cong 40°$, to coincidence with the class C_0 minimum at $\phi \cong 83°$. The intermediate variation of this ridge is depicted on Figure 5. The results have been inferred from Section V, and information provided by the second-order derivatives of U.

With this picture of the extended energy surface in mind it is finally easy to understand how the hysteresis cycle of the four-magnet system arises. As the hysteresis coordinate ϕ is varied, the system is impelled to move along the bottom of a trough on the extended energy surface. This trough becomes increasingly shallower and narrower as we approach one of the singular points associated with an irreversible jump. Precisely at the singular point the trough terminates in a saddle point. From a physical point of view, this means that the system abruptly becomes unstable. It is then easy to see that it must "roll" down the (extended) energy surface into a valley of (relative) stability. This rolling corresponds exactly to the observed twitch of the magnet patterns. Physically this is brought about by small disturbances such as vibration in the magnet bearings, and fluctuations in the ambient fields. It is clear that the influence of these perturbations will be most important in distorting the flat regions of the energy surface which characterize the neighborhood of instability points. In our experimental setup these minute noise sources roll the system over the edge with less than a one-degree variation of ϕ.

During the jump the individual magnets acquire a maximum kinetic energy equal to the difference in potential energy between the initial and terminal states of the "roll". It is observed that they eventually come to rest in one of the stable configurations available at this value of ϕ. The damping time is of the order of one to two seconds. It is known from previous studies that the magnets dissipate this kinetic energy principally by means of mutual magnetic work and not in pivot friction.[6] Thus the macroscopic jump of the system is accompanied by an agitation of the microstructure of the individual magnets. In principle, therefore, the hysteresis jump ultimately results in an irreversible warming of the system.

Once the array has settled in a new equilibrium state, further variation of the hysteresis coordinate ϕ will impel it along the bottom of another stability valley. As Figures 2, 4, and 5 show, additional lattice deformations may then lead it to a second encounter with an irreversible instability. In the present instance the subsequent jump will bring the system back into the initial stability valley, and thereby close the hysteresis loop.

These irreversible transitions between two configurations of different symmetry' of course represent discontinuities in the potential energy of the system. The exact energy lost per cycle in these irreversible transitions can be computed from the expressions for the respective potential energies (4.1) and (4.2):

$$\begin{aligned} \Delta U_c &= |U_1(\phi = 39.84^\circ) - U_0(\phi = 39.84^\circ)| \\ &\quad + |U_1(\phi = 83.34^\circ) - U_0(\phi = 83.34^\circ)| \\ &= (1.17 + 3.80)\frac{\mu^2}{a^3} = 4.97\frac{\mu^2}{a^3} \end{aligned} \tag{4.3}$$

The various experimental conditions listed in Table I then yield the specific values

$$\Delta U_c(\mu = 14.5 \text{ G-cm}^3, a = 5.2 \text{ cm}) = 7.5 \text{ erg} \tag{4.4a}$$

$$\Delta U_c(\mu = 18.4 \text{ G-cm}^3, a = 5.2 \text{ cm}) = 11.2 \text{ erg} \tag{4.4b}$$

$$\Delta U_c(\mu = 14.5 \text{ G-cm}^3, a = 3.2 \text{ cm}) = 32.0 \text{ erg} \tag{4.4c}$$

From this discussion it is evident that the hysteresis cycle will show a definite sense of direction (time's arrow!). This is due to the fact that the problem is asymmetric with respect to the boundary conditions; that is, if we increase ϕ from 0° to 90° the system does not follow the time-reversed path of the evolution beginning at $\phi = 90°$ and proceeding to 0°. In this sense the origin of hysteresis in the four-magnet system can be traced to the existence of complementary instability points that cause irreversible transitions between different lines of evolution.

B. Quasi-Reversible Symmetry Breaking

We have seen that in addition to irreversible jump discontinuities, the four-magnet system also has instability points that are associated with the onset of continuous symmetry-breaking solutions at $\phi = 45.63°$, 37.94°, 68.72°, and 80.37°. As shown in Figure 4, a fork originates at each of these points. It is observed that the system crosses these instabilities continuously without any visible agitation of the magnets. Obviously this corresponds to a smooth variation of the energy during the crossing. Analytically we can verify the continuity in a precise mathematical sense by evaluating the energy expressions (4.1a) and (4.1b) and showing that they coincide exactly at $\phi = 45.63°$. The same arguments can be carried through for (4.2a) and (4.2b) at $\phi = 37.94°$, and (4.2b) and (4.2c) at $\phi = 68.72°$ and 80.37°. Although these transitions are perfectly smooth with respect to energy variations it is nevertheless appropriate to characterize them as being quasi-reversible since crossings are uniquely determined in *one* direction only! Specifically, if the linkage is deformed across this type of instability point there is a unique transition in the direction of a confluence; a deformation in the opposite sense will lead to a dichotomic ambiguity at the fork. Thus more information is required to distinguish on which branch of the fork the system is located, for example, the direction of the magnetic moments of the respective patterns. This discontinuity in the amount of information may be associated with the Kolmogorov interpretation of entropy jumps.[10] However the proper definition of entropy in a wider sense for systems exhibiting hysteresis is still an open question.[35]

These arguments can be extended by considering the magnetization which characterizes the patterns of the array. As usual we define this in terms of the magnetic moments of the individual magnets ($\boldsymbol{\mu}_i$) by the expression:

$$M = \frac{|\sum_{i=1}^{4} \boldsymbol{\mu}_i|}{\sum_{i=1}^{4} |\boldsymbol{\mu}_i|} \tag{4.5}$$

It is then a straightforward matter to compute the magnetization for all patterns of the hysteresis cycle. As in the analogous macroscopic case, it is useful to display the results on a two-dimensional graph. One axis will then correspond to the magnetization, the other to the hysteresis coordinate, that is, the linkage deformation angle ϕ. The results are summarized in Table III and on Figure 6a. We see that at the quasi-reversible transitions the magnetization is continuous but the slope of the magnetization (susceptibility as a function of ϕ) is discontinuous. In this sense the quasi-reversible junctions correspond to phase transitions of the second order.

TABLE III

Variation of Magnetization and Energy During Hysteresis Cycle[a]

Descending branch (C_1)				Ascending branch (C_0)			
Pattern	ϕ (degrees)	M	$-U$ (ergs)	Pattern	ϕ (degrees)	M	$-U$ (ergs)
	90	0	10.06	Symmetry breaking	83.35	0.988	4.32
	80	0	9.99		80.37	1	4.65
	70	0	9.79	Ferromagnetic	70	1	5.77
Neutral	60	0	9.58		68.72	1	5.89
	50	0	9.63		60	0.930	6.75
	45.63	0	9.89	Symmetry breaking	50	0.753	8.48
	45	0.155	9.96		40	0.339	13.07
Symmetry breaking	43	0.296	10.32		37.94	0	14.79
	41	0.368	10.96	Antiferromagnetic	30	0	26.22
	39.84	0.398	11.46		20	0	76.88

[a] Values computed for an array with $\mu = 14.5$ G-cm^3, $a = 5.2$ cm ("wide" spacing).

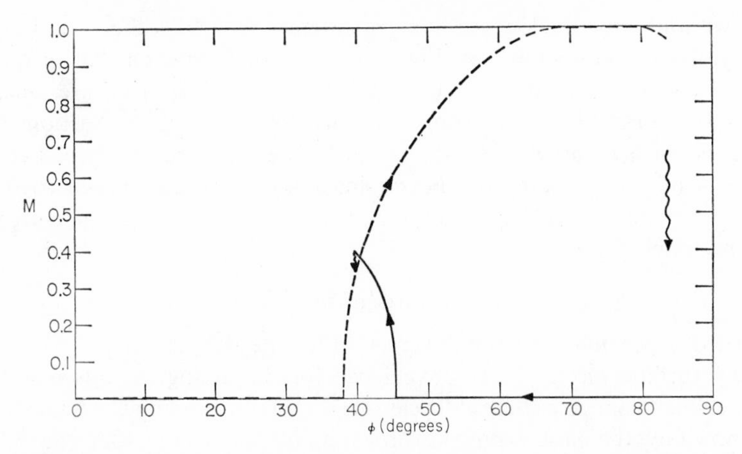

Fig. 6a. Variation of magnetization for hysteresis cycle. The magnetization (4.5) of the two classes of minima is shown as a function of ϕ. The solid line represents class C_1, the dashed line represents class C_0 [compare (3.9a) and (3.9b)]. The slopes are discontinuous at the symmetry-breaking transitions $\phi = 37.9°, 39.8°, 45.6°, 68.7°, 80.4°, 83.4°$.

On the other hand, at the irreversible jumps the magnetization, susceptibility, energy, "entropy", and configuration coordinates are all discontinuous. This suggests an analogy to a first- or "zero"-order phase transition.

The simple loop shown on Figure 2 suggests that it is in principle possible to construct a hysteresis cycle including at least two jump discontinuities

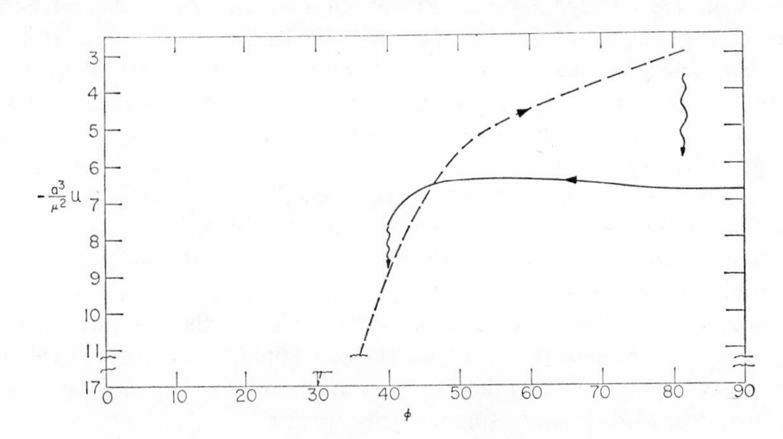

Fig. 6b. Variation of energy during hysteresis cycle. Note the cross-over at $\phi \simeq 46°$ [compare (4.1) and (4.2)].

but without any quasi-reversible transitions. On the other hand, the unabridged version shown on Figure 4 raises the suspicion that in more complex systems there may be loops all of whose vertices are quasi-reversible forks. These questions are discussed further in Section VI. Here we wish to draw attention to the interesting circumstance that in the four-magnet cycle both types of instabilities are interwoven so that quasi-reversible transitions always appear as the immediate precursors of irreversible jumps.

C. Qualitative Features of the Energy Surface

In the preceding discussions we saw that the concept of an extended energy surface was a great convenience for explaining various features of the four-magnet hysteresis cycle. It is therefore extremely important to know how the basic expressions for the potential energy (3.3) [or (3.5)] are altered by the influence of disturbances which may be present in the experimental arrangement. From the comparisons between theory and experiment which have already been given in Tables I and II we know that in any case these perturbations will be small. Nevertheless, we shall discuss these in more detail.

First of all we recall that the magnets are actually not ideal dipoles, as assumed in the derivation of (3.3), but have more complex fields which engender octopole admixtures in higher-order multipole expansions. This problem has been studied in detail for $\phi = 90°$ (square tessellation[1]), and $\phi = 60°$ (triangular tessellation[4]) with the following results: the octopole contribution to the potential energy is scaled by the factor $\varepsilon = d^2/2a^2$, where $2d$ is effectively the physical length of a magnet ($2d = 0.89$ cm), and a is the lattice constant. For all values of $\varepsilon \gtrsim 0.01$ the octopole admixtures can only affect the shape of the higher regions of the energy surface, for example, the location of saddle points and maxima. In the present instance we find that for the wide-spaced linkage ($a = 5.2$ cm), the octopole contributions are certainly negligible since $\varepsilon \simeq 0.004$. In the case of the close-spaced linkage ($a = 3.2$ cm) the octopole coupling constant increases to $\varepsilon \cong 0.01$, and therefore we may expect detectable alterations in the upper portions of the energy surface. From Figure 6b, and the corresponding projection on the $U - \phi$ plane of Figure 5, it is clear that this sensitive "highland" is precisely the ferromagnetic portion of the hysteresis loop. This is consistent with the results displayed in the last column of Table I for $\phi \lesssim 60°$. Further corroboration of this point is provided by the results of the state-area investigation (Section V).

The mechanical inaccuracies of the experimental setup are additional sources of error: first of all there is an irregularity and scatter in the lengths of the sides of the deformable rhombus shown on Figure 3. However these geometrical variations are at most of the order of 1% and produce negligible effects. Great care has also been taken in leveling the array and plexiglass board (Fig. 1) to better than two degrees, thereby minimizing the bias of gravity. In addition one can argue that for the deformations involved in the hysteresis cycle the center of gravity of the array remains unchanged to first order with respect to the plane of the linkage. Therefore a tilt of this plane could affect the interactions of the magnets only through a possible enhancement of friction in the magnet supports, and also change the field patterns. It can, however, be verified explicitly that even if the level of the array is tilted by as much as two degrees (see above) this last effect is less than one percent.

A thorough study of the problems associated with friction in the pivots has been carried out in Reference 6. The essential result is that the coefficient of sliding friction is of the order of 0.1, and therefore the magnets will respond to torques as small as 0.1 dyne-cm or field gradients of 0.01 G/cm. This ensures that the magnets will in fact trace out the calculated minimum trajectories on the energy surface and not be hung up by friction. However, near the instability points the energy surface becomes flat, the corresponding field gradients dwindle, and we might expect to see a "supercooling" effect; that is, friction retains the magnets in a pattern which should already have become unstable. Observationally, this amounts to less than a one-degree scatter in the values of ϕ associated with the various instability points.

Finally we recall that the shielding arrangement shown on Figure 1 is vulnerable to penetration by external magnetic fields. These contribute an additional potential energy to (3.3) which is given by

$$\Delta U_H = \sum_{i=1}^{4} (\boldsymbol{\mu}_i \cdot \mathbf{H}) \tag{4.6}$$

where H is the residual field inside the shield. The magnitude of this energy perturbation is at most of the order of 1 erg for our experimental values, that is, $\mu = 14.5$ G-cm^3 and $H \cong 0.02$ G, compared with the potential energy of about 6 ergs that prevails at the higher portions of the hysteresis cycle (see Table III). From this it is clear that magnetic shielding is an absolute necessity, since the unattenuated earth's field would produce effects larger by a factor of about 25, and dominate the magnet interactions in this region.

V. STATE-AREA AND HYSTERESIS

A. Introduction

At any fixed value of the hysteresis coordinate ϕ the four-magnet system may be animated to produce phase transitions. Experimentally this is arranged by agitating the array with external magnetic noise fields produced by "boiling" coils underneath the linkage (Fig. 1). When these perturbations are diminished, the system "congeals" into one of the local minima of the energy surface. Since metastable transitions are involved, the relative probabilities for condensing into these states cannot be computed within the framework of equilibrium statical mechanics. It has been shown previously that it is useful to regard these transitions as branching processes in phase space.[1] The quiescence of the system then corresponds to a contraction of the phase space image to the energy (hyper) surface, whose topography completely determines the probability for populating any particular final state. In Figure 5 the excerpted section shows the energy surface at a fixed value of ϕ. According to the state-area principle developed in Reference 1 the relative probabilities p_1 and p_0 for populating the states C_1 and C_0 are simply proportional to the (hyper) areas A_1 and A_0, that is,

$$\frac{p_1}{p_0} = \frac{A_1}{A_0} \tag{5.1}$$

The continuously deformable four-magnet system is ideally suited for testing this principle since each of the relative probabilities must vary through the entire range from 0% to 100% as the linkage angle is deformed from 0° to 90°. Occasionally these metastable condensations are also described by a Boltzmann-type relation such as

$$\frac{p_1}{p_0} = \exp\{-(U_1 - U_0)/kT_f\} \tag{5.2}$$

where U_1 and U_0 are the respective energies (Fig. 5), and T_f, the "fictive" temperature, is an adjustable parameter.[11] Since we know U_1 and U_0 exactly as functions of ϕ, in virtue of (4.1) and (4.2), we can check the Boltzmann approach (5.2) in a parameter-independent way by plotting the logarithm of the experimental probability ratios versus the potential energy differences.

B. Results

The experimental findings are summarized in Table IV and Figure 7. An outstanding feature of these results is the rapidity of the variation of the relative populations in the interval $45° \gtrsim \phi \gtrsim 55°$. This is in marked

TABLE IV

Population Probabilities for Hysteresis Cycle[a]

ϕ (degrees)	Descending branch (C_1)		Ascending branch (C_0)		p_1/p_0		$-\frac{a^3}{\mu^2}(U_1-U_0)$[b]
	$a = 5.2$ cm ("wide")	$a = 3.2$ cm ("close")	$a = 5.2$ cm ("wide")	$a = 3.2$ cm ("close")	$a = 5.2$ cm ("wide")	$a = 3.2$ cm ("close")	
90	100	100	0	0	∞	∞	—
↓	↓	↓	↓	↓	↓	↓	
70	100	100	0	0	∞	∞	2.684
65	96	97	4	3	24.00	32.33	2.293
60	95	94	5	6	19.00	15.67	1.889
55	84	95	16	5	5.25	19.00	1.410
52.5	75	91	25	9	3.00	10.10	1.105
50	57	61	43	39	1.33	1.56	0.767
47.5	44	31	56	69	0.786	0.450	0.375
45	29	2	71	98	0.408	0.020	−0.160
40	0	0	100	100	0	0	−1.129
↓	↓	↓	↓	↓	↓	↓	
20	0	0	100	100	0	0	—

[a] One hundred trials run at each value of ϕ. For both linkages, the magnetic moments were 14.5 G-cm^3.

[b] Compare (4.1) and (4.2).

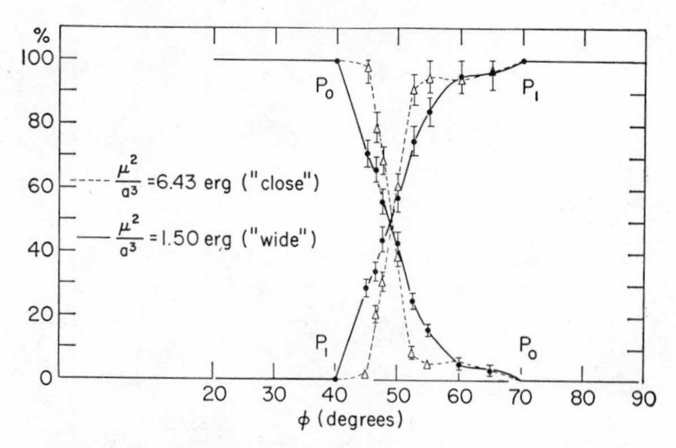

Fig. 7. Relative population probabilities of the four-magnet system. The variation of the probabilities for minima of class C_1 and C_0 is shown both for the wide ($a = 5.2$ cm) and close ($a = 3.2$ cm) spaced lattices. The error bars indicate $\pm 5\%$ uncertainties due to the statistics (100 trials at each point).

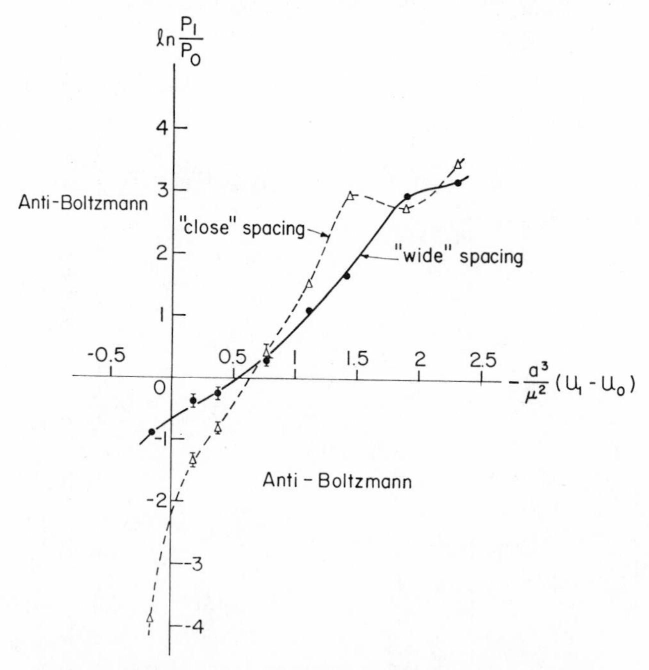

Fig. 8. Plot of $\ln p_1/p_0$ versus $U_1 - U_0$. Experiment indicates a population inversion in one "anti-Boltzmann" quadrant [see (5.2)].

contrast to the gentle variation of the energy—and energy differences—which is displayed on Figure 6b. It is obvious from (5.1) in conjunction with Figure 5 that this distinction arises from the fact that the state-area principle is sensitive to the global structure of the energy surface, and does not merely measure the differences in the levels of the stability troughs.

The results are displayed in a different way on Figure 8. This also provides a direct check for (5.2). We note that experimentally the relative probabilities become equal at $\phi \cong 49°$, while the respective energies coincide at $\phi \cong 46°$. Although these values are close, it is decisive that they are not identical. This gives rise to experimental points in the "anti-Boltzmann" quadrants which are indicated on Figure 8. It is interesting to observe that in the band of rapid transition, $45° \gtrsim \phi \gtrsim 55°$, there is in fact an approximately exponential behavior. Nevertheless, the state-area principle is more comprehensive in accounting for the behavior in the entire range. It is apparent from Figure 5 that the variation in the state-area ratio is a manifestation of the shift in the saddle point from the vicinity of the C_0 track ($\phi \sim 83°$) to the C_1 track ($\phi \sim 40°$). Finally, we note that the differences between the wide-space and close-space arrays shown on Figures 7 and 8 can be attributed to octopole admixtures. This is discussed further in Reference 4.

C. Magnetization (Statistical)

The statistical results obtained in the preceding section (Table IV) can be combined with the magnetization of the respective configurations (Fig. 6 and Table III) to yield an ensemble magnetization. This is defined by

$$\mathcal{M} = \sum_{i=0}^{1} M_i p_i \tag{5.3}$$

which for any ϕ represents the gross magnetization of the four-magnet system as observed in many boiling-condensation cycles. The graphs for each term in the summation (5.3) and the resulting ensemble magnetization as functions of ϕ are shown in Figure 9. From a comparison with Figure 6 it is clear that the discontinuities in magnetization associated with the irreversible jumps at $\phi \cong 40°$ and 83° have been obliterated. At $\phi \cong 68°$, the variation of p_0 has smoothed away the discontinuity in susceptibility. Furthermore at $\phi \cong 80°$, we have $p_0 \approx 0$, and so this transition is effectively invisible. Therefore, the only surviving "second-order" transitions are those corresponding to $\phi \cong 38°$ and 45°.

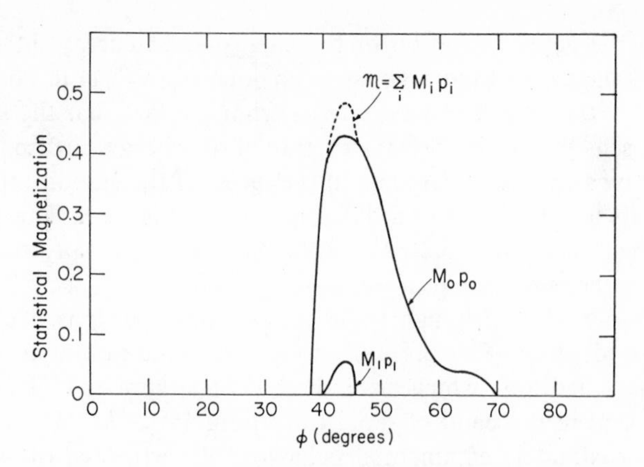

Fig. 9. Statistical ensemble magnetization (5.3). Note the contrast with Figure 6a.

VI. TOPOLOGICAL CONSIDERATIONS

A. The Circulation

The division of various minimum configurations of the four-magnet system into two classes C_1 and C_0 was guided by the observation that within each class there exist continuous transformations connecting all the patterns (Fig. 4). It would obviously be desirable to find a topological invariant characterizing each class. An immediate candidate for such an invariant is the magnetization. From Figure 6, however, it is clear that this parameter varies from 0 to 1 in class C_0, and 0 to 0.4 in class C_1 and therefore is unsuitable. It turns out that the appropriate object is the "circulation" vector which is defined by

$$\mathscr{C} = \frac{\sum_{i=1}^{4} \boldsymbol{\mu}_i \times \mathbf{r}_i}{\sum_{i=1}^{4} |\boldsymbol{\mu}_i \times \mathbf{r}_i|} \tag{6.1}$$

where $\mathbf{r}_i$ is the radius vector from the center of the lattice to the center of the ith magnet, and $\boldsymbol{\mu}_i$ is the corresponding magnetic moment. The neutral state is characterized by $\mathscr{C} = \pm 1$ ("up" or "down") according to the direction of the rotation indicated on Figure 4. The ferro- and antiferromagnetic patterns both have $\mathscr{C} = 0$. A crucial test of whether the circulation (6.1) indeed represents an invariant for each class is provided by considering the respective symmetry-breaking configurations. One can easily check that invariant circulation for C_0 implies parallel moments across the diagonals [see (3.9b)], and for C_1 antiparallel projection of moments perpendicular to the connecting diagonals [see (3.15)]. Hence circulation is also invariant for these states. This proof is of course based on the theoretically predicted configurations. The observed patterns, listed in Table II, also yield the same result within the experimental errors.

In topological terms the phase space of the four-magnet system can be regarded as a four-torus, since obviously the potential energy is periodic in all four coordinates θ_i. The smooth transformations of the system with changing ϕ can then be considered as continuous mappings of the torus onto itself. The topological invariant $\mathscr{C}$, however, divides this torus into three disjoint regions corresponding to $\mathscr{C} = +1$, $\mathscr{C} = -1$, $\mathscr{C} = 0$, and consequently there exist no continuous mappings between these sections: a transition from a pattern with one value of the circulation to a configuration with a different circulation must necessarily involve at least one discontinuous jump.

In this connection it is worth mentioning that in the zero-circulation region one can go from the antiferromagnetic state with zero magnetization to the ferromagnetic state with unit magnetization by means of a smooth transformation. If we take into account the vector nature of the magnetization, that is

$$\boldsymbol{M} = \frac{\sum_{i=1}^{4} \boldsymbol{\mu}_i}{\sum_{i=1}^{4} |\boldsymbol{\mu}_i|} \tag{6.2}$$

this leads to the possibility of a hysteresis cycle in $\boldsymbol{M}$ as ϕ is varied. Such a magnetization cycle can indeed be constructed for the $\mathscr{C} = 0$ patterns in virtue of the existence of the quasi-reversible instability points (Fig. 4). This emphasizes that there is not a unique correspondence between the magnetization (6.2) and the hysteresis coordinate. It should be stressed, however, that since this magnetization loop does not involve irreversible energy losses it is completely distinct from the other hysteresis cycle which includes jump discontinuities.

The expression (6.1) for the circulation can in principle be generalized to more complex magnet systems. Preliminary studies on larger arrays indicate that the invariance of the circulation continues to have a significance.[4]

B. A Lemma on Magnet Revolutions

The concept of circulation as a topological invariant provides us with a global characterization of the four-magnet system, expressing the fact that transitions between patterns of different circulation can only occur in association with at least one discontinuity. A local characterization which leads to a similar, although weaker, conclusion can be obtained from

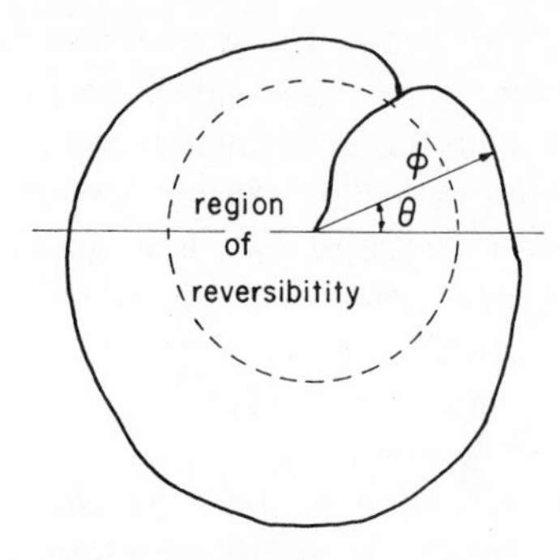

Fig. 10. Polar plot of magnet orientation on $\phi - \theta$ plane (Section VI B).

the experimental observation that in a hysteresis cycle magnets may turn through 360 degrees with respect to a fixed reference direction.

Consider a polar plot of the variation of the angle θ (magnet orientation with respect to a fixed direction) as a function of the hysteresis coordinate ϕ. This is shown in Figure 10. We assume that there exists a certain maximum value of ϕ below which the system is completely reversible; that is, for any given ϕ within the dashed circle in Figure 10 the angle θ is uniquely determined. If the magnet now revolves through 360 degrees, while ϕ goes from 0 to $\pi/2$ and back to 0, then the corresponding trajectory in the $\phi - \theta$ plane will be a continuous closed curve for continuous transformations of the system. The construction in Figure 10 makes it evident that the system necessarily has to encounter at least a fork discontinuity. Therefore transformations of the system, in which a magnet is observed to spin around completely, must include at least one quasi-reversible fork. This lemma, in virtue of the construction applies also to magnet arrays of arbitrary size. It is obvious that dropping the assumption of a reversible region of ϕ in the neighborhood of zero does not alter the conclusion of the lemma since $\phi = 0$ now becomes a branching point.

C. Formal Stability Theory and Topological Estimates for n-Magnet Systems

The four-magnet configuration discussed in the preceding sections has an obvious heuristic interest insofar as it may be regarded as a convenient experimental realization of the Nix-Shockley model[12] for hysteresis. The significance of this model, however, is, purely phenomenological since it simply represents an empirical interpolation scheme which may be adjusted to correspond to various experimental situations. A well-known example is Borelius' analysis of temperature hysteresis in order-disorder transitions of alloys.[13] In order to progress beyond the phenomenological level—particularly in the case of magnetic systems—it is first necessary to settle the question of whether the "compulsory" instabilities which give rise to the four-magnet hysteresis cycle are accidental, and presumably rare occurrences, or are in fact manifestations of some more pervasive and deeper feature of cooperative systems. From an experimental point of view this question can be approached in a straighforward way simply by extending the linkages to include more magnets, and to check whether these larger systems also exhibit irreversible jump discontinuities as the lattice is cyclically deformed. Experimental studies of this kind have already been carried out, and the principal results are described in Section VII A. For purposes of the present discussion it will be sufficient to anticipate the essential conclusion, which is that *as the systems become more complex the incidence of irreversible jump discontinuties rapidly increases.* A necessary consequence is that the average interval—in the hysteresis coordinate—between discontinuities rapidly decreases. If these trends can be extrapolated to really large systems, that is, configurations having of the order of 10^5 (domains) or even 10^{23} constituents, then an interesting new aspect of the irreversibility of macroscopic systems can be envisaged. Specifically, we could surmise that irreversible behavior is generally associated with the existence of dense sets of instability points which the system *must* traverse if its evolution is governed by a hysteresis coordinate such as temperature, strain, or magnetic field. We will show elsewhere that results derived from complex magnetic models[4] and structural "shake-down" models[36] indicate that repeated traversals of these instability sets can lead either to a quenching of hysteresis or a continuous dissipation of energy into internal degrees of freedom.

In this section we prepare the formal structure for interpreting these results. We begin by introducing the n-magnet Hamiltonian

$$H = H(\theta_1, \theta_2, \ldots, \theta_n, \phi) \tag{6.3}$$

where $\theta_1, \ldots, \theta_n$ as usual denote the configuration coordinates of the magnets, and ϕ is the hysteresis (deformation) coordinate. Since only the "zero-temperature" limit will matter, all other variables such as momenta and internal coordinates have been suppressed. The physically realizable states are a subset of the extremals determined by the intersection of the solutions of the equations

$$\frac{\partial H}{\partial \theta_i} = 0; \qquad i = 1, \ldots, n \tag{6.4}$$

It is convenient to write these solutions in the form

$$\theta_i^{(\sigma)} = \theta_i^{(\sigma)}(\phi) \tag{6.5}$$

which emphasizes the explicit dependence on ϕ, and also introduces the index σ as a label for each distinct family of extremals. The physical states correspond to solutions which are local minima. As before (compare the discussion in Section III B) we can select these by constructing the Hessian matrix $\mathscr{H}$. In terms of the Hamiltonian this may be written as

$$\mathscr{H}(\phi \mid \sigma) = \left(\frac{\partial^2 H}{\partial \theta_i^{(\sigma)} \, \partial \theta_j^{(\sigma)}} \right) \tag{6.6}$$

which displays the explicit ϕ-dependence. The matrix is of course also a functional of the extremal sets, and this connection is notationally represented by the index σ. The stable states are then determined by an obvious generalization of the sequence of inequalities corresponding to the set of minors listed in (3.26a–d). It is well known that these are necessary and sufficient conditions for the Hessian matrix to be positive definite.[14] It is convenient to denote this property by

$$\mathscr{H}(\phi \mid \sigma) \succ 0 \tag{6.7}$$

which is an intuitive shorthand for the *sufficient* conditions guaranteeing the local stability of the physical system on the extremal set $\theta_i^{(\sigma)}(\phi)$. We now recall that a transition point at which the Hessian becomes singular, that is,

$$\det | \mathscr{H}(\phi \mid \sigma) | = 0 \tag{6.8}$$

signals the onset of instability of the system. In the spirit of the notation (6.7) this is written simply as

$$\mathscr{H}(\phi \mid \sigma) = 0 \tag{6.9}$$

Finally, in case any of the minors in the sequence $\{D_{n-k}\}_{k=0}^{n-1}$ are negative, it is natural to introduce the convention

$$\mathscr{H}(\phi \mid \sigma) \prec 0 \tag{6.10}$$

which is a *sufficient* condition for the instability of the system.

Suppose now ϕ_σ is a solution of (6.8) in a physically accessible region, and that as ϕ varies monotonically across some interval including ϕ_σ, the Hessian goes through the sequence (6.7) → (6.9) → (6.10). It is natural to call this a "downward zero-crossing" of the Hessian at the singular point ϕ_σ. Clearly this marks a transition from stability to instability of the physical system. Similarly an "upward zero-crossing" of the Hessian at the singular point ϕ_σ corresponds to the sequence (6.10) → (6.9) → (6.7). In practically all cases this is associated with the transformation of the extremal set $\theta_i^{(\sigma)}$ into a local minimum.* If ϕ is repeatedly varied back and forth across the hysteresis interval including ϕ_σ, the physical system will in general *not* retrace the states corresponding

* Note that $\theta_i^{(\sigma)}$ may fail to exist below a downward zero-crossing.

to the extremal set $\theta_i^{(\sigma)}$. This is the essential connection between instability and irreversibility which has already been worked out in the specific case of the four-magnet linkage. The general situation can now be formally described as follows: let the downward zero-crossing sequence

$$(6.7) \quad \rightarrow \quad (6.9) \tag{6.11a}$$

correspond to the hysteresis coordinate variation

$$\phi \quad \rightarrow \quad \phi_\sigma + \quad \text{(limit from "above")} \tag{6.11b}$$

Then the evolution of the physical system is described by

$$\lim_{\phi \to \phi_\sigma +} \theta_i^{(\sigma)}(\phi) \rightarrow \theta_i^{(\tau)}(\phi_\sigma)$$

where τ is the index of the nearest (not necessarily unique!) locally stable extremal set into which the system "falls" at the singular point ϕ_σ. If the hysteresis coordinate is varied in the opposite sense, there is an upward zero-crossing of the Hessian,

$$(6.10) \quad \rightarrow \quad (6.9) \tag{6.12a}$$

and, in analogy with (6.11b), we write

$$\phi \quad \rightarrow \quad \phi_\sigma - \tag{6.12b}$$

However, in this case the evolution of the physical system is described by

$$\lim_{\phi \to \phi_\sigma -} \theta_i^{(\tau)}(\phi) \rightarrow \theta_i^{(\tau)}(\phi_\sigma); \qquad \tau \neq \sigma \tag{6.12c}$$

since "below" the singular point the system must be on a locally stable extremal. In fact, even the singular point then becomes harmless since for the $\theta_i^{(\tau)}$ configuration we generally have the stability condition

$$\mathscr{H}(\phi_\sigma \mid \tau) \succ 0 \tag{6.13}$$

which corresponds to a positive definite regime of the Hessian. In a cyclic variation across the point ϕ_σ, that is, $(\phi_\sigma + \rightarrow \phi_\sigma - \rightarrow \phi_\sigma +)$, we therefore encounter a discontinuity in the trajectories which is given by

$$\sum_{i=1}^{n} |\theta_i^{(\sigma)}(\phi_\sigma +) - \theta_i^{(\tau)}(\phi_\sigma -)| = \Delta(\sigma, \phi_\sigma +; \tau, \phi_\sigma -) \tag{6.14}$$

If this increment doesn't vanish, that is,

$$\Delta(\sigma, \phi_\sigma; \tau, \phi_\sigma) > 0, \sigma \neq \tau \tag{6.15a}$$

then the singular point ϕ_σ gives rise to a jump discontinuity. In the four-magnet example we have already found two specific cases, namely

$$\phi_\sigma \simeq 83° \Rightarrow \frac{\Delta}{4} \simeq 90° \tag{6.15b}$$

and

$$\phi_\sigma \simeq 40° \Rightarrow \frac{\Delta}{4} \simeq 70° \tag{6.15c}$$

which constitute the unstable junctions of the hysteresis cycle.

It may of course also occur that in a single pass across ϕ_σ, the configuration increment Δ vanishes. If, however, the cycle is repeated several times, then the stable terminal states may vary in virtue of a branching of extremals at the singular point. This implies that even if there is a smooth behavior at ϕ_σ, that is,

$$\Delta(\sigma, \phi_\sigma +; \tau', \phi_\sigma -) = \Delta(\sigma, \phi_\sigma +; \tau'', \phi_\sigma -) = 0 \tag{6.16a}$$

where τ' and τ'' denote two locally stable extremals with a confluence at ϕ_σ, there will nevertheless be a finite divergence

$$\Delta(\tau', \phi; \tau'', \phi) > 0 \tag{6.16b}$$

for ϕ values in the range "below" ϕ_σ. Clearly in this case there is also a link between instability and irreversibility, although as pointed out in our earlier discussion of forked extremals (Section IV B) it seems to be more suitable to distinguish this type of directional discontinuity as "quasi-reversible."

With all this formal machinery in hand, we can now return to the original conjecture that macroscopic irreversibility arises from dense sets of microscopic instabilities and assert that none of the arguments (6.6) through (6.16b) depends in any way on special time-reversal properties of the Hamiltonian or the equations of motion. Rather, the nub of the problem has been shifted to defining the nature of instability. Naturally we cannot pull any rabbits out of the hat that may have escaped from Boltzmann, Loschmidt, or Zermelo (see, for instance, Ref. 15), but at least we can be very precise in pin-pointing where the Hessian stability concept goes beyond (6.3) and (6.4). First we note that by hypothesis the Hamiltonian (6.3) *must* be incomplete, since an auxiliary set of "noise forces" is required to continually sense whether an extremal is locally stable or not. Secondly, the equations of motion (6.4) must necessarily be incomplete, since after the system has been nudged from an unstable extremal by "noise," it is assumed that ultimately there is a re-equilibration at another configuration determined by (6.4) and (6.7). This of course implies dissipative interactions, and is the *deus ex machina* required for a complete stability theory. In the case of (idealized) n-magnet systems, internal magnetic work and radiation are two dissipative options. The coupling to these degrees of freedom, however, is only important at the points of instability.

The total number ($\mathcal{N}$) of downward zero-crossings of the Hessian is in principle completely determined by the Hamiltonian, $H(\theta_1, \ldots \theta_n, \phi)$. What we would really like to establish is a lower bound for $\mathcal{N}$ as a function of n. It would of course also be a gain in plausibility if such a growth estimate were not tied exclusively to Hamiltonians with simple dipole interactions. Even in the context of our model systems this restriction is too severe, since it is known that the addition of octopole forces tends to *increase* $\mathcal{N}$ while n is kept fixed. An obvious line of attack is through probabilistic methods. Since the Hessian is real and symmetric, the extensive machinery already developed for statistical theories of spectra of complex systems can be brought to bear.[16] However, it is more in the spirit of the present approach to refer to topological estimates which yield exact, albeit very weak, results. We begin with the following theorem.

Theorem (Lusternik-Schnirelmann). Let $H(\theta_1, \ldots, \theta_n, \phi)$ be a singly differentiable real valued function on a closed manifold $\mathcal{M}$. Let the number of geometrically distinct solutions (extremals) of

$$dH = \sum_{i=1}^{n} \frac{\partial H}{\partial \theta_i} d\theta_i = 0 \tag{6.17}$$

be n_{LS}. Then

$$n_{\mathrm{LS}} \geqslant cat\ \mathcal{M} \tag{6.18a}$$

where $cat\ \mathcal{M}$—the category of $\mathcal{M}$—is the topological invariant of Lusternik and Schnirelmann.[17]

This result can easily be adapted to our situation by noting that the Hamiltonian is periodic in each of the configuration coordinates. The manifold $\mathcal{M}$ is therefore equivalent to the n-torus, and it is known that in this case $cat\ \mathcal{M} = n + 1$. Equation (6.18a) can therefore be sharpened to

$$n_{\mathrm{LS}} \geqslant n + 1 \tag{6.18b}$$

As long as nothing is specified about second derivatives, there is of course no Hessian stability theory and one cannot identify any of the extremals with local minima. Nevertheless it is interesting to see that *any* smooth Hamiltonian with appropriate periodicities will yield an energy surface whose extremal corrugations are bounded from below by n even in the limit $n \to 10^{23}$. We can now add more structural information with the next theorem.

Theorem (Morse). Let $H(\theta_1, \ldots, \theta_n, \phi)$ be a real valued function on a closed manifold $\mathcal{M}$. Suppose the Hessian associated with the θ_i variables exists and is nowhere singular. Then the number (n_M) of geometrically distinct solutions of (6.17) has the lower bound

$$n_M \geqslant \sum_{j=0}^{n} B_j \tag{6.19a}$$

where B_j is the jth Betti number of $\mathscr{M}$.[18,19]

In the special case where $\mathscr{M}$ is the n-torus, the Betti invariants are the binomial coefficients,[20,21]

$$B_j = \binom{n}{j}$$

and (6.19a) can be sharpened to

$$n_M \geqslant 2^n \tag{6.19b}$$

This estimate also indicates that the number of extremals of the Hamiltonian increases with enormous rapidity as n approaches macroscopic magnitudes. The gap between (6.18b) and (6.19b) can to some extent be narrowed in special circumstances by an enumeration of "fluted" singularities.[22] The essential point, however, is that the category and Betti number characterizations of the extremals of the energy surface differ precisely because of the occurrence of singularities of the Hessian. Unfortunately we cannot simply invert the argument and deduce lower bounds for $\mathscr{N}$ in terms of the difference of (6.18b) and (6.19b). In general $\mathscr{N}$ will depend on the hysteresis coordinate ϕ as well as some other properties of the Hamiltonian. It is an open question whether these features are also associated with some (computable) invariant of $\mathscr{M}$. For the moment, therefore, despite experimental indications, probabilistic likelihood, and topological lower bounds, the existence of dense sets of instability points in complex systems remains a conjecture.

VII. EXTENSIONS TO MORE COMPLEX SYSTEMS

A. Experimental Results

In Section VI C we raised the question as to whether the hysteresis exhibited by the four-magnet array is an accidental occurrence or represents the simplest form of a general feature of complex magnetic systems. Experimentally, the answer may be determined in a straightforward way by studying increasingly larger magnet arrays. As indicated on Figure 1, provision was made at the outset to accommodate these systems on our linkages. The analysis has so far been carried out in detail for 3×3 and 4×4 magnet arrays, and qualitatively extended to 6×6 systems. The main results pertinent to the present discussion are given in Table V and Figure 11. A more detailed account of this work will be published elsewhere.

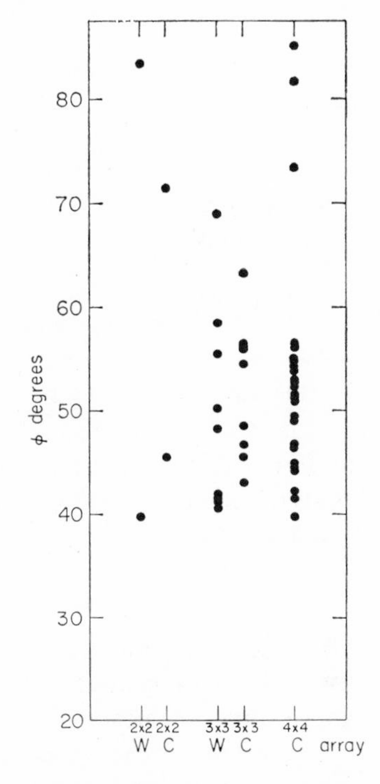

Fig. 11. Irreversible jump discontinuities in complex arrays. Results for 2 × 2 to 4 × 4 systems. The subscripts indicate wide ($a = 5.2$ cm) and close ($a = 3.2$ cm) spaced linkages.

Due to the simple structure of these systems at $\phi = 0°$, the patterns representing local minima in the vicinity of this point can be enumerated explicitly without having to solve for the extremals of the Hamiltonian. This feature can be exploited in mapping the extended energy surfaces of these arrays. The experimental procedure consists of setting up a particular pattern at $\phi \cong 20°$ ($\Rightarrow \phi \sim 0°$), following the systems' evolution to $\phi = 90°$, and then tracing the development of all subsequent configurations back to $\phi \cong 20°$. The essential result is that the larger systems also exhibit hysteresis cycles. In fact these cycles are not unique, and numerous major and minor hysteresis loops appear. Table V summarizes this information. Also shown is the total number of irreversible jump discontinuities ("breaks") encountered during the variation of ϕ along all paths. Clearly, increases in complexity are associated with a drastic rise in the occurrence of breaks. A complementary way of appreciating these trends is indicated by the average number of breaks per path, as well as the average distance between breaks. This distance is defined by

$$\Delta\phi \equiv (\text{number of paths})^{-1} \sum_{\text{paths}} \frac{|\phi_f - \phi_i|}{(1 + \text{number of breaks})} \tag{7.1}$$

where ϕ_i and ϕ_f are the starting and end points of each path. These indices are listed in the last rows of Table V.

Figure 11 reflects the increasing complexity of hysteresis in larger systems by showing the variation of the density of breaks. It is apparent that the density indeed increases rapidly, and we also begin to see the remarkable feature that the hysteresis structure tends to be concentrated in definite bands. In particular, the results obtained from arrays extending

TABLE V

Comparison Between Magnet Systems of Increasing Complexity

System	2×2	3×3	4×4
Number of hysteresis loops[a]	2	12	>20
Number of paths	2	6	19
Number of breaks	2	17	66
Average number of breaks per path	1	2.8	3.5
Average distance between breaks $\Delta\phi$ (degrees)	35.0	18.9	16.5

[a] The sequences generated by the pole-reversed patterns are included.

up to 6 × 6 show that no irreversible jumps exist for $\phi \gtrsim 40°$. This is the basis for the extrapolations represented by the band edge ϕ_L' on Figures 12a and 12b.

B. General Hysteresis Theory

On the basis of the preceeding results it is possible to make some quantitative extrapolations to the hysteresis behavior of very large systems. It will appear that it is most convenient to do so in terms of statistical properties of the extended energy surface. In order to have a specific example in mind we consider a hypothetical magnetic dipole array mounted on a deformable linkage where the number of components has been increased to macroscopic magnitudes, that is, $n \lesssim 10^8$. The simple extended-energy surface shown in Figure 5 will in this case become tremendously crinkled and convoluted. In fact we know from topological estimates (Section VI C) that for each value of the hysteresis coordinate ϕ there must be at least $n + 1$ extremal points. A schematic plan view of the extended-energy surface for this large system is given on Figure 12a. No attempt has been made to simulate the actual topography since this would require graying out the entire diagram with a dense network of extremal curves. Rather, to aid the imagination only a few representative hysteresis trajectories have been indicated.

Strictly speaking, the extended-energy surface has been defined in a space of $n + 2$ dimensions comprising the energy, hysteresis parameter, and n configuration coordinates (compare Section IV A). In the diagram we have simplified matters drastically by showing only a two-dimensional projection. However, this is exactly what is needed to establish a correspondence with the usual macroscopic descriptions of hysteresis. In the magnetic case, it is convenient to consider that this reduction is equivalent

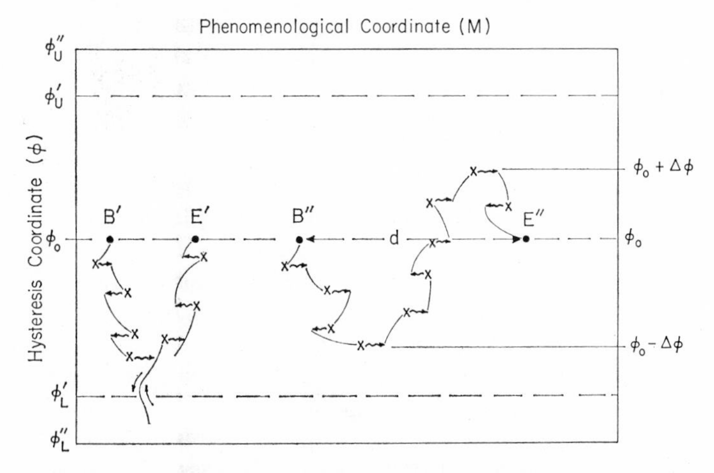

Fig. 12a. Plan view of the extended-energy surface of a complex system. The physically accessible range of the hysteresis coordinates is $\phi_L'' \leqslant \phi \leqslant \phi_U''$. The bands $\phi_L'' \leqslant \phi \leqslant \phi_L'$, and $\phi_U' \leqslant \phi \leqslant \phi_U''$ are free of jump discontinuities. Two hysteresis cycles are indicated: The smooth segments correspond to locally stable states. The x-ᨓ portions represent the irreversible jumps; note that these are always parallel to the abscissa.

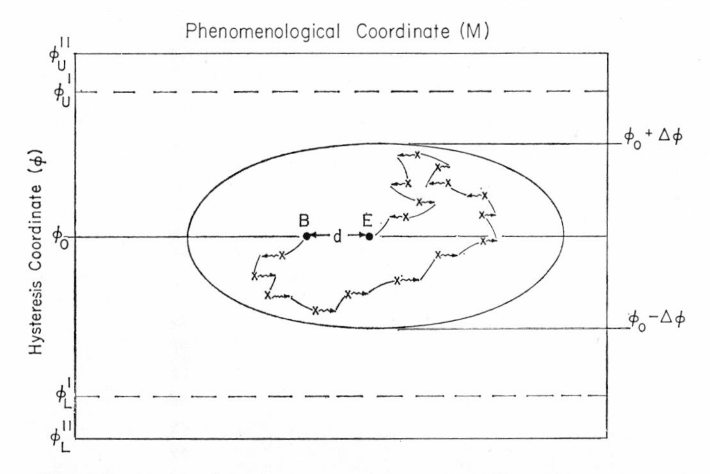

Fig. 12b. Hysteresis cycle on extended-energy surface. The superposed ellipse has foci at B and E. This construction leads to the ellipse metric (7.6).

to telescoping all the configuration coordinates into a single phenomenological parameter, the magnetization. For the present, all we require is that this mapping reflect the complexity of the extremal structure of the full (hyper) surface. Other details won't be essential for our discussion.

In order to get down to specific cases, let us begin with the hysteresis locus connecting B' and E' on Figure 12a. We suppose that the system is initially in the state B', and that the hysteresis coordinate is subsequently varied uniformly in the direction of ϕ_L'. The point B' is assumed to be a local minimum, and as ϕ is altered, the system moves along a locally stable trough until it reaches the first x-point marked on the figure. This is a downward zero-crossing of the Hessian, and therefore a point of instability. As indicated by the arrow, we suppose that at this point the system "falls" down the energy surface until it is intercepted by the trough of another locally stable extremal. The system re-equilibrates in this new minimum, and as ϕ is varied further, it is constrained to move along this extremal until the next instability point is reached. In very complex systems this alternation between sidling and stumbling occurs with great frequency on regions of the energy surfaces where downward zero-crossings of the Hessian are dense. The hysteresis trajectories shown on Figures 12a and 12b may therefore be identified with a kind of drunken walk—aggravated by numerous falls—which is prodded over the extended-energy surface by the variation of a hysteresis coordinate.

Both Figures 12a and 12b show surfaces in which the physically accessible range of the hysteresis coordinate is bounded by $\phi_U'' \geqslant \phi \geqslant \phi_L''$. We have also extrapolated the trend of the results shown on Figure 11, and indicated bands $\phi_U'' \geqslant \phi \geqslant \phi_U'$ and $\phi_L' \geqslant \phi \geqslant \phi_L''$ in which no jump discontinuities are assumed to occur. This implies that the portion of the trajectory $B' \to E'$ for which $\phi < \phi_L'$ will be strictly reversible, as shown on the figure, or at most may pass over quasi-reversible forks (compare Section VII A). The trajectory $B'' \to E''$ illustrates a case in which the hysteresis coordinate has gone through a complete cycle, that is, $\phi_0 \to \phi_0 - \Delta\phi \to \phi_0 \to \phi_0 + \Delta\phi \to \phi_0$, with a corresponding sequence of shifts in the states of the system but no further systematic relation between the beginning and end points. In particular, if the ϕ-cycle were repeated, the system would be dragged along on another tour of random stumbles, and come to rest somewhere on the ϕ_0 line at an arbitrary state E'''. This is exactly the behavior found in small magnet systems ($n \sim 36$) during the first few cycles of the hysteresis coordinate. In macroscopic systems the obvious analogy is the virgin hysteresis regime where the loops have not yet closed or settled down to an asymptotic regularity.

It is a striking feature of many hysteresis systems that repeated cycles of the hysteresis coordinate ultimately become correlated with cycles in a phenomenological coordinate. This transition from the virgin regime ($B'' \rightarrow E''$, Fig. 12a) to an asymptotic periodicity ($B \rightarrow E$, Fig. 12b) can easily be demonstrated with our experimental dipole arrays. For the small systems this evolution can be interpreted very simply in terms of the state-areas of the energy surface. In particular if the initial states are prepared randomly, then all the troughs of local equilibrium will be populated in proportion to their state-areas. The cycling of the hysteresis coordinate then induces a decanting from trough to trough each time an instability point is reached. The energy surface cross-section on Figure 13 makes it intuitively clear that this repeated decanting will ultimately result in a concentration of the system in an asymptotic hysteresis regime. This can also be verified in detail from the combinational structure of the hysteresis maps of small magnetic systems. Of course, in general we cannot presume that there is a simple and unique "up" and "down" direction on the energy surface as indicated on Figure 13, but it still seems plausible to suppose that the existence of asymptotic hysteresis loops is associated with preferred regions of the extended-energy surface into which the system is prodded by repeated cycling of the hysteresis coordinate.

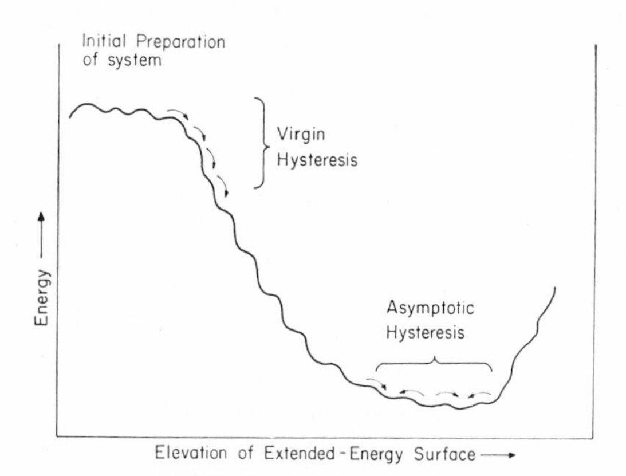

Fig. 13. Profile view of extended-energy surface. The arrows indicate the wanderings of the system in various regimes, for example, virgin hysteresis and asymptotic hysteresis. The approach to the asymptotic region is not unique and is exploited in metallurgic "training" techniques to toughen materials. Chain reaction cascades down the surface may be associated with "rapid" processes such as the deflagration of explosives.

If we now restrict our attention to the hysteresis loop $B \rightarrow E$ of Figure 12b, and suppose that it corresponds to the asymptotic cycle of a very large system, then there should be enough statistical regularity to lead to some quantitative conclusions. The simplest relations of this kind are the Rayleigh cube law of ferromagnetic hysteresis[23, 24] and Dorey's rule of stress-strain hysteresis.[32,34] In order to make contact with these results, it is convenient to introduce a metric on the extended-energy surface projection in the $\phi - M$ plane. We can then speak of a distance d between the points B and E, and also assign a length l to the hysteresis path. The notion of a closed loop then corresponds to the condition

$$d \ll l \tag{7.2}$$

since we suppose that the initial and final states are "close" to each other in a statistical sense. Just as in the analogous case of Boltzmann statistics where the partitioning of phase space is crucial but the actual scale of length is irrelevant,* the course of argument here will not depend on a unique choice for this $\phi - M$ metric.

The essence of the Rayleigh and Dorey laws is a correspondence between the average energy lost per cycle in a hysteresis loop and the average maximum excursion of the hysteresis coordinate. For the loop $B \rightarrow E$ it is easy to write down a result which is in principle exact. If ΔW is the energy lost per cycle, then obviously

$$\Delta W = \sum_i e_i \tag{7.3}$$

where e_i is the energy dissipated at the ith jump discontinuity, and the sum is extended over the entire path. In order to compute the average values of these quantities we make the simple ansatz

$$\left\langle \sum_i e_i \right\rangle_{\text{Av}} = \langle l \times \rho \times e \rangle_{\text{Av}} \tag{7.4}$$

where ρ is the inverse frequency of jump-discontinuities along $B \rightarrow E$, and e is the energy lost at each discontinuity. Since $B \rightarrow E$ is supposed to be bounded away from the hysteresis edges at ϕ_L', ϕ_U', and the system is very complex, it is reasonable to assume that both ρ and e are essentially constant along the entire trajectory. These trends are also supported by experimental evidence from small arrays. Equation (7.4) can therefore be simplified still further by factoring

$$\left\langle \sum_i e_i \right\rangle_{\text{Av}} = \rho e \langle l \rangle_{\text{Av}} \tag{7.5}$$

* With all due respect to Sackur-Tetrode.

and this reduces the problem to calculating the average length of the hysteresis trajectory. In order to do this it is convenient to assign a statistical metric to the hysteresis paths linking B with E:[25]

Consider the set of all continuous curves of length σ, where $\sigma < l$, which connect points B and E. Clearly these must all lie within an ellipse of semi-major axis $l/2$, and semi-minor axis $(l^2 - d^2)^{1/2}/2$, with B and E as focal points. We shall assume that the probability, $P(l)$, that a path linking B and E have length not exceeding l is proportional to the area of this ellipse. Specifically,

$$P(l) = \begin{cases} 0, & l < d \\ \dfrac{\pi}{4L^2}\, l(l^2 - d^2)^{1/2}, & l \geqslant d \end{cases} \tag{7.6}$$

where the normalizing factor L^2 reflects the finite extent of the hysteresis band ($\phi'_U \geqslant \phi \geqslant \phi'_L$). In terms of this statistical metric it is easy to derive an average length. The result is

$$\begin{aligned} \langle l \rangle_{\text{Av}} &= \int_0^l \sigma \, dP(\sigma) \\ &= \frac{\pi}{4L^2}\left\{ l^2(l^2 - d^2)^{1/2} - \frac{1}{3}(l^2 - d^2)^{3/2} \right\}, \qquad l \geqslant d \end{aligned} \tag{7.7}$$

This simple expression can be reduced still further for hysteresis cycles where we also have the inequality (7.2). In this case

$$\langle l \rangle_{\text{Av}} \simeq \frac{\pi}{4L^2}\, l^3 \tag{7.8}$$

which already exhibits the cubic dependence. If we finally recall that the semi-minor axis is equal to $\Delta\phi$—modulo the choice of metric scale—then Equations (7.3), (7.5), and (7.8) lead to the proportionality

$$\langle \Delta W \rangle_{\text{Av}} \propto (\Delta\phi)^3 \tag{7.9}$$

which is equivalent to both the Rayleigh and Dorey results.

The ellipse metric can of course also be applied to the noncyclic trajectory $B'' \to E''$ of Figure 12a. We again have to stipulate that $\phi_0 \pm \Delta\phi$ is bounded away from the hysteresis band edges, and that the statistical (7.4) and uniformity (7.5) estimates are valid. In this case Equation (7.2) is replaced by the approximate equality $l \approx d$. The general expression (7.7) then implies the linear relation

$$\Delta W \propto \Delta\phi \tag{7.10}$$

which is the appropriate form of the Rayleigh law in the virgin hysteresis regime. For practical applications it is advisable to summarize the sequence (7.3) → (7.7) in the form

$$\langle \Delta W \rangle_{\text{Av}} = \frac{4\pi}{3} \frac{\rho e}{L^2} \left\{ \frac{3}{8} d^2 \Delta\phi + (\Delta\phi)^3 \right\} \tag{7.11}$$

since this shows explicitly the relation between the energy losses and the magnitude of the hysteresis coordinate. Because our approach has been essentially model independent, Equation (7.11) should apply to a wide variety of systems exhibiting hysteresis. Indeed, experimental evidence indicates that the linear/cubic damping behavior predicted by (7.11) occurs in the following systems:

(i) magnetic cooperative model systems (Section VII A, Ref. 4),
(ii) structural "shake-down",[36]
(iii) ferromagnetic hysteresis,[26]
(iv) stress-strain hysteresis,[37] and
(v) superconducting hysteresis.[38]

Although the underlying mechanisms vary widely in each case, the weak assumption that with each system we can associate an extended-energy surface densely sprinkled with instability points, is already sufficient to reach quantitative conclusions regarding the range of the damping exponent (e.g., Ref. 37). In Table VI we present a detailed comparison of hysteresis for stress-strain and ferromagnetism. It is clear that at least in a formal sense the abstract hysteresis theory stands in good correspondence with the observations.

The parallels indicated in Table VI may be elaborated further on more speculative levels. For example in the case of stress-strain hysteresis it has been surmised that the work per cycle, that is, the area of the loop, is associated with the rate of approach to fatigue failure.[29, 30, 31] This has a simple interpretation in terms of the extended-energy surface: hysteresis circuits necessarily include many jump discontinuities, and at each one energy is irreversibly pumped into the microstructure of the material. The magnetic analogy for this process is obvious. In fact we can even illustrate it with our original four-magnet system. During each hysteresis cycle a certain amount of energy ($\sim$15 ergs) is irreversibly lost to internal degrees of freedom. If radiation and thermal conduction in the supports are neglected, it is clear that this will result in a gradual warming of the individual magnets. The magnetic moments will therefore tend to decrease, but the cycle will still persist since the elementary dipole interactions are homogeneous in the moments. In principle this could be continued until

TABLE VI

Comparison of Magnetic and Stress-Strain Hysteresis

	Specific systems	
Hysteresis Theory	Ferromagnetic hysteresis	Stress-strain hysteresis
Hysteresis coordinate	Applied magnetic field	Stress
Phenomenological coordinate	Magnetization	Strain
Band without hysteresis ($\phi_U'' - \phi_U'; \phi_L' - \phi_L''$)	Reversible magnetization	Hooke's law region
Hysteresis band edge	Threshold of nonlinear magnetization[a]	Proportional limit (σ_p) (endurance limit)
Jump discontinuities	Barkhausen jumps (domain wall motion)	Portevin-le Chatelier effect[b] (dislocation drag[c])
Quasi-reversible forks	"Reversible" hysteresis contribution[d]	Elastic hysteresis[e]
$\Delta W \sim \Delta\phi$	Barkhausen energy loss[d] (small fields)	Linear loss regime ($\sigma \lesssim \sigma_p$)[f]
$\Delta W \sim (\Delta\phi)^3$	Rayleigh law (iron, cobalt, nickel)	Dorey rule[g] (steel, iron, lucite, plywood, bakelite, monel metal)

[a] Initial Barkhausen jump.
[b] Ref. 27.
[c] Ref. 28.
[d] Ref. 24, Chap. 25.
[e] Ref. 33.
[f] S. A. Guralnick, private communication.
[g] Refs. 32, 34, and 37.

the magnets reached the Curie temperature. At this point we could imagine an abrupt internal phase transition; the moments would vanish; and the hysteresis engine would suddenly cease to function at the 10^8 cycle!

Acknowledgments

Some of the equipment utilized in this work was constructed by G. R. Marousek and G. K. Forsberg. G. B. Baumgartner, R. J. Fontana, B. Halphen, J. E. Nuti, and B. Tauber carried out numerous experimental checks. We would like to acknowledge many helpful conversations with Professor T. J. Neubert, Professor O. Levenspiel, and P. Everett. We are especially grateful to Professor S. A. Guralnick for detailed advice regarding stress-strain hysteresis; Professor B. Schweizer for directing our attention to the topological theorems; and Professor A. Sklar who proposed the ellipse metric.

This work has benefited from sustained support from the Research Corporation. Additional assistance was provided by Sigma Xi, and grants from the U.S. Army Research Office (Durham), and the National Science Foundation.

APPENDIX A. STABILITY RANGE FOR MINIMA OF THE FOUR-MAGNET SYSTEM

We first apply the stability conditions (3.25a–d) to the neutral state. Inserting (3.12) into (3.7a–j) we find the following expressions for the respective second-order derivatives:

$$U_{\alpha_1\alpha_1} = \frac{\mu^2}{a^3}\left[3\,\frac{\cos^3\dfrac{\phi}{2} + \sin^3\dfrac{\phi}{2}}{2\sin^3\phi}\right] \tag{A.1a}$$

$$U_{\beta_2\beta_2} = \frac{\mu^2}{a^3}\left[3\,\frac{\cos^3\dfrac{\phi}{2} + \sin^3\dfrac{\phi}{2}}{2\sin^3\phi} + 6\sin\phi\right] \tag{A.1b}$$

$$U_{\alpha_2\alpha_2} = U_{\beta_1\beta_1} = \frac{\mu^2}{a^3}\left[3\sin\phi - \frac{\cos^3\dfrac{\phi}{2} + \sin^3\dfrac{\phi}{2}}{2\sin^3\phi}\right] \tag{A.1c}$$

$$U_{\alpha_1\alpha_2} = U_{\alpha_1\beta_1} = U_{\alpha_2\beta_2} = U_{\beta_1\beta_2} = 0 \tag{A.1d}$$

$$U_{\alpha_1\beta_2} = \frac{\mu^2}{a^3}\left[3\,\frac{\cos^3\dfrac{\phi}{2} - \sin^3\dfrac{\phi}{2}}{2\sin^3\phi}\right] \tag{A.1e}$$

$$U_{\alpha_2\beta_1} = \frac{\mu^2}{a^3}\left[-\frac{\cos^3\dfrac{\phi}{2} - \sin^3\dfrac{\phi}{2}}{2\sin^3\phi}\right] \tag{A.1f}$$

With these the stability criteria for the neutral state are

$$\cos^3\frac{\phi}{2} + \sin^3\frac{\phi}{2} > 0 \tag{A.2a}$$

$$1 + 8\left(\cos^3\frac{\phi}{2} + \sin^3\frac{\phi}{2}\right)\sin\phi > 0 \tag{A.2b}$$

$$\left[6\sin^4\phi - \cos^3\frac{\phi}{2} - \sin^3\frac{\phi}{2}\right]\left[1 + 8\left(\cos^3\frac{\phi}{2} + \sin^3\frac{\phi}{2}\right)\sin\phi\right] > 0 \tag{A.2c}$$

$$\left[48 \sin\frac{\phi}{2}\cos^4\frac{\phi}{2} - 1\right]\left[48 \sin^4\frac{\phi}{2}\cos\frac{\phi}{2} - 1\right] \times \left[1 + 8\left(\cos^3\frac{\phi}{2} + \sin^3\frac{\phi}{2}\right)\sin\phi\right] > 0 \quad \text{(A.2d)}$$

The first two of these inequalities are obviously satisfied for any deformation in the interval $0° < \phi \leqslant 90°$. The third, (A.2c), holds only for $90° \geqslant \phi > 38.22°$, and finally, (A.2d) restricts the range to $90° \geqslant \phi > 45.63°$. This lower bound coincides exactly with the appearance of the symmetry breaking solution of class C_1 represented by (3.15) and (3.17). For this state the second-order derivatives are as follows:

$$U_{\alpha_1\alpha_1} = \frac{\mu^2}{a^3}\left[3\frac{\cos^3\frac{\phi}{2} + \sin^3\frac{\phi}{2}}{2\sin^3\phi}\right] \quad \text{(A.3a)}$$

$$U_{\beta_2\beta_2} = \frac{\mu^2}{a^3}\left[144\sin^2\phi\sin^3\frac{\phi}{2} + 3\frac{\cos^3\frac{\phi}{2} + \sin^3\frac{\phi}{2}}{2\sin^3\phi}\right] \quad \text{(A.3b)}$$

$$U_{\alpha_2\alpha_2} = U_{\beta_1\beta_1} = \frac{\mu^2}{a^3}\left[\frac{\cos^3\frac{\phi}{2} - \sin^3\frac{\phi}{2}}{2\sin^3\phi}\right] \quad \text{(A.3c)}$$

$$U_{\alpha_1\alpha_2} = -U_{\alpha_1\beta_1} = \frac{\mu^2}{a^3}[-\sin\alpha_2] \quad \text{(A.3d)}$$

$$U_{\alpha_1\beta_2} = \frac{\mu^2}{a^3}\left[3\frac{\cos^3\frac{\phi}{2} + \sin^3\frac{\phi}{2}}{2\sin^3\phi}\right] \quad \text{(A.3e)}$$

$$U_{\alpha_2\beta_1} = \frac{\mu^2}{a^3}\left[\frac{\cos^3\frac{\phi}{2} + \sin^3\frac{\phi}{2}}{2\sin^3\phi} - 72\sin^2\phi\sin^3\frac{\phi}{2}\right] \quad \text{(A.3f)}$$

$$U_{\alpha_2\beta_2} = -U_{\beta_1\beta_2} = \frac{\mu^2}{a^3}[-3\cos\phi\sin\alpha_2] \quad \text{(A.3g)}$$

The factor $\sin \alpha_2$ may be obtained explicitly from (3.17). These lead to the stability conditions

$$\cos^3 \frac{\phi}{2} + \sin^3 \frac{\phi}{2} > 0 \tag{A.4a}$$

$$1 + 192\left(\cos^3 \frac{\phi}{2} + \sin^3 \frac{\phi}{2}\right) \sin^2 \phi \sin^3 \frac{\phi}{2} > 0 \tag{A.4b}$$

$$\begin{aligned} &3\left(\cos^3 \frac{\phi}{2} - \sin^3 \frac{\phi}{2}\right)\left[1 + 192\left(\cos^3 \frac{\phi}{2} + \sin^3 \frac{\phi}{2}\right)\sin^2 \phi \sin^3 \frac{\phi}{2}\right] \\ &- 8 \sin^3 \phi\left[1 - 576 \sin^2 \phi \sin^6 \frac{\phi}{2}\right] \\ &\times \left[\cos^3 \frac{\phi}{2}(3 \cos \phi - 1)^2 + \sin^3 \frac{\phi}{2}(3 \cos \phi + 1)^2 + 96 \sin^5 \phi \sin^3 \frac{\phi}{2}\right] > 0 \end{aligned} \tag{A.4c}$$

$$\begin{aligned} &\left[1 - 576 \sin^2 \phi \sin^6 \frac{\phi}{2}\right]\left\{3\left[1 + 192\left(\cos^3 \frac{\phi}{2} + \sin^3 \frac{\phi}{2}\right) \sin^2 \phi \sin^3 \frac{\phi}{2}\right]\right. \\ &\quad \times [72 \sin^5 \phi - 1] - 64 \cos^3 \frac{\phi}{2} \\ &\quad \times \left(1 - 576 \sin^2 \phi \sin^6 \frac{\phi}{2}\right)\left[\cos^3 \frac{\phi}{2}(3 \cos \phi - 1)^2\right. \\ &\quad \left.\left. + \sin^3 \frac{\phi}{2}(3 \cos \phi + 1)^2 + 96 \sin^5 \phi \sin^3 \frac{\phi}{2}\right]\right\} > 0 \end{aligned} \tag{A.4d}$$

The first three inequalities are satisfied for $0° < \phi \leqslant 90°$. The fourth, (A.4d), is fulfilled only in the interval $45.63° > \phi > 39.84°$. The upper limit coincides exactly with the onset of the symmetry-breaking solution (3.15).

Summarizing these results, we see that the two solutions of class C_1 smoothly merge at $\phi = 45.63°$, as is also indicated by the energy expressions (4.1a) and (4.1b). Above $\phi = 45.63°$, the neutral state represents the minimum configuration; below this point the symmetry-breaking solution takes over, and below $\phi = 39.84°$ neither one of these is a local minimum. Therefore minima of class C_1 are restricted to the range

$$C_1 : 90° \geqslant \phi > 39.84° \tag{A.5}$$

In order to carry out an analogous investigation for class C_0 it is convenient to begin with the antiferromagnetic state (3.23a): the corresponding second-order derivatives are

$$U_{\alpha_1\alpha_1} = \frac{\mu^2}{a^3}\left[-2 + 3\,\frac{\cos^3\frac{\phi}{2} - \sin^3\frac{\phi}{2}}{2\sin^3\phi}\right] \tag{A.6a}$$

$$U_{\beta_2\beta_2} = \frac{\mu^2}{a^3}\left[6\cos\phi + 3\,\frac{\cos^3\frac{\phi}{2} - \sin^3\frac{\phi}{2}}{2\sin^3\phi}\right] \tag{A.6b}$$

$$U_{\alpha_2\alpha_2} = U_{\beta_1\beta_1} = \frac{\mu^2}{a^3}\left[3\cos\phi - 1 + \frac{\cos^3\frac{\phi}{2} + \sin^3\frac{\phi}{2}}{2\sin^3\phi}\right] \tag{A.6c}$$

$$U_{\alpha_1\alpha_2} = U_{\alpha_1\beta_1} = U_{\alpha_2\beta_2} = U_{\beta_1\beta_2} = 0 \tag{A.6d}$$

$$U_{\alpha_1\beta_2} = \frac{\mu^2}{a^3}\left[3\,\frac{\cos^3\frac{\phi}{2} + \sin^3\frac{\phi}{2}}{2\sin^3\phi}\right] \tag{A.6e}$$

$$U_{\alpha_2\beta_1} = \frac{\mu^2}{a^3}\left[\frac{\cos^3\frac{\phi}{2} - \sin^3\frac{\phi}{2}}{2\sin^3\phi}\right] \tag{A.6f}$$

These lead to the stability conditions

$$3\left(\cos^3\frac{\phi}{2} - \sin^3\frac{\phi}{2}\right) - 4\sin^3\phi > 0 \tag{A.7a}$$

$$8\left(\cos^3\frac{\phi}{2} - \sin^3\frac{\phi}{2}\right)(3\cos\phi - 1) - 32\sin^3\phi\cos\phi - 3 > 0 \tag{A.7b}$$

$$\left[8\left(\cos^3\frac{\phi}{2} - \sin^3\frac{\phi}{2}\right)(3\cos\phi - 1) - 32\sin^3\phi\cos\phi - 3\right] \times \left[6\sin^3\phi\cos\phi + \cos^3\frac{\phi}{2} + \sin^3\frac{\phi}{2} - 2\sin^3\phi\right] > 0 \tag{A.7c}$$

$$\left[8\left(\cos^3\frac{\phi}{2}-\sin^3\frac{\phi}{2}\right)(3\cos\phi-1)-32\sin^3\phi\cos\phi-3\right]$$
$$\times\left[8\sin^3\frac{\phi}{2}(3\cos\phi-1)+1\right]\left[8\cos^3\frac{\phi}{2}(3\cos\phi-1)+1\right]>0 \tag{A.7d}$$

The first of these, (A.7a), holds for $0° < \phi < 51.74°$. The last three are restricted to the range $0° < \phi < 37.94°$. This upper limit for the antiferromagnetic minimum coincides with the onset of the symmetry-breaking solution (3.22b, c). For this symmetry-breaking solution the required second-order derivatives are given by

$$U_{\alpha_1\alpha_1}=\frac{\mu^2}{a^3}\left[3c\,\frac{\cos^3\dfrac{\phi}{2}+\sin^3\dfrac{\phi}{2}}{2\sin^3\phi}\right] \tag{A.8a}$$

$$U_{\beta_2\beta_2}=\frac{\mu^2}{a^3}\left[3\,\frac{\cos^3\dfrac{\phi}{2}+\sin^3\dfrac{\phi}{2}}{2c\sin^3\phi}\right] \tag{A.8b}$$

$$U_{\alpha_2\alpha_2}=\frac{\mu^2}{a^3}\left[24c\left(\cos^3\frac{\phi}{2}+\sin^3\frac{\phi}{2}\right)\cos^2\phi-16\left(\cos^3\frac{\phi}{2}-\sin^3\frac{\phi}{2}\right)\cos\phi\right.$$
$$\left.+\left(\cos^3\frac{\phi}{2}+\sin^3\frac{\phi}{2}\right)\left(\frac{8}{3c}+\frac{1}{2\sin^3\phi}\right)-3\sin\phi\sin(\beta_2-\phi)\right] \tag{A.8c}$$

$$U_{\beta_1\beta_1}=\frac{\mu^2}{a^3}\left[24c\left(\cos^3\frac{\phi}{2}+\sin^3\frac{\phi}{2}\right)\cos^2\phi-16\left(\cos^3\frac{\phi}{2}-\sin^3\frac{\phi}{2}\right)\cos\phi\right.$$
$$\left.+\left(\cos^3\frac{\phi}{2}+\sin^3\frac{\phi}{2}\right)\left(\frac{8}{3c}+\frac{1}{2\sin^3\phi}\right)+3\sin\phi\sin(\beta_2-\phi)\right] \tag{A.8d}$$

$$U_{\alpha_1\alpha_2}=U_{\alpha_1\beta_1}=U_{\alpha_2\beta_2}=U_{\beta_1\beta_2}=0 \tag{A.8e}$$

$$U_{\alpha_1\beta_2}=\frac{\mu^2}{a^3}\left\{-16\cos\phi\left[\cos^3\frac{\phi}{2}+\sin^3\frac{\phi}{2}-3c\left(\cos^3\frac{\phi}{2}-\sin^3\frac{\phi}{2}\right)\cos\phi\right]\right.$$
$$\left.+3c\,\frac{\cos^3\dfrac{\phi}{2}-\sin^3\dfrac{\phi}{2}}{2\sin^3\phi}\right\} \tag{A.8f}$$

$$U_{\alpha_2\beta_1} = \frac{\mu^2}{a^3}\left[\frac{\cos^3\dfrac{\phi}{2} - \sin^3\dfrac{\phi}{2}}{2\sin^3\phi}\right] \tag{A.8g}$$

The factor $\sin(\beta_2 - \phi)$ may be obtained explicitly from (3.22c); c is defined in (3.20b). The complete stability conditions for the symmetry-breaking solution then are

$$\left(\cos^3\frac{\phi}{2} + \sin^2\frac{\phi}{2}\right)c > 0 \tag{A.9a}$$

$$\begin{aligned}
&\left[\left(\cos^3\frac{\phi}{2} + \sin^3\frac{\phi}{2}\right)(3 - 32\sin^3\phi\cos\phi)\right.\\
&\qquad \left. + 3c\left(\cos^3\frac{\phi}{2} - \sin^3\frac{\phi}{2}\right)(1 + 32\sin^3\phi\cos^2\phi)\right]\\
&\qquad \times\left[\left(\cos^3\frac{\phi}{2} + \sin^3\frac{\phi}{2}\right)(3 + 32\sin^3\phi\cos\phi)\right.\\
&\qquad \left. - 3c\left(\cos^3\frac{\phi}{2} - \sin^3\frac{\phi}{2}\right)(1 + 32\sin^3\phi\cos^2\phi)\right] > 0
\end{aligned} \tag{A.9b}$$

$$\begin{aligned}
&\left[24c\left(\cos^3\frac{\phi}{2} + \sin^3\frac{\phi}{2}\right)\cos^2\phi - 16\cos\phi\left(\cos^3\frac{\phi}{2} - \sin^3\frac{\phi}{2}\right)\right.\\
&\qquad \left. + \left(\cos^3\frac{\phi}{2} + \sin^3\frac{\phi}{2}\right)\left(\frac{8}{3c} + \frac{1}{2\sin^3\phi}\right) - 3\sin\phi\sin(\beta_2 - \phi)\right]\\
&\qquad \times\left[\left(\cos^3\frac{\phi}{2} + \sin^3\frac{\phi}{2}\right)(3 - 32\sin^3\phi\cos\phi)\right.\\
&\qquad \left. + 3c\left(\cos^3\frac{\phi}{2} - \sin^3\frac{\phi}{2}\right)(1 + 32\sin^3\phi\cos^2\phi)\right]\\
&\qquad \times\left[\left(\cos^3\frac{\phi}{2} + \sin^3\frac{\phi}{2}\right)(3 + 32\sin^3\phi\cos\phi)\right.\\
&\qquad \left. - 3c\left(\cos^3\frac{\phi}{2} - \sin^3\frac{\phi}{2}\right)(1 + 32\sin^3\phi\cos^2\phi)\right] > 0
\end{aligned} \tag{A.9c}$$

$$\left\{\left[24c\left(\cos^3\frac{\phi}{2}+\sin^3\frac{\phi}{2}\right)\cos^2\phi-16\left(\cos^3\frac{\phi}{2}-\sin^3\frac{\phi}{2}\right)\cos\phi\right.\right.$$
$$\left.+\left(\cos^3\frac{\phi}{2}+\sin^3\frac{\phi}{2}\right)\left(\frac{8}{3c}+\frac{1}{2\sin^3\phi}\right)\right]^2-9\sin^2\phi+\frac{64\sin^2\phi}{c^2}$$
$$\left.\times\left[\cos^3\frac{\phi}{2}+\sin^3\frac{\phi}{2}-3c\left(\cos^3\frac{\phi}{2}-\sin^3\frac{\phi}{2}\right)\cos\phi\right]^2\right\}$$
$$\times\left[\left(\cos^3\frac{\phi}{2}+\sin^3\frac{\phi}{2}\right)(3-32\sin^3\phi\cos\phi)\right.$$
$$\left.+3c\left(\cos^3\frac{\phi}{2}-\sin^3\frac{\phi}{2}\right)(1+32\sin^3\phi\cos^2\phi)\right]$$
$$\times\left[\left(\cos^3\frac{\phi}{3}+\sin^3\frac{\phi}{2}\right)(3+32\sin^3\phi\cos\phi)\right.$$
$$\left.-3c\left(\cos^3\frac{\phi}{2}-\sin^3\frac{\phi}{2}\right)(1+32\sin^3\phi\cos^2\phi)\right]>0 \qquad \text{(A.9d)}$$

The first of these inequalities, (A.9a), is fulfilled identically in virtue of the choice of the sign of c (3.20b). The second, (A.9b), and third, (A.9c), conditions both hold for the intervals $37.94^\circ < \phi < 68.72^\circ$ and $80.37^\circ < \phi \leqslant 90^\circ$. These coincide exactly with the range of existence of this symmetry breaking extremum. The fourth inequality (A.9d) limits the stability range to the disjoint regions $37.94^\circ < \phi < 68.72^\circ$ and $80.37^\circ < \phi < 83.35^\circ$.

Finally, the second-order derivatives for the ferromagnetic state (3.23b) are

$$U_{\alpha_1\alpha_1}=\frac{\mu^2}{a^3}\left[2-3\,\frac{\cos^3\dfrac{\phi}{2}-\sin^3\dfrac{\phi}{2}}{2\sin^3\phi}\right] \qquad \text{(A.10a)}$$

$$U_{\beta_2\beta_2}=\frac{\mu^2}{a^3}\left[6\cos\phi-3\,\frac{\cos^3\dfrac{\phi}{2}-\sin^3\dfrac{}{2}}{2\sin^3\phi}\right] \qquad \text{(A.10b)}$$

$$U_{\alpha_2\alpha_2}=U_{\beta_1\beta_1}=\frac{\mu^2}{a^3}\left[1+3\cos\phi+\frac{\cos^3\dfrac{\phi}{2}+\sin^3\dfrac{\phi}{2}}{2\sin^3\phi}\right] \qquad \text{(A.10c)}$$

$$U_{\alpha_1\alpha_2}=U_{\alpha_1\beta_1}=U_{\alpha_2\beta_2}=U_{\beta_1\beta_2}=0 \qquad \text{(A.10d)}$$

$$U_{\alpha_1\beta_2} = \frac{\mu^2}{a^3}\left[-3\,\frac{\cos^3\dfrac{\phi}{2} + \sin^3\dfrac{\phi}{2}}{2\sin^3\phi}\right] \tag{A.10e}$$

$$U_{\alpha_2\beta_1} = \frac{\mu^2}{a^3}\left[\frac{\cos^3\dfrac{\phi}{2} - \sin^3\dfrac{\phi}{2}}{2\sin^3\phi}\right] \tag{A.10f}$$

These lead to the stability conditions

$$4\sin^3\phi - 3\left(\cos^3\frac{\phi}{2} - \sin^3\frac{\phi}{2}\right) > 0 \tag{A.11a}$$

$$32\sin^3\phi\cos\phi - 8(3\cos\phi + 1)\left(\cos^3\frac{\phi}{2} - \sin^3\frac{\phi}{2}\right) - 3 > 0 \tag{A.11b}$$

$$\left[2\sin^3\phi(3\cos\phi + 1) + \cos^3\frac{\phi}{2} + \sin^3\frac{\phi}{2}\right] \times \left[32\sin^3\phi\cos\phi - 8(3\cos\phi + 1)\left(\cos^3\frac{\phi}{2} - \sin^3\frac{\phi}{2}\right) - 3\right] > 0 \tag{A.11c}$$

$$\left[8\sin^3\frac{\phi}{2}(3\cos\phi + 1) + 1\right]\left[8\cos^3\frac{\phi}{2}(3\cos\phi + 1) + 1\right] \times \left[32\sin^3\phi\cos\phi - 8(3\cos\phi + 1)\left(\cos^3\frac{\phi}{2} - \sin^3\frac{\phi}{2}\right) - 3\right] > 0 \tag{A.11d}$$

The first of these inequalities, (A.11a), is satisfied in the range $90° \geqslant \phi > 51.74°$. The second, (A.11b), however, reduces this interval to $80.37° > \phi > 68.72°$. The last two inequalities, (A.11c, d), do not restrict this range any further. Thus the ferromagnetic state precisely fills the gap between the intervals in ϕ where the symmetry-breaking extremum represents a local minimum. Summarizing the preceding discussion, we see that class C_0 consists of the following sequence: an antiferromagnetic configuration from 0° to 37.94°; a symmetry-breaking pattern between 37.94° and 68.72°; a ferromagnetic state from 68.72° to 80.37°; and finally another symmetry-breaking portion between 80.37° and 83.35°. Stability ceases at 83.35°, which terminates the sequence.

TABLE VII

Summary of Existence and Stability Intervals

Descending branch (C_1)			Ascending branch (C_0)		
Extremum	Existence	Stability	Extremum	Existence	Stability
Neutral (3.12)	$90° \geqslant \phi > 0°$	$90° \geqslant \phi > 45.63°$	Symmetry-breaking (3.9b) (3.22)	$90° \geqslant \phi \geqslant 80.37°$	$83.35° > \phi > 80.37°$
			Ferromagnetic (3.23b)	$90° \geqslant \phi > 0°$	$80.37° > \phi > 68.72°$
Symmetry-breaking (3.15) (3.17)	$45.63° \geqslant \phi > 0°$	$45.63° > \phi > 39.84°$	Symmetry-breaking (3.9b) (3.22)	$68.72° \geqslant \phi \geqslant 37.94°$	$68.72° > \phi > 37.94°$
			Antiferromagnetic (3.23a)	$90° \geqslant \phi > 0°$	$37.94° > \phi > 0°$

It can be shown that the patterns given by the various solutions of class C_0 merge smoothly at the instability points $\phi = 37.94°$, $68.72°$, and $80.37°$. A similar continuity is exhibited by the corresponding energy expressions (4.2a, b, c). The class C_0 therefore represents local minima in the entire interval

$$C_0 : 0° < \phi < 83.35° \tag{A.12}$$

References

1. T. Erber and H. G. Latal, *Bull. Acad. Roy. Belg. Cl. Sc.*, **53,** 1019 (1967).
2. T. Erber and H. G. Latal, *Physica*, **37,** 489 (1967); *Bull. Am. Phys. Soc.*, **12,** 86 (1967).
3. H. G. Latal, *Magnet Laboratory Internal Report*, Illinois Institute of Technology, Chicago, 1967.
4. T. Erber and H. G. Latal, in preparation.
5. T. Erber, G. R. Marousek, and G. K. Forsberg, *Bull. Am. Phys. Soc.*, **12,** 374 (1967); *Acta Phys. Austriaca*, **30**, 271 (1969); see also *Prospects for Simulation and Simulators of Dynamic Systems*, Spartan Books, New York, 1967, pp. 201–226.
6. G. R. Marousek, M. S. Thesis, Illinois Institute of Technology, Chicago, 1967.
7. G. K. Forsberg, M. S. Thesis, Illinois Institute of Technology, Chicago, 1966.
8. T. M. Apostol, *Mathematical Analysis*, Addison-Wesley, Reading, Mass., 1958, p. 151.
9. A. B. Pippard, *Elements of Classical Thermodynamics*, Cambridge University Press, London, 1957, p. 122.
10. A. N. Kolmogorov, *Selec. Transl. Math. Statist. Probab.*, **7,** 293 (1968).
11. A. Q. Tool, *J. Research Natl. Bur. Standards*, **37,** 73 (1946).
12. F. C. Nix and W. Shockley, *Rev. Mod. Phys.*, **10,** 1 (1938).
13. G. Borelius, *Ann. Physik*, **20,** 57 (1934).
14. R. Bellman, *Introduction to Matrix Analysis*, McGraw-Hill, New York, 1960.
15. M. Kac, *Probability and Related Topics in Physical Sciences*, Interscience, New York, 1959.
16. C. E. Porter, *Statistical Theories of Spectra: Fluctuations*, Academic Press, New York, 1965.
17. L. Lusternik and L. Schnirelmann, "Méthodes Topologiques dans les Problémes Variationnels," *Actualités Scientifique et Industrielles*, Vol. 188, Hermann & Cie., Paris, 1934.
18. M. Morse, *Calculus of Variations in the Large*, Vol. 18 of American Mathematical Society Colloquium Publications, Providence, R.I., 1934.
19. H. Seifert and W. Threlfall, *Variationsrechnung im Grossen*, Chelsea, New York, 1948.
20. H. Künneth, *Math. Ann.*, **90,** 65 (1923).
21. V. I. Arnold, *Ergodic Problems of Classical Mechanics*, W. A. Benjamin, New York, 1968.
22. J. C. Phillips, *Phys. Rev.*, **104,** 1263 (1956).
23. Lord Rayleigh, *Phil. Mag.*, **23,** 225 (1887).
24. E. Kneller, *Ferromagnetismus*, Springer, Berlin, 1962.
25. B. Schweizer and A. Sklar, *Pacific J. Math.*, **10,** 313 (1960).
26. L. Néel, *Cahiers Phys.*, **12,** 1 (1942).

27. J. F. Bell, *The Physics of Large Deformation of Crystalline Solids*, Springer, New York, 1968.
28. J. P. Hirth and J. Lothe, *Theory of Dislocations*, McGraw-Hill, New York, 1968.
29. F. R. Shanley, *Strength of Materials*, McGraw-Hill, New York, 1957.
30. J. Morrow, *Cyclic Plastic Strain Energy and Fatigue of Metals*, Spec. Tech. Publ. 738, American Society for Testing Materials, Philadelphia, 1965.
31. S. A. Guralnick, unpublished.
32. S. F. Dorey, *Proc. Inst. Mech. Engrs.* (*London*), **123,** 479 (1932).
33. O. Föppl, *J. Iron Steel Inst.* (*London*), **134,** 393 (1936).
34. J. M. Robertson and A. J. Yorgiadis, *J. Appl. Mechanics*, **13,** A-173 (1946).
35. P. W. Bridgman, *Rev. Mod. Phys.*, **22,** 56 (1950).
36. S. A. Guralnick, T. Erber, and H. G. Latal, in preparation.
37. B. J. Lazan, *Damping of Materials and Members in Structural Mechanics*, Pergamon, Oxford, 1968.
38. W. S. Gilbert, R. B. Meuser, and F. Voelker, *Losses in Pulsed Superconducting Magnets*, UCRL-18885, 1969.

THE LINEAR GAS

M. R. HOARE

Department of Physics, Bedford College, Regent's Park, London, England

I. INTRODUCTION

In 1891 Lord Rayleigh proposed and began to elucidate what is perhaps the simplest conceivable, nontrivial model in statistical dynamics—the "Rayleigh piston" or, as it might now be called, the "one-dimensional test-particle gas problem."[1] One considers an ensemble of frictionless pistons, mass m_1, subject to random collisions with a one-dimensional heat bath of particles, mass m_2, and asks for the evolution of the velocity distribution function $P(\mathbf{V}, t)$ from specified initial conditions (Fig. 1). Rayleigh solved this initial-value problem under the special assumption $m_1 \gg m_2$ but did not obtain a formulation for more general mass ratio. His reason for treating such an idealized case was "... a conviction that the present rather unsatisfactory state in the Theory of Gases is due in some degree to a want of preparation in the mind of readers who are confronted suddenly with ideas and processes of no ordinary difficulty..." This excuse seems as pertinent as ever in the still unsatisfactory state of the theory of gases eighty years later.

For all his insight, Rayleigh, who worked in one dimension "In order to bring out fundamental statistical questions unencumbered with other difficulties ...," is unlikely to have foreseen the degree of real mathematical subtlety behind this simple-looking model, a subtlety which is still distinctly under-appreciated among statistical physicists. The comparative neglect of Rayleigh's problem, and to an extent, linear transport theory as a whole, in the half-century following his original paper is a curious aspect of the history of statistical mechanics. On the one hand, the successors of Boltzmann were preoccupied with nonlinear aspects of kinetic theory, while on the other hand, the school of Brownian motion theorists carried to an extreme the implications of the approximation $m_1 \gg m_2$; somewhere between these immense and fruitful fields the general linear problem for $m_1 \lesseqgtr m_2$ seemed to be lost. When it finally came into focus again, it was by a devious route which owed little either to Rayleigh's simplistic approach or to the rest of orthodox statistical mechanics.

The growth of neutron transport theory, with its obvious technological stimuli, gave sudden urgency to the whole class of linear transport problems, of which Rayleigh's piston is the prototype. Much of the work that has flowed from this direction since the late 1940's deals with spatial solutions in special geometries, with initial pulses of very high energy and other conditions of less interest in "pure" statistical mechanics. A large part is

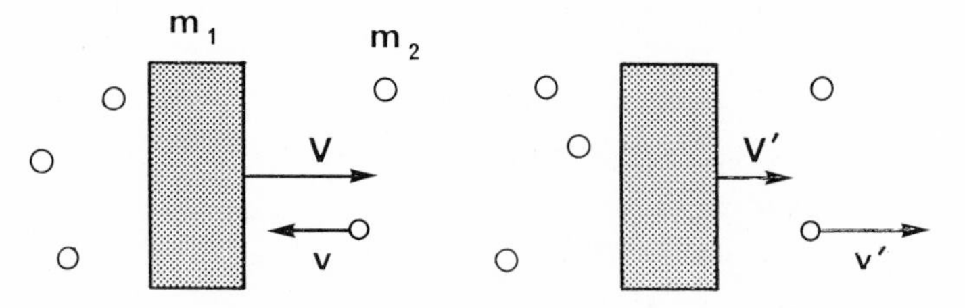

Fig. 1. The Rayleigh piston.

written in language and mathematical conventions virtually private to the field of reactor theory and published in limited-circulation reports, some originally "classified." Nevertheless, out of this emerges a coherent body of work centered around the initial-value problem for particles in simple, spatially homogeneous, gaslike moderators, which is of clear relevance to "ordinary" statistical mechanics and can be turned at once toward the Rayleigh problem and its ramifications.

The main concern of this article will be to reconsider the classic test-particle problem in the light of this work and other recent developments in the theory of linear integro-differential equations of "Master Equation" type, and their corresponding Markov processes. Although we shall be forced to concentrate on the more realistic three-dimensional hard-sphere gas, it will quickly be apparent that the one-dimensional model, in precisely Rayleigh's original terms, is by no means a trivial simplification of this, but embodies the crux of the mathematical problem, which is the handling of linear integral operators with mixed continuous and discrete spectra and singular eigenfunctions. Since the basic theory, computational mathematics, and physical interpretation of these operators all remain imperfectly understood, the same limitations must be accepted in speaking of the Rayleigh model itself, which seems likely to remain a source of both mathematical and physical interest going considerably beyond what it will be possible to cover in this article.

We shall begin by discussing the transport equation for the linear relaxation problem and the kernels that arise, particularly for the case of "hard" interactions between system and heat bath. The eigenvalue problem for scalar relaxation in a homogeneous system will then be considered with special reference to the qualitative theory of the spectra involved. We shall then review practical methods of computation and present the results of some of these with some attention to the light they throw on the accuracy of the Rayleigh-Fokker-Planck equation for the case of heavy test particles. It is planned to treat the more difficult spatial relaxation problem in a separate article. The aim throughout will be to emphasize those parts of the theory that are less well-known in conventional statistical mechanics, e.g. the spectral theory of the Master equation, the WKB method, and so on, and give less space to topics that are equally important but have been better publicized. For this reason we shall concentrate on the true linear test-particle gas throughout and not take up the Linearized Boltzmann Equation (LBE) itself in any detail. Several excellent reviews of the latter already exist.[2,3] In the matter of notation, it has been very difficult to extract a convenient set of symbols from either side of the boundary between "neutron transport" and "kinetic theory" and apply this consistently throughout. Where a good notation exists, some attempt has been made to preserve it; otherwise new symbols have been introduced rather than risk confusion with different usages.

Although no injustice is intended to several notable papers on the atomic relaxation problem, it would not be inaccurate to say that much of what follows is a translation into molecular terms, with appropriate shifts of emphasis, of a few key papers in the theory of neutron thermalization. This task would have been considerably more difficult without the definitive text by M. M. R. Williams.[4] The author's debt to this work is too great to be acknowledged only in postscript. Nevertheless, certain shifts of viewpoint will be apparent. The general statistical mechanical problem is open to a much greater variety of interactions than the neutron case and is obviously free from the condition $m_1 \ll m_2$ which always applies in a nuclear reactor. In fact, particular interest will attach to the "Rayleigh regime" $m_1 > m_2$ and the approach to the Brownian condition $m_1 \gg m_2$, about which very little has been written. In presenting practical methods we shall also deviate somewhat from the conventions of neutron transport theory, particularly by working in probability distributions rather than fluxes and avoiding where possible the use of unsymmetric or non-self-adjoint equations.

In referring to original work in the neutron field no attempt has been made to cite the vast report and conference literature in which so much of this material has appeared. Where such a reference is given, it is an indication that no publication in the open literature exists. Extensive references to these sources will be found in the monographs by Williams,[4] Zweifel and Case,[5] and Ferzinger and Zweifel.[6]

II. TRANSPORT EQUATIONS

Let us generalize the Rayleigh problem somewhat by considering the motion of an ensemble of test-particles mass m_1, in a three-dimensional heat bath of particles, mass m_2, with temperature T. Both the position $\mathbf{r}$ and velocity $\mathbf{V}$ of the test-particles are random variables which are only stationary in the special case of equilibrium at temperature T. In general, we may define a distribution function $P(\mathbf{V}, \mathbf{r}; t)$ whose evolution from some initial condition $P(\mathbf{V}, \mathbf{r}; 0)$ is the subject of interest, though averages over this will be sufficient for the solution of many physical problems.

To obtain a linear transport equation for $P(\mathbf{V}, \mathbf{r}; t)$ a number of assumptions must be stated, though they are really implicit in the very concept of a *system* and *heat bath* as distinct entities interacting by "collisions."

If the ensemble is physically realized as an actual set of many test particles distributed in a gaslike moderator, then they must be present in sufficient dilution that the consequences of collisions between them can be neglected in comparison with those with the heat-bath itself. There is no difficulty in fulfilling this condition with neutrons, but it is a more severe limitation to the treatment of physicochemical systems. Secondly, the heat bath must present an aspect of molecular chaos to the system; that is, there should be no correlations either in position or velocity between successive collisions. For strict fulfillment this must require an overall dilution of both system and heat bath to dilute-gas densities; when this is not the case all that follows is reduced to the level of an approximation which must be treated with some reserve, particularly where short time intervals are considered. Finally, as is implicit in the above, collisions should be well-defined in the sense that a total scattering cross section for system-heat bath interactions exists. In practice, only "hard" interactions are amenable computationally, and most of what we shall consider here will be limited to the extreme case of "hard-sphere" scattering.

If we further exclude the possibility of interactions with external fields, then the distribution $P(\mathbf{V}, \mathbf{r}; t)$ will evolve spatially by free streaming, and in velocity by collisional scattering with the heat-bath particles alone. This leads straightforwardly to the most general type of transport equation we shall need to consider here,

$$\frac{\partial}{\partial t} P(\mathbf{V}, \mathbf{r}, t) + \mathbf{V} \cdot \nabla P(\mathbf{V}, \mathbf{r}, t) = \mathscr{A} P(\mathbf{V}, \mathbf{r}, t) \tag{2.1}$$

In this we have written $\mathscr{A}$ for the scattering operator governing collisional transitions, which, under the assumptions described, will be a linear integral operator whose kernel represents the statistical outcome of collisions with specified incoming and outgoing velocity. By assuming classical mechanics throughout we may take Equation (2.1) to be a self-evident balance equation for the flux into and out of the state $(\mathbf{V}, \mathbf{r})$ with the stochastic contribution isolated on the right-hand side.

It is now natural to write the scattering operator in terms of a *transition kernel* $K(\mathbf{V}, \mathbf{V}')$ giving the probability per unit time that a system particle with velocity $\mathbf{V}$ will scatter to a velocity range $d\mathbf{V}'$ about $\mathbf{V}'$. The right-hand side of (2.1) then becomes a difference of "scattering-in" and "scattering-out" terms as follows:

$$\mathscr{A} P(\mathbf{V}, \mathbf{r}, t) = \int \left\{ K(\mathbf{V}', \mathbf{V}) P(\mathbf{V}', \mathbf{r}, t) - K(\mathbf{V}, \mathbf{V}') P(\mathbf{V}, \mathbf{r}, t) \right\} d\mathbf{V}' \tag{2.2}$$

and we may rewrite this as

$$\mathscr{A}P(\mathbf{V}, \mathbf{r}, t) = \int K(\mathbf{V}', \mathbf{V})P(\mathbf{V}', \mathbf{r}, t)\, d\mathbf{V}' - Z(V)P(\mathbf{V}, \mathbf{r}, t) \tag{2.3}$$

where the scalar function $Z(V)$ is simply the velocity(speed)-dependent collision number:

$$Z(V) = \int K(\mathbf{V}, \mathbf{V}')\, d\mathbf{V}' \tag{2.4}$$

More symbolically, it will be convenient to write

$$\mathscr{A} \equiv \mathscr{K} - Z(V)\delta(\mathbf{V} - \mathbf{V}') \tag{2.5}$$

where $\mathscr{K}$ is the integral operator associated with the kernel K and the second term is also to be understood as a symbolic integral operator.

A number of reduced forms of Equation (2.1) are possible on integrating over angles or removing the spatial dependence altogether, or by introducing initial conditions of special symmetry. One may sometimes also multiply by various quantities of interest and integrate to give moment equations, again for a variety of special initial conditions. Thus, in appropriate cases one may write transport equations to attempt solutions for distributions such as $P(V, r, t)$, $P(\mathbf{V}, t)$, $P(V, t)$, $P(\mathbf{r}, t)$, and $P(r, t)$ and moments such as $\langle V(t)\rangle$ and $\langle r^2(t)\rangle$.*

Most of this article will be concerned with the spatially homogeneous relaxation problem in a scalar variable.

* We can only pause to mention some of the various extensions to Equation (2.1) which are possible still within the framework of linear transport theory. One may add source and absorption terms to the left-hand side to represent injection of particles into the system or their removal, for example by chemical reaction, or one can add an acceleration to the streaming term for the case of charged particles in an external field. All these cases are of ample interest in physicochemical systems. Homogeneous sources occur, for example, in the chemical kinetics of recoil tritium atoms[7] and local sources in the case of "chemical activation" by infusion of hydrogen atoms.[8] One may also bear in mind the possibility of producing velocity transients in molecular systems by secondary collisions with charged particles and the use of pulsed beams from shock waves. Where both source and absorption terms, or external field terms are present, the steady-state problem with $\partial P/\partial t = 0$ becomes an interesting and difficult case in its own right. The charged-particle equation would refer not to plasma conditions but to the transport of dilute ions in a neutral heat bath. This case is distinguished from the others in having very little in common with neutron transport theory. Although it will not be possible to go into details of these properties here, it need hardly be emphasized that an understanding of the simple, homogeneous, infinite medium problem is a necessary first step in approaching them.

For this it will be convenient to write the basic transport equation as

$$\begin{aligned}\frac{\partial P(x,t)}{\partial t} &= \int K(y,x)P(y,t)\,dy - Z(x)P(x,t) \\ &= \mathscr{A}P(x,t) \qquad (2.6)\end{aligned}$$

with the understanding that the variable x may stand for either the speed V or (more usually) the energy of the test particle, preferably expressed in dimensionless form.

We shall first establish some of the most general properties which must be shown by all kernels of physical interest and have important consequences for the solutions of the "Master Equation" (2.6). The first is that, for the infinite, closed, homogeneous case, an equilibrium solution $P(x, \infty)$ should exist corresponding to the Maxwellian distribution for the appropriate variable. Denoting this by $M(x)$ and putting $\partial P(x,t)/\partial t = 0$, we observe that

$$\int K(y,x)M(y)\,dy = Z(x)M(x) \qquad (2.7)$$

This can be seen to be just a statement that the scattering operator $\mathscr{A}$ has one eigenfunction equal to $M(x)$ with corresponding eigenvalue zero, that is, $\mathscr{A}M(x) = 0$.

A second property is that the transition kernel K should satisfy the detailed-balance condition in the form

$$M(x)K(x,y) = M(y)K(y,x) \qquad (2.8)$$

This is particularly important because it guarantees that all eigenvalues of the scattering operator $\mathscr{A}$ are real, and makes it possible to write the Master Equation (2.6) in a conveniently symmetric form.

To see this we have only to introduce the new function

$$h(x,t) = P(x,t)/[M(x)]^{1/2} \qquad (2.9)$$

and kernel

$$G(x,y) = K(x,y)[M(x)/M(y)]^{1/2} \qquad (2.10)$$

which, by (2.8), must be symmetric. This leads immediately to the new, self-adjoint transport equation

$$\begin{aligned}\frac{\partial h(x,t)}{\partial t} &= \int_0^\infty G(y,t)h(y,t)\,dy - Z(x)h(x,t) \\ &= \int_0^\infty B(x,y)h(y,t)\,dy = \mathscr{B}h(x,t) \qquad (2.11)\end{aligned}$$

This is the form we shall usually discuss.* Concerning the properties (2.7) and (2.8), it should be emphasized that they are invariably "built in" to the functional form of the transistion kernel whenever this can be derived by systematic averaging over all possible collisions with the heat-bath particles, and they do not have to be supplied in any form of additional information. The equilibrium condition is $\mathscr{B}\,[M(x)]^{1/2} = 0$.

Finally, and still at this quite general level, we can prove that the properties (2.7) and (2.8) are sufficient for the scattering operators $\mathscr{A}$ and $\mathscr{B}$ to be negative semi-definite and posess a unique equilibrium distribution. The former is a necessary condition for conservation of probability which we express by the normalization

$$\int P(x, t)\, dx = 1 \qquad \text{for all } t \tag{2.12}$$

To see this we examine the quadratic form of the operator with an arbitrary square-integrable function $f(x)$ on the required range. Writing $[M(x)]^{1/2} = N_0(x)$ and using (2.7) and (2.8), we have successively

$$(f, \mathscr{B}f) = \int f(x) \int G(x, y) f(y)\, dy\, dx - \int Z(x)[f(x)]^2\, dx$$

$$\begin{aligned}(f, \mathscr{B}f) &= \int f(x) \int G(x, y) f(y)\, dy\, dx \\ &\quad - \int [f(x)]^2 \int G(x, y)[N_0(y)/N_0(x)]\, dy\, dx \\ &= -\frac{1}{2} \iint G(x, y) N_0(x) N_0(y) \left\{\frac{f(x)}{N_0(x)} - \frac{f(y)}{N_0(y)}\right\}^2 dy\, dx \\ &\leqslant 0\end{aligned} \tag{2.13}$$

Since $G(x, y)$ and $N_0(x)$ are positive functions, this shows the operator $\mathscr{B}$ to be negative semi-definite. The form of the last integral confirms that $N_0(x)$ is the equilibrium eigenfunction of $\mathscr{B}$ and that this must be non-degenerate if x is scalar quantity.

* We shall occasionally write $\mathscr{B} = \mathscr{G} + z(x)\delta(x - y)$.

III. TRANSITION KERNELS

A. The Rayleigh Kernel

Before we consider solutions of the Master Equation (2.6) it will be well to illustrate some of the above ideas with reference to specific kernels which can be written explicitly. Of these the simplest is probably that which governs the original Rayleigh problem in one dimension.

This is very easily derived via the Maxwellian distribution and the impact equation for the piston with a heat-bath particle. Let a piston with velocity $\mathbf{V}$ collide with a heat-bath particle, velocity $\mathbf{v}$, and acquire final velocity $\mathbf{V}'$. The conservation of momentum and energy lead immediately to the relationship

$$\mathbf{V}' = \frac{2\mathbf{v}\gamma + \mathbf{V}(1-\gamma)}{1+\gamma} \tag{3.1}$$

where $\gamma = m_2/m_1$ is the mass ratio. We note in passing the special case $\gamma = 1$ where the two particles appear to change places with each other. To derive the transition kernel $K(\mathbf{V}, \mathbf{V}')$ we must write the collision frequency $z(\mathbf{v}, \mathbf{V})$ and then average the outcomes, according to Equation (3.1) with this as weighting factor. The required frequency is simply

$$z(\mathbf{V}, \mathbf{v}) = n\sigma\,|\mathbf{V} - \mathbf{v}|\,f_2(\mathbf{v}) \tag{3.2}$$

where $n =$ the number density of heat-bath particles and $\sigma =$ the geometrical cross section of the piston. $f_2(\mathbf{v})$ is the Maxwellian distribution for mass m_2, which in one dimension is

$$f_2(\mathbf{v}) = (m_2/2\pi kT)^{1/2} \exp\left(-m_2\mathbf{v}^2/2kT\right) \tag{3.3}$$

A simple accounting of collisions leads to the vector kernel

$$K(\mathbf{V}, \mathbf{V}') = \left(\frac{n\sigma}{4\gamma^2}\right)\left(\frac{m_2}{2\pi kT}\right)^{1/2} (1+\gamma)^2\,|\mathbf{V} - \mathbf{V}'| \times \exp\left[-\frac{m_1}{8\gamma kT}\left(\mathbf{V}(\gamma-1) + \mathbf{V}'(\gamma+1)\right)^2\right] \tag{3.4}$$

This can also be written in the compact form

$$K(\mathbf{V}, \mathbf{V}') = \tfrac{1}{4}n\sigma(1+\gamma)^2\,|\mathbf{V} - \mathbf{V}'|\,f_2\{(1/2\gamma)[\mathbf{V}(\gamma-1) + \mathbf{V}'(\gamma+1)]\} \tag{3.5}$$

Although a commonsense derivation is adequate in one dimension, it is interesting to write all the above steps in a single symbolic expression thus:

$$K(\mathbf{V}, \mathbf{V}') = n\sigma \iint f_2(\mathbf{v})\, \delta(\tfrac{1}{2}m_1(\mathbf{V}^2 - \mathbf{V}'^2) + \tfrac{1}{2}m_2(\mathbf{v}^2 - \mathbf{v}'^2))$$
$$\times\, \delta(m_1(\mathbf{V} - \mathbf{V}') - m_2(\mathbf{v} - \mathbf{v}'))\, d\mathbf{v}\, d\mathbf{v}' \quad (3.6)$$

The use of the delta function in this way is a device due to Waldmann,[2] which considerably simplifies the integrations in higher dimensions.*

The kernel (3.4) was not given by Rayleigh himself and was probably written for the first time by Lebowitz and Bergmann, who used it in their study of the thermodynamics of nonstationary ensembles.[9] An application to simple chemical kinetic models has also been given by Bak and Lebowitz.[10]

It is very easy to see the limiting behavior of the kernel in the two cases $\gamma \to 0$ and $\gamma \to \infty$. The effect of the Gaussian factor is as follows:

$$m_2 \ll m_1: \quad \mathbf{V}' \approx \mathbf{V} \text{ (but } \mathbf{V}' \neq \mathbf{V}\text{) almost always}$$
$$m_2 \gg m_1: \quad \mathbf{V}' \approx -\mathbf{V} \text{ (}\mathbf{V}' = -\mathbf{V} \text{ possible) almost always}$$
$$m_2 = m_1: \quad K(\mathbf{V}, \mathbf{V}') = n\sigma\, |\mathbf{V} - \mathbf{V}'|\, f_2(\mathbf{V}')$$

The first two cases are the Rayleigh and Lorentz limits, respectively; the third reflects the behavior already referred to where, for $\gamma = 1$, the probability of reaching $\mathbf{V}'$ is simply the probability of meeting a heat-bath atom with minus that velocity.

The velocity-dependent collision number can now be obtained according to Equation (2.4).

$$Z(V) = n\sigma\{V \operatorname{erf}[(m_2/2kT)^{1/2} V] + (2kT/\pi m_2)^{1/2} \exp-(m_2 V^2/2kT)\} \quad (3.7)$$

From this we note that $Z(V)$ has the following properties:

(i) $Z(0) = n\sigma(2kT/\pi m_2)^{1/2}$

(ii) $Z(V) = Z(0) + bV^2 + 0(V^3)$; $b = n\sigma(m_2/2\pi kT)^{1/2}$

(iii) $Z(V) \to n\sigma V$ as $V \to \infty$

(iv) $Z(V)$ is a monotonically increasing function of V.

All these features are of importance in the spectral theory of the operator $\mathscr{B}$.

Finally, we may use the equilibrium distribution

$$f_1(V) = (m_1/2\pi kT)^{1/2} \exp(-m_1 V^2/2kT) \quad (3.8)$$

to obtain the symmetric kernel $G(\mathbf{V}, \mathbf{V}')$ according to Equation (2.10).

* The vector notation $\mathbf{V}$ has been retained in the above equations to emphasize that K is defined on the range $(-\infty, +\infty)$. The "speed-kernel" $K(|\mathbf{V}|, |\mathbf{V}'|)$ would clearly be different and contain less information.

We find

$$G(\mathbf{V}, \mathbf{V}') = A(1+\gamma)^2 |\mathbf{V} - \mathbf{V}'| \times \exp\left\{-\frac{M}{8\gamma kT}[V^2(1+\gamma^2) + 2VV'(\gamma^2 - 1) + V'^2(1+\gamma^2)]\right\} \quad (3.9)$$

where $A = (n\sigma/4\gamma^2)(m_2/2\pi kT)^{1/2}$.

The symmetry of G is self-evident; moreover the quadratic form in the Gaussian exponent is easily shown to be positive definite. The nonvanishing of this term for finite V and V' is a sufficient condition for G to be square integrable over the whole $V - V'$ plane.

B. The Wigner–Wilkins Kernel

The three-dimensional analog of Rayleigh's problem is the statistical dynamics of an ensemble of hard-sphere test particles embedded in a similar heat-bath gas. Although this retains some of the character of the one-dimensional problem there are important differences, notably the *persistence of velocity* effect occurring through oblique collisions.

The transition kernel for this problem has a somewhat curious history. Most surprisingly, it appears not to have been known to the Hilbert-Enskog school of kinetic theory and does not appear anywhere in Chapman and Cowling's book.[11]*

The first derivation of a scalar kernel for energy transitions is usually credited to Wigner and Wilkins, who obtained it by straightforward kinetic-theory arguments in a classified 1944 report on the neutron-moderator problem.[12] Later, independent and somewhat neater derivations were given by Andersen and Shuler[13] and Nielsen and Bak,[14] who obtained both scalar and vector forms. The latter derivation, based on the device of Waldmann (Eq. (3.6)), is perhaps the clearest. They obtain a number of equivalent forms, of which the following, in velocities, is the most expressive.

$$K(\mathbf{V}, \mathbf{V}') = \frac{A}{|\mathbf{V}' - \mathbf{V}|} \exp\left\{-\frac{m_1 \mathbf{V}'^2}{2kT} + \frac{\gamma}{2kT}\frac{(\mathbf{V}' \times \mathbf{V})^2}{(\mathbf{V}' - \mathbf{V})^2} - \frac{(\gamma - 1)m_1}{8\gamma kT}[\gamma(\mathbf{V}' + \mathbf{V})^2 - (\mathbf{V}' - \mathbf{V})^2]\right\}$$

$$A = n\sigma^2(1+\gamma)^2/4m_2(2\pi m_2 kT)^{1/2} \quad (3.10)$$

Although the Gaussian term is quite similar to that in the Rayleigh kernel, the factor $|\mathbf{V} - \mathbf{V}'|^{-1}$ tends to give a large contribution when $\mathbf{V}'$

* The corresponding collision-number expression $Z(V)$ (Eq. (3.16) of this article) does appear, however. See Ref. 11, p. 95, Eq. (5.4.5).

and $\mathbf{V}$ are similar and in the same direction, that is, the persistence of velocity effect.

Introducing the quantity

$$\cos\theta = (\mathbf{V}\cdot\mathbf{V}')/|\mathbf{V}|\cdot|\mathbf{V}'| \tag{3.11}$$

we see that the kernel can be written in the functional form $K(V, V', \cos\theta)$, that is, as a function of the speeds and the angle between velocities before and after collision.

The original Wigner-Wilkins kernel is the scalar version of the above expressed in initial and final energies. On integrating over angles and using the reduced energies $x = (m_1 V^2/kT)$ the result is

$$\begin{aligned} K(x, y) = \frac{1}{2} AQ^2 \frac{\pi^{1/2}}{x^{1/2}} \{&\operatorname{erf}(Qy^{1/2} + Rx^{1/2}) + e^{x-y}\operatorname{erf}(Ry^{1/2} + Qx^{1/2}) \\ &\pm [\operatorname{erf}(Qy^{1/2} - Rx^{1/2}) + e^{x-y}\operatorname{erf}(Ry^{1/2} - Qx^{1/2})]\} \end{aligned} \tag{3.12}$$

($+$ for $y < x$, $-$ for $y > x$).

The quantities Q and R are related to the mass ratio as follows:

$$Q = \tfrac{1}{2}(\gamma^{-1/2} + \gamma^{1/2}); \quad R = \tfrac{1}{2}(\gamma^{-1/2} - \gamma^{1/2}) \tag{3.13}$$

A is a collision number, $A = n\pi\sigma^2(kT/2m_1)^{1/2}$, n is the number density as before, and σ is now the sum of the radii of the two interacting hard spheres. We use the function erf (x) in the form:

$$\operatorname{erf}(x) = (2/\pi^{1/2})\int_0^x e^{-s^2}\,ds \tag{3.14}$$

The structure of the terms in (3.10) is again such that the initial and final velocities become either nearly identical or nearly reversed in the Rayleigh and Lorentz limits $\gamma \to 0$ and $\gamma \to \infty$, respectively.

For the special case $\gamma = 1$ (the uniform gas), the kernel is considerably simplified. We have

$$\begin{aligned} K(x, y) &= A\pi^{1/2}x^{-1/2}\operatorname{erf}(y^{1/2})\,;\ y < x \\ &= A\pi^{1/2}x^{-1/2}e^{x-y}\operatorname{erf}(x^{1/2})\,;\ y > x \end{aligned} \tag{3.15}$$

The symmetric form of this has the character of a Green function for a second-order differential equation, a property which can be put to use in a number of approximations, as we shall see later (Section VID).

The integration of the kernel can be carried out explicitly and leads to the energy-dependent collision number:

$$Z(x) = A\gamma^{-1/2}\left[(2(\gamma x)^{1/2} + (\gamma x)^{-1/2})\frac{\pi^{1/2}}{2}\operatorname{erf}(\gamma x)^{1/2} + e^{-\gamma x}\right] \quad (3.16)$$

From this we obtain ($b = \frac{2}{3}A\gamma^{1/2}$)

(i) $$Z(0) = 2A/\gamma^{1/2} \quad (3.17)$$

(ii) $$Z(x) = Z(0) + bx + O(x^2) \quad (3.18)$$

(iii) $$Z(x) \to 2Ax^{1/2} = n\pi\sigma^2 V \quad \text{as} \quad x \to \infty \quad (3.19)$$

Integration over the Maxwellian distribution leads to the mean equilibrium collision number:

$$Z_{00} = 2A(1 + \gamma^{-1})^{1/2} \quad (3.20)$$

an elementary result.

Just as with $Z(x)$, the error functions appearing in the kernel $K(x, y)$ can be expanded to lowest order giving the low-velocity approximation

$$\begin{aligned} K(x, y) &= Z(0)Q^2(y/x)^{1/2}\ ; \quad y < x \\ &= Z(0)Q^2\ ; \qquad\qquad\quad y > x \end{aligned} \quad (3.21)$$

This is of considerable importance in the theory of the spectrum of the operator $\mathscr{B}$ (see Section VB). We note that the symmetric form of the above can be written quite simply as

$$G(x, y) = Z(0)Q^2(y/x)^{\pm 1/4} \quad (3.22)$$

($+$ for $y < x$; $-$ for $y > x$).

Finally, we can consider the question of the square integrability of the kernels, which is of considerable importance both in the theory of spectra and in making practical approximations. It is possible to argue informally[43] that a whole class of kernels derivable from the Van Hove scattering theory, including the Wigner-Wilkins one, are square integrable. However a perfectly explicit proof can also be constructed without difficulty.[25] The approximation (3.22) guarantees good behavior at the origin, while the asymptotic expansion of erf (x) for large x can be shown to lead to exponential boundedness along the diagonal. It does not seem possible to derive the actual value of the square integral in closed form, though this would be a useful quantity.

Note on the Linearized Boltzmann Equation

A danger of confusion arises in comparing the test-particle relaxation problem for $\gamma = 1$ with the linearized solution of the Boltzmann equation for a system near equilibrium. Insofar as the latter describes the probability of finding *any* particle of the closed system with velocity $d\mathbf{V}$ about $\mathbf{V}$, rather than simply one of an ensemble of test particles, the scattering kernel for the problem can be expected to differ essentially from the one just considered (Fig. 2). In fact the vector kernel for the LBE is as follows (putting $V_0 = V(2kT/m_1)^{-1/2}$):

$$K_B(\mathbf{V}_0, \mathbf{V}_0') = Ae^{-V_0'^2}\left[\frac{2}{|\mathbf{V}_0 - \mathbf{V}_0'|}\right]\exp\left\{\left(\frac{|\mathbf{V}_0 \times \mathbf{V}_0|}{|\mathbf{V}_0' - \mathbf{V}_0|}\right)^2\right\} - |\mathbf{V}_0' - \mathbf{V}_0|\Bigg] \tag{3.23}$$

This differs significantly from the $\gamma = 1$ form of Equation (3.10), which reads

$$K(\mathbf{V}_0, \mathbf{V}_0') = \frac{A}{|\mathbf{V}_0 - \mathbf{V}_0'|}e^{-V_0'^2}\exp\left\{\left(\frac{|\mathbf{V}_0' \times \mathbf{V}_0|}{|\mathbf{V}_0 - \mathbf{V}_0'|}\right)^2\right\} \tag{3.24}$$

The second term in the expression for K_B may be explained as representing the decreased probability of large changes in velocity when the average taken also includes the *recoil* atom affected in the collision. However, even this comparison is confusing since the kernel (3.23) is not a stochastic transition probability at all; it is certainly not non-negative and has never been shown to be positive definite. The energy kernel corresponding to Equation (3.23) likewise includes an additional term and reads

$$\begin{aligned} K_B(x, y) &= A\pi^{1/2}[x^{-1/2}\,\mathrm{erf}\,(y^{1/2}) - (\tfrac{1}{3}y^{3/2} + xy^{1/2})e^{-y}]\,;\ y < x \\ &= A\pi^{1/2}[x^{-1/2}e^{x-y}\,\mathrm{erf}\,(x^{1/2}) - (\tfrac{1}{3}x^{3/2} + yx^{1/2})e^{-y}]\,;\ y > x \end{aligned} \tag{3.25}$$

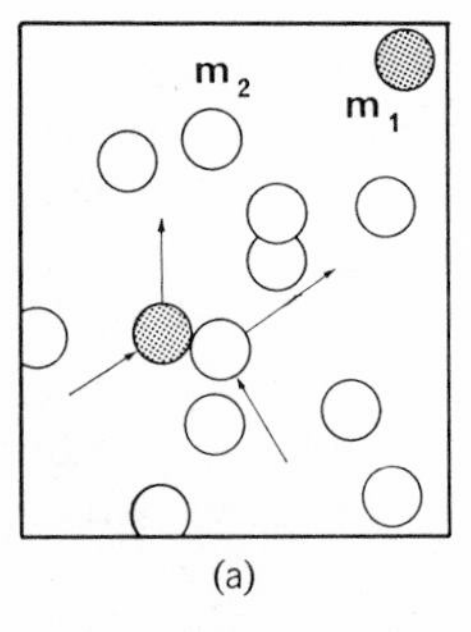

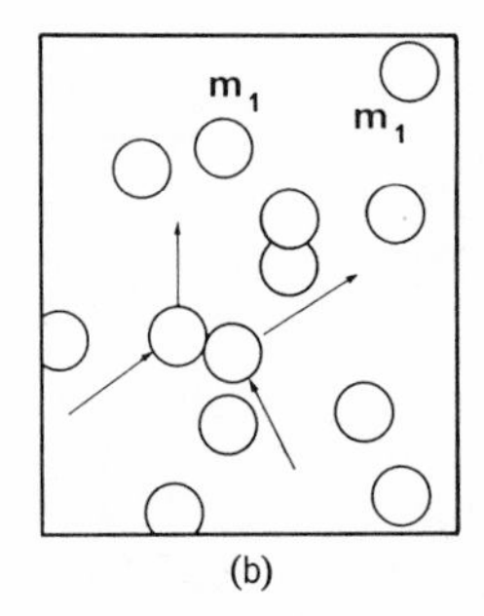

Fig. 2. The three-dimensional Rayleigh problem. Members of a dilute ensemble of labeled atoms, mass m_1, undergo random collisions with heat-bath atoms, mass m_2, preserved at an equilibrium temperature T (a). In the linearized Boltzmann problem (b), identical atoms exchange energy in a near-equilibrium state. The evolution is measured by a distribution function taken over *all* atoms and not just a labeled subset. Relaxation is therefore slower than for case (a) with $m_1 = m_2$.

The first clear distinction between these two cases seems to have been drawn by Kuščer and Williams,[15] though a detailed account of the LBE kernel is to be found in Grad.[3,16]

Note on the Van Hove-Glauber Theory of Neutron Scattering

There is an important, though slightly fortuitous, connection between the Wigner-Wilkins kernel for hard-sphere scattering and the Van Hove-Glauber theory of scattering for neutrons in a moderator.[17,18] Although the present theory can be developed quite independently of this, some appreciation of the implications and scope of the Van Hove theory is essential in translating from the neutron literature into the language of molecular systems. Without going into details (see, for example, Williams,[4] and Eggelstaff[19]), we shall simply state the extent of and limitations to this parallel.

The Van Hove theory relates the differential cross section for a statistical assembly of scatters to their space-time correlation function $G(\mathbf{r}, t)$ through its Fourier transform. It is entirely conditional upon the validity of the Born approximation, which, in the case of neutrons, may be trivially satisfied by assuming interaction through the Fermi pseudo-potential. When the correlation function for an ideal gas moderator (heat bath) is introduced, a form essentially the same as the Wigner-Wilkins kernel results and this, rather than the original kinetic-theory argument, is now usually taken as the starting point for discussion of the ideal gas moderator.[4] Although the formal similarity between the results for the Fermi pseudo-potential and classical hard-sphere dynamics is not surprising, it must be emphasized that the former is a quantum-mechanical case and that the taking of its classical limit is not without complications.[4,19]

The practical effect of these considerations is that, while any neutron-scattering treatment may be translated immediately into the corresponding atomic problem for *hard-spherical* systems (which can be in liquid or crystalline as well as ideal-gas form), the lack of validity of the Born approximation for thermal scattering by softer potentials rules out any extension to more general cases. This concentration on the hard-sphere gas is unfortunate but at the same time very fruitful. A considerable number of papers, particularly in the journal *Nuclear Science and Engineering*, might have been written as investigations of the hard-sphere gas with minor changes of wording, and numerical results for the gaseous moderator can be accepted without change under this heading. Although these parallels are not widely appreciated, they have been recognized and clearly stated by a few authors, particularly Desai and Nelkin[20] and Kuščer and Williams[15] (see also Williams[21]).

C. Approximate and Synthetic Kernels

Before we can take up the question of practical methods of solution in any detail it will be necessary to discuss some of the derived properties of the transition kernel which go into the various approximations of use in numerical work.

1. *Moments of the Transition Kernel*

We shall refer to the following quantities as the *moments* of the transition kernel:

$$k_n(x) = \int_0^\infty K(x, y)y^n \, dy \tag{3.26}$$

These are related in turn to the *transfer moments* which can be written

$$a_n(x) = \int_0^\infty K(x, y)(y - x)^n \, dy \tag{3.27}$$

Equivalent relationships in terms of the symmetric kernel $G(x, y)$ may also be written. Direct evaluation of the integrals for the hard-sphere kernel is extremely tedious but presents no difficulties of principle. Some of the lower transfer moments of the Wigner-Wilkins kernel were obtained in this way by Andersen and Shuler.[13] A number of shortcuts are possible. One is to define a parametric quantity

$$F(x, \mu) = \int_0^\infty K(x, y)e^{-\mu(x-y)} \, dy \tag{3.28}$$

in terms of which the transfer moments are

$$a_n(x) = (-)^n \left[\frac{\partial^n}{\partial \mu^n} F(x, \mu) \right]_{\mu=0} \tag{3.29}$$

We shall also be needing related quantities defined from the symmetric kernel. Writing, as before, $M(x) = [N_0(x)]^2$, we define

$$g_i(x) = \int_0^\infty G(x, y)N_0(y)y^i \, dy = k_i(x)N_0(x) \tag{3.30}$$

2. *Matrix Elements of the Kernel*

Further quantities of interest are obtained on integrating the moments again with the equilibrium distribution as a convergence factor. These are usually called the "matrix elements" of the kernel, though we shall use other types of derived matrices as well. We have the alternative expressions

$$(\mathbf{S})_{ij} = \int_0^\infty M(x)x^i k_j(x)\,dx \tag{3.31}$$

$$= \int_0^\infty N_0(x)x^i g_j(x)\,dx = (\mathbf{S})_{ji} \tag{3.32}$$

Various possibilities exist for obtaining asymptotic expansions of the transfer moments for large x and of both these and the elements $\mathbf{S}_{ij}$ for the two extremes of mass ratio.[4]

3. *Expansion in Orthogonal Functions*

The expansion of the transition kernel in a double series of orthogonal functions is important both as a computational and theoretical tool. Suppose we have an orthonormal basis $\{\phi_i\}$ on the interval $(0, \infty)$. Then the following representation of the symmetric kernel may be constructed:

$$G^N(x, y) = \sum_{i=1}^{N} \sum_{j=1}^{N} G_{ij}\,\phi_i(x)\phi_j(y) \tag{3.33}$$

where

$$G_{ij} = \langle \phi_i | \mathscr{G} | \phi_j \rangle = G_{ji}$$

$$= \int_0^\infty \phi_i(x) \int_0^\infty G(x, y)\phi_j(y)\,dy\,dx \tag{3.34}$$

The normal conditions for generalized Fourier expansions apply (Ref. 22, p. 83), namely that, for G_N to converge in the mean to G as $N \to \infty$, $G(x, y)$ should be square integrable. This is the case for the hard-sphere kernel[25,43] but is not necessarily so for other transition kernels (see discussion below). It should be emphasized here that expansions of this type carried out with a modest number of basis functions (say, $N \approx 20$) are not particularly accurate. Moreover they are liable to give oscillating character with negative-going regions which violate the positive kernel condition $G(x, y) > 0$ for all x, y. (We note, however, that the *positive definite* character of the kernel *is* preserved in the representation.)

4. *The Bohm-Gross Kernel*

A natural way to make an orthogonal expansion of a transition kernel is in terms of functions orthogonal to $N_0(x)$, which is the *exact* eigenfunction corresponding to equilibrium. If now as an extreme case we take only the first term in the expansion (3.33), the approximate kernel becomes $G^0(x, y) = G_{00}N_0(x)N_0(Y)$. Working out the meaning of G_{00} with the help of Equation (2.10), we find that, in fact, $G_{00} = Z_{00}$ with Z_{00} the mean collision number in the heat bath at equilibrium. Translating back into the unsymmetric kernel, this becomes equivalent to the approximation

$$K^0(x, y) = Z_{00}M(y) \tag{3.35}$$

This simple kernel, which clearly satisfies the detailed-balance condition (2.8), has a ready physical explanation. Each system particle, of whatever velocity, is given a chance of collision Z_{00}^{-1} per second and the outcome is precisely the equilibrium distribution—that is, system particles undergo "instant thermalization" on their first collision. This clearly has the effect of collapsing the whole detailed spectrum of the process into the single relaxation time Z_{00}^{-1}. Although so simple, the vector form of this kernel has played a part in relaxation theory, having been introduced by Bohm and Gross[23] in a study of damped plasma oscillations. It also provides a certain link with the single relaxation-time approximation for the full Boltzmann equation (Ref. 73, Section 13).

5. *Degenerate Kernel Approximations*

As we have indicated, the approximations which can be achieved with the usual basis sets constructed from orthogonal polynomials are likely to be poor, mainly because of their unsuitability for fitting the cusp at $G(x, x)$ and preserving the positive character of the kernel. A very powerful alternative exists which avoids some of this difficulty and has other computational advantages. This is to expand the kernel in a series of degenerate kernels where the factors in each term are not the orthogonal basis appearing in Equation (3.33) but other functions more suitably tailored to the profile of $G(x, y)$. As before, we shall work with the symmetric form, since this simplifies labor considerably in practical calculations. The method described is due to Shapiro and Corngold.[24]

Consider the approximation

$$G_N(x, y) = \sum_{i=1}^{N} \sum_{j=1}^{N} B_{ij}\, g_i(x) g_j(y) \tag{3.36}$$

where the functions $g_i(x)$ are as defined in Equation (3.30) and the matrix $\mathbf{B}$ is to be determined.

If we now require consistency in the sense that both sides multiplied by $N_0(y)y^i$ and integrated should yield $g_i(x)$, then the following condition is obtained:

$$g_k(x) = \sum_{i=0}^{N} \sum_{j=0}^{N} \mathbf{S}_{kj} B_{ij}\, g_i(x) \tag{3.37}$$

This clearly requires that $\mathbf{B} = \mathbf{S}^{-1}$. The degenerate-kernel expansion of $G(x, y)$ thus reads

$$G_N(x, y) = \sum_{i=0}^{N} \sum_{j=0}^{N} (\mathbf{S}^{-1})_{ij} g_i(x) g_j(y) \tag{3.38}$$

where the matrix $\mathbf{S}$ is that given in Equation (3.27). For convenience in later calculations we shall rewrite this in the form

$$G_N(x, y) = \sum_{j=0}^{N} f_j(x) g_j(y) \tag{3.39}$$

with $f_i(x)$ defined by

$$f_j(x) = \sum_{i=0}^{N} (\mathbf{S}^{-1})_{ij} g_i(x) \tag{3.40}$$

The greater efficiency of this type of expansion can be traced to the fact that *the Nth order degenerate-kernel expansion correctly reproduces all the moments of the transition kernel* (*and hence all the transfer moments*) *up to order N*. This contrasts with the orthogonal function expansion where the separate terms relate only to Fourier components in the particular basis set and do not embody any clear statistical property of the kernel. The statement in italics is readily proved on converting the representation (3.38) back into unsymmetric form and inserting the result into the definition (3.26) for the moments $k_n(x)$.

Shapiro and Corngold[24] demonstrate convincingly how a five-by-five degenerate kernel representation can reproduce the shape of $G(x, y)$ to an accuracy that would probably require nearer one hundred squared polynomial terms.

6. *The "Amnesia" Kernel*

Just as with the orthogonal function representation it is interesting to see what happens in the "one-by-one" degenerate-kernel case. Interpreting the necessary quantities we find from Equations (3.26) and (3.30) that

$$g_0(x) = N_0(x) Z(x) \tag{3.41}$$

and

$$S_{00} = Z_{00} \tag{3.42}$$

Thus the representation of the symmetric kernel becomes

$$G_0(x, y) = Z_{00}{}^{-1} N_0(x) N_0(y) Z(x) Z(y) \tag{3.43}$$

and correspondingly,

$$K_0(x, y) = Z_{00}{}^{-1} M(y) Z(x) Z(y) \tag{3.44}$$

Since the transition probability from a state x is proportional only to the collision number, a system atom scattered to state y has no "knowledge" of where it originated—hence the above name. This approximate kernel has been used for some time in neutron transport theory. Its properties are by no means as trivial as those of the Bohm-Gross kernel just described; in fact, as we shall see later, its spectral behavior is highly interesting.

It should be added in conclusion of this section that, although the various matrix elements for polynomial and degenerate-kernel expansions are conceptually quite straightforward integrals, very little progress can be made with calculations so long as numerical quadrature is required for their evaluation. To obtain as many as several hundred double integrals at each mass ratio with difficulties arising from the cusp at $x = y$ is hardly a practical undertaking unless explicit algebraic expressions can be found. In all the cases where extensive calculations have been made[24,25] the key step has been to obtain these algebraic forms. The necessary expressions for the hard-sphere kernel, which are extremely complicated and tedious to write, will be found in the papers cited.

IV. THE EIGENVALUE PROBLEM

Although certain direct methods of solution are possible, the whole character of the transport equation (2.1) and the initial-value problem is intimately bound up with the eigenvalue properties of the collision operator appearing on the right-hand side. In this article we shall consider only the spatially homogeneous case governed by the symmetric Master Equation (2.11).

As a first move the equation is reduced by separation of variables:

$$h(x, t) = N(x)\Theta(t) \tag{4.1}$$

which leads immediately to time-dependent factors of the form

$$\Theta(t) = \text{const} \times e^{-\lambda t} \tag{4.2}$$

and an integral-operator eigenvalue problem*

$$[Z(x) - \lambda]N(x) = \int_0^\infty G(y, x)N(y)\, dy \tag{4.3}$$

or

$$\mathscr{B}N(x) = -\lambda N(x) \tag{4.4}$$

* The use of the minus sign with λ implied as positive is conventional. Since $\mathscr{B}$ is negative definite we can take it in all following equations that $\lambda \geqslant 0$ with equality giving the equilibrium condition.

If the collision number Z were independent of x as, for example, in the relaxation of internal degrees of freedom, the solution of the eigenvalue problem would be relatively straightforward and could be expected to yield a complete set of eigenfunctions $N_k(x)$ with a finite or infinite point spectrum λ_k.

Let us suppose for the moment that this simple picture suffices for the more complicated Master Equation with the "scattering-out" term containing the state-dependence $Z(x)$. The solution of the initial-value problem for $P(x, t)$ or $h(x, t)$ is then a fairly routine matter.[26] The symmetry of the kernel guarantees real eigenvalues and orthogonal eigenfunctions $N_k(x)$, which we shall assume normalized to unity:

$$\int_0^\infty N_i(x)N_j(x)\,dx = \delta_{ij} \tag{4.5}$$

On solving the eigenvalue problem for the $N_k(x)$ and λ_k solutions can be constructed in the form

$$h(x, t) = N_0(x) + \sum_{k=1}^{\infty} a_k N_k(x)e^{-\lambda_k t} \tag{4.6}$$

where the expansion coefficients are to be determined by the orthogonality property as

$$a_k = \int_0^\infty h(x, 0)N_k(x)\,dx \tag{4.7}$$

In taking Equation (4.6) as a general solution it is tacitly assumed that the spectrum is infinite and the set $N_k(x)$ *complete* for L_2-initial conditions.

The realization, some ten years ago, that this simple outcome is not the case was a landmark in transport theory. Arguing from simple examples in the spatial diffusion problem, Case[27–29] and others pointed out that the presence of the function $Z(x)$ as a "multiplicative operator" introduced a singular character into the operator $\mathscr{B}$ with the effect that its spectrum must contain a bounded continuous region in addition to any discrete eigenvalues that might occur. The point, which had previously been made in the context of plasma-theory by Van Kampen[30] and had connections going back to Dirac,[31] was quickly taken up and its implications for the spatially homogeneous problem explored.[32–36]

It is hardly an overstatement to say that the consequences of this to any simple relaxation theory are quite traumatic. Even though an infinite point spectrum may still be obtained, a simple sum of the type (4.6) is inadequate since the set $N_k(x)$, though possibly infinite, is no longer *complete* and thus cannot form a basis for the solution of the initial-value problem. Furthermore, although the domain of the continuous spectrum may be easy to determine, the continuous eigenfunctions $N(x, \lambda)$ will be (Schwartz-) distributions and will require very special handling as regards normalization, the fitting of initial conditions, and so on. Even worse, the possibility exists that, for some kernels at least, the discretum might be empty (except for $\lambda_0 = 0$) and the evolution of the system governed entirely by the continuous spectrum. In this case the relaxation would be in no sense exponential and the continuum part of the solutions might carry a very special physical interpretation. Although certain aspects of these problems are still incompletely understood and we are still a long way from having explicit computations of the continuum eigenfunctions for a useful kernel, the main features of the theory of the "singular Master Equation" (2.6) have been made clear and prescriptions have been given for the calculation of solutions, both discrete and continuous.

We shall describe this work in the next section. Here it will be sufficient to emphasize that Equation (4.6) must be replaced by one of the form

$$h(x, t) = N_0(x) + \mathsf{S}_{\lambda > 0}\, a(\lambda) N(x, \lambda) e^{-\lambda t} \tag{4.8}$$

where the S symbol indicates both summation over the discretum and integration over the continuum as appropriate. Even so, proof is required that this form of expression is a valid representation of the solution for all reasonable initial conditions $h(x, 0)$. We shall return to this point later.

V. SPECTRAL THEORY OF THE MASTER EQUATION

A. General

Although even the semi-rigorous theory of singular operators such as $\mathscr{B}$ can hardly be expected to be easy, the existence proofs for the complex types of spectra involved are probably of more practical concern than the corresponding ones for more well-behaved operators.[22] We shall not attempt here to discuss the more formal side of this theory (for this, see Shizuta[37]), but will outline some of the qualitative treatments which have been introduced in neutron transport studies and, more recently, applied to the linearized Boltzmann equation. It will be seen that these are to some extent constructive; they lead to certain definite predictions about the numerical values of the discrete λ_k and have an intimate connection with practical methods such as the Rayleigh-Ritz and WKB procedures.

To see the origin of the singularity of the integral operator $\mathscr{B}$ it is only necessary to rewrite Equation (4.3) with the term $Z(x) - \lambda$ as a denominator on the right-hand side:

$$N(x) = \int_0^\infty \frac{G(x, y)}{Z(x) - \lambda} N(y)\, dy \tag{5.1}$$

Clearly, for any value of λ to which $Z(x)$ can become equal for some x in the range of the integral, there will be an infinity, which must be compensated by a delta-function term in the eigenfunction $N(x)$ and the solution $h(x, t)$. The continuum region of the spectrum (C) thus fills the whole set of real λ for which $Z(x_\lambda) = \lambda$ has a root x_λ. Since $Z(x)$ is usually continuous and monotonic with a bound at $Z(0)$, the spectrum can be expected to be of one of the two forms shown in Figures 3a and 3b, depending on whether $(\partial Z/\partial x)$ is positive or negative at the origin. These complications may be incorporated into the eigenvalue equation by the formal device of writing

$$N(x, \lambda) = \frac{\mathfrak{P}}{Z(x) - \lambda} \int_0^\infty G(x, y) N(y, \lambda)\, dy + \omega(\lambda)\, \delta(x - x_\lambda) \tag{5.2}$$

The continuum weighting function $\omega(\lambda)$ is to be found and the $\mathfrak{P}$ indicates a principal-value prescription which must be used in any integral involving $N(x, \lambda)$ where a singularity occurs. In practice $\omega(\lambda)$ may be determined by defining a normalization for the functions $N(x, \lambda)$. An obvious choice is

$$\int N(x, \lambda)\, dx = 1 \tag{5.3}$$

and in this case we have, operating on Equation (5.2),

$$\omega(\lambda) = \mathfrak{P} \int_0^\infty [\lambda - Z(x)]^{-1} \int_0^\infty G(x, y) N(y, \lambda)\, dy\, dx. \tag{5.4}$$

The form of solution (5.2) is unfamiliar and probably mystifying at first sight. In practice it is implemented by obtaining a finite resolution of the integral on the right in terms of coefficients and variational functions and then entering this into the expression as written along with the delta function. (We note that the latter can equally well be written $\delta(\lambda - Z(x))$ provided that it is correctly interpreted.)

The crucial step of writing the solution of the equation

$$(x - \lambda) g(x) = f(x) \tag{5.5}$$

in the symbolic form

$$g(x) = \mathfrak{P} \frac{f(x)}{x - \lambda} + \omega(\lambda)\delta(x - \lambda) \tag{5.6}$$

was taken by Van Kampen,[30] though Dirac[31] had previously used this expression without the arbitrary function $\omega(\lambda)$. The quantity x in the above can be replaced by a function, for example, $Z(x)$, through a simple change of variables, and in this context is said to be a "*multiplicative operator*." This type of operator equation is slowly establishing itself in the literature outside the confines of quantum mechanics and formal distribution theory.*

In explanation of the above, we cannot do better than quote from Van Kampen's original paper: "Since [the term $f(x)/(x - \lambda)$] is a distribution function, it is sufficient to know how to use it for calculating averages, for example, how to integrate it after multiplication with other functions. Hence [it] is a good distribution as soon as a prescription is given how to integrate across the pole. This prescription cannot be determined *a priori*, for instance by decreeing that the Cauchy principal value is to be taken. On the contrary, every different way of dealing with the pole gives rise to a different distribution function. All these different functions are comprised in [the result given.]"

As we have seen, rather more is known about the possible forms of the collision number $Z(x)$ than about the transition kernels themselves. Using the results of Grad[16] it follows that, for "hard" inverse-power interaction laws ($V(r) = r^{-s}$ with $s > 5$), the spectrum must be of the form Figure 3a, while for "soft" potentials ($s < 5$), it must be of the form Figure 3b. The intermediate case $s = 5$ is the Maxwell gas, which can be shown to give the single point spectrum $\lambda = Z(0)$, = the constant collision number (Fig. 3c). We see again how totally unrepresentative this case is as an indicator of behavior for less special force laws. For want of further knowledge of the scattering behavior of potentials more general than the inverse-power type, it can only be conjectured that the spectrum is primarily determined by the core-interaction, with "hard" repulsion giving increase of collision number with velocity and hence a type 3a spectrum, and "soft" repulsion giving decrease of collision number and hence type 3b.[16] (The hard-sphere case may be considered as a limiting form of the r^{-s} potential for $s \to \infty$.) However, against this it has been pointed out[16] that all the known results except the hard-sphere case are heavily dependent on the introduction of a suitable cutoff, for which there is no unique procedure. *Were it possible* to treat a kernel systematically for

* The standard introductions to distribution theory are Schwartz[38a] and Lighthill.[38b] More recent expositions are Zemanian[39a] and Donoghue.[39b] Cercignani[32] gives an elementary account of the kinetic-theory aspects of Schwartz-distributions, and some useful background to singular equations and the Cauchy principal value will be found in Tricomi (Ref. 22, Chap. 4). For a recent application of multiplicative operators, see Tokizawa.[40]

infinite-range interactions, its spectrum would presumably be totally discrete since it would give $Z(0) = \infty$.

The physical manifestations of these various models can be no less varied than the spectra themselves. Models of type 3a will give non-exponential behavior at short times, going over to exponential relaxation as the system is "aged," with long-time behavior dominated by the smallest nonzero eigenvalue λ_1. By contrast, systems of type 3b will begin to relax exponentially but with progressive domination by the "continuum modes." This breakdown of exponential behavior is clearly connected with the slow relaxation of very fast particles, and this can sometimes be observed experimentally as with the "electron runaway" phenomenon in plasma physics.[41] The occurrence of such radical changes in the evolution of a system for relatively minor variations in the scattering law is widely admitted to be one of the most worrying features of contemporary kinetic theory, whether linear or nonlinear.[16]

Turning to the discrete part of the spectrum we can see that, so long as λ is not in the continuum range, Equation (5.1) is nonsingular and of Fredholm type; it can thus be expected to yield a discrete spectrum $\lambda_0 < \lambda_1 < \lambda_2 \ldots \lambda_N$ where N can be finite or infinite. The possibility of discriminating between finite and infinite discrete spectra without first arriving at solutions is taken up in the next section.

As we have emphasized, the occurrence of an infinite set of discrete eigenfunctions $N_k(x)$ $(k = 0, 1, 2, \ldots \infty)$ is no guarantee that they will form a *complete* set for the expansion of any wider class of functions on the range of x; neither can it be assumed without detailed proof that the set $N_k(x)$ supplemented by the continuum set $N(x, \lambda)$ $(\lambda > Z(0))$ is complete for the class of all "reasonable" initial distributions $h(x, 0)$.*

That this is in fact so was first demonstrated by Koppel,[34] writing on the neutron thermalization problem. His proof was later elaborated somewhat in the context of the linearized Boltzmann Equation by Ferzinger.[35] The result is subject to the condition of monotonicity of the function $Z(x)$, but this may be relaxed with slight weakening of the class of functions which may be expanded.[36] The proofs are constructive and consist in showing that an expansion function $A(\lambda)$ can always be found for the representation of any function orthogonal to the discrete $N_k(x)$ as a weighted integral over the continuum eigenfunctions $N(x, \lambda)$. It follows that any "reasonable" function $f(x)$ can then be expressed in the form

* "Reasonable" may be taken in the present context to describe any function which possesses a Laplace transform. This is altogether less restrictive than the condition of square integrability, though the simple integrability of the initial distribution $P(x, 0) = N_0(x)h(x, 0)$ remains necessary for conservation of probability.

$$f(x) = \sum_{k=0}^{M} a_k N_k(x) + \int_{\lambda \in c} A(\lambda) N(x, \lambda)\, d\lambda \tag{5.7}$$

where M can be finite or infinity and the integral is over whatever interval is filled by the continuum. The solution of the initial-value problem in the form (4.8) is then valid.

Two final complications may be mentioned, but will not be pursued. The first is that, if the continuum weight function $\omega(\lambda)$ were to vanish for some $\lambda > Z(0)$, then discrete eigenvalues might appear embedded in the continuum. It is conjectured,[34] though so far unproven, that this type of behavior cannot occur with physically realizable kernels; and no actual or synthetic kernels with such a peculiarity appear to have been found. The second possibility is that a contribution to the continuum might arise, not from the "multiplicative operator" $Z(x)$, but through the kernel $G(x, y)$ itself in the case where the latter is non-square integrable.[22] Again, this has not been observed but is something to be watched for if new kernels are discovered.

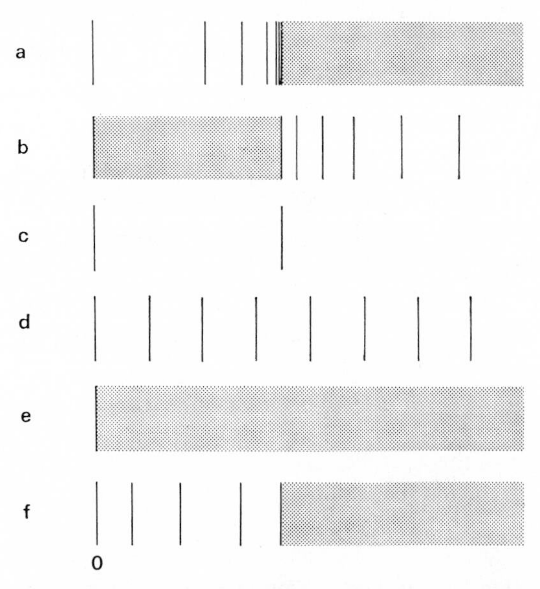

Fig. 3. Types of spectra in linear relaxation processes. (a) Spectrum for "hard" interactions with point of accumulation below the continuum threshold. (b) Spectrum for "soft" interactions. (The structure of the discrete part is unknown in this case.) (c) The Maxwell gas. (d) "Hard" system in the Rayleigh limit ($m_1 \gg m_2$). (e) "Hard" system in the Lorentz limit ($m_1 \ll m_2$). (f) Spectrum for crystal moderators.

This overall picture of the eigenvalue spectra cannot be complete without some consideration of the behavior in the two important limits $\gamma \to 0$ (the Rayleigh gas) and $\gamma \to \infty$ (the Lorentz gas). Since, in three dimensions we have $Z(0) = 2A/\gamma^{1/2}$ for the hard-sphere gas, it is clear that, in this case, for $\gamma \to 0$, the continuum should disappear to infinity leaving a totally discrete spectrum (Fig. 3d). This will be demonstrated explicitly in Section VIIIB. The case $\gamma \to \infty$ is slightly more problematic. Since the collision frequency A contains a factor $m_1^{-1/2}$ the net effect is to make $Z(0)$ tend to zero on a real time scale as the mass of the heat-bath particles m_2 is increased. Thus, in these terms the continuum threshold converges down onto the zero eigenvalue $\lambda_0 = 0$ as the Lorentz limit is approached (Fig. 3e).

B. Kuščer-Corngold Theory

Although the problem arose quite independently in transport theory, the methods that have been applied to the analysis of the discrete spectrum of operators such as $\mathscr{B}$ have much in common with the techniques used in quantum mechanics for the investigation of the number of bound states supported by a given potential.[42] The object in both cases is to obtain information about the qualitative character of the discrete spectrum—for example, its finite or infinite nature and its points of accumulation—without the necessity of finding explicit solutions to the eigenvalue problem itself.

The starting point is to consider the *conjugate eigenvalue problem* for the operator of interest. One introduces a scaling factor c into the eigenvalue equation and absorbs the true eigenvalue λ as an implicit parameter in the operator itself. The eigenvalue problem for c is then considered and, in favorable cases, can lead to at least a qualitative picture of the behavior of the eigenvalues $c_n(\lambda)$ treated as functions of λ. Evidently, the true spectrum corresponds to the values of λ for which the family of curves $c_n(\lambda)$ take on the value unity. The main problem in analyzing the discretum for the integral operators of transport theory is to determine whether an infinite or simply a finite number of such values exist.

This method was first applied to the scalar Master Equation by Corngold and Kuščer.[43] In an earlier paper Corngold, Michael, and Wollman[44] had succeeded in exploiting the analogy between the Wigner-Wilkins differential equation (6.44) and Schrödinger's equation in order to prove the infinity of discrete eigenvalues for the special case of the hard-sphere gas with mass ratio $\gamma = 1$. This was possible by a transformation which put the equation into simple relationship with the Schrödinger equation for the one-dimensional x^{-2} potential which is known to have an infinite number of bound states. A later development by Williams[45,46] showed that this procedure can be made quantitative in terms of the WKB method (see Section VID).

In Kuščer and Corngold's method the eigenvalue equation (5.1) is rewritten in the form

$$c[Z(x)-\lambda] = \int_0^\infty G(x, y)\, N(y)\, dy \tag{5.8}$$

and a new variable is defined as

$$\Psi(x, \lambda) = [Z(x) - \lambda]^{1/2} N(x) \tag{5.9}$$

If now a new, singular kernel, $G(x, y; \lambda)$, is constructed as

$$G(x, y; \lambda) = G(x, y)[(Z(x) - \lambda)(Z(y) - \lambda)]^{-1/2} \tag{5.10}$$

we obtain an eigenvalue equation for $c(\lambda)$ in the form

$$c(\lambda)\psi(x, \lambda) = \int_0^\infty G(x, y; \lambda)\psi(y, \lambda)\, dy \tag{5.11}$$

A number of quite general properties of the $c_n(\lambda)$ can now be deduced. First, it is observed that

$$c_0(0) = 1 \tag{5.12}$$

This follows on translating Equation (5.8) back into the unsymmetrical form and integrating both sides over all x.

Secondly, Feymann's theorem[47] can be applied to the kernel (5.10) to obtain an expression for $dc_n(\lambda)/d\lambda$, which is then shown to be positive for all $0 < \lambda < \lambda$.* Thus, by some simple manipulations,

$$\frac{dc_n(\lambda)}{d\lambda} = \int_0^\infty \psi_n(x, \lambda) \int_0^\infty (\partial/\partial\lambda) G(x, y; \lambda)\psi_n(y, \lambda)\, dy\, dx \tag{5.13}$$

$$= c_n(\lambda) \int_0^\infty \frac{(\psi_n(x, \lambda))^2}{Z(x) - \lambda}\, dx \geqslant 0 \tag{5.14}$$

The square integral $\|G(x, y; \lambda)\|$ now plays a special part since this quantity is known to be a bound to the squares of the eigenvalues.[48] We have

$$\int_0^\infty \int_0^\infty [G(x, y; \lambda)]^2\, dx\, dy \geqslant c_0^2 > c_1^2 > \cdots > 0. \tag{5.15}$$

Kuščer and Corngold continue this analysis for various types of moderator—gas, liquid, and solid—using very general arguments made possible by the Van Hove scattering theory. Here the application must be specialized to the hard-sphere case, though the essential mathematics remains unchanged. A number of subtle steps follow before it is possible to establish the complete picture of the curves $c_n(\lambda)$. Attention is first turned to the range of λ very near to the continuum threshold λ^*. Since the transients due to these are necessarily governed by collisions in the small velocity range, the spectrum will be qualitatively unaffected by truncation of the kernel at some low velocity with use of the approximations (3.18) and (3.22) for the part remaining. A change of both independent and dependent variables leads to a differential equation of hypergeometric type whose boundary conditions can be recovered from the original integral equation. The crucial step is to show that this equation and boundary conditions correspond to an infinite set of complex eigenvalues and that equivalently all the $c_n(\lambda)$ are bounded by a finite value c^* to which number they tend as $\lambda \to \lambda^*$. The value of this bound is

$$c^* = 6(1 + \gamma)^2 \tag{5.16}$$

(Note a superfluous factor $\pi^{1/2}$ in Kuščer and Corngold's article, pointed out by Williams.[45])

Since $c^* \geqslant 6$ for the whole range of possible mass ratios, it is thus established that all the infinite family of curves cut the line $c = 1$ at some point, indicating the existence of an infinite point spectrum for $\lambda < \lambda^*$. A closer investigation shows that the point of accumulation is, in fact, at λ^*. Since this essential property is unaffected by the assumption of a cutoff in the kernel, the same behavior can be inferred for the original integral operator. The resulting picture is therefore that shown in Figure 4, with the actual spectrum $\lambda_n = c_n(1)$ of the type illustrated in Figure 3a.

The development just described is not entirely qualitative and a useful estimate of the rate of accumulation of the eigenvalues in the near-continuum region can also be derived from the differential equation. This is that

$$(\lambda^* - \lambda_n)/(\lambda^* - \lambda_{n+1}) \approx e^{4\pi/\sqrt{23}} \tag{5.17}$$

Numerical results we shall describe later show that, for the uniform gas ($\gamma = 1$) this estimate is remarkably good for all eigenvalues other than λ_1, though its reliability weakens in the Lorentz regime ($\gamma > 1$) along with that of the truncation approximation.

As indicated earlier, this whole spectral theory has been taken up in rigorous form by Shizuta,[37] who comes to the same conclusions in an abstract operator language.

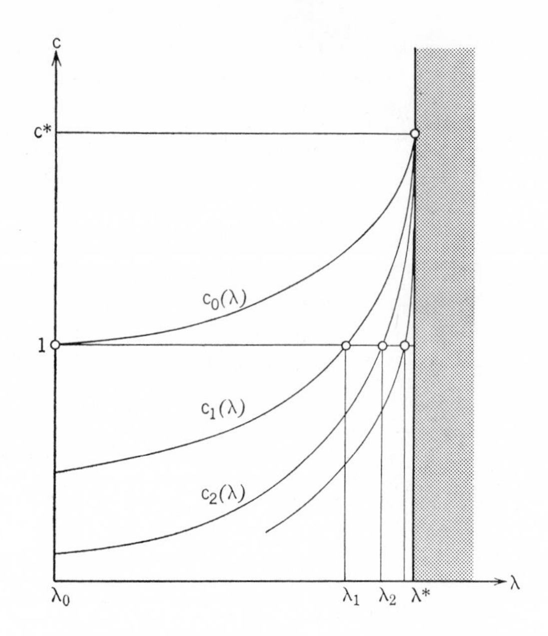

Fig. 4. Kuščer-Corngold diagram for the hard-sphere gas.

VI. PRACTICAL METHODS FOR THE DISCRETUM

As with most eigenvalue problems there is here considerable interplay between operations that are considered in the theory of the spectra and those that are of practical value in carrying out computations. Although this is perhaps less true in the rigorous operator-theoretic approach, the more accessible proofs we shall concentrate on here tend to be constructive and, with slight modification, lead to definite algorithms for obtaining information about at least part of the eigenvalue spectrum or, more directly, the initial-value problem itself. In doing this certain themes recur constantly such as the expansion of kernels in orthogonal functions, their expression in terms of transfer moments, and degenerate forms. The mechanics of deriving these expressions has been considered in Section IIIC and will not be discussed further here. We shall continue to concentrate on the scalar Master Equation (2.6) in energy or speed, though several of the results will retain validity or have bearing on the vector transport problem (2.1).

Although nonspectral methods can play some part, it will be clear that any method flexible enough to yield solutions for arbitrary initial conditions is likely to be directed to the determination of some "representative set" of eigenvalues and eigenfunctions for the operators $\mathscr{A}$ or $\mathscr{B}$. The term "representative set" conceals a certain dilemma which should be explained at the beginning. With a variety of methods available there may well be a choice between computations which yield a few good eigenvalues and functions, and others which yield more, but to less accuracy. No general rule can be given as to which alternative is most useful in a physical problem since, even with given initial conditions, different value may be placed on the representation of short-time behavior or behavior near equilibrium. The main point which must be emphasized is that the "true" eigenvalues for the transport problem are only in very exceptional cases physical observables; the exact numerical values of some of them may be sought in a Platonistic spirit, but (quite unlike the case in vibration theory or quantum mechanics, for example) their physical role is inseparable from the solutions into which they enter. Stated differently, for practical purposes a good *spectral resolution* of the operator concerned will have priority over the tabulation of particular eigenvalues λ_k.

Since $B(x, y)$ is a positive kernel we are allowed to write the spectral resolution in the form (Ref. 48, Chap. 3, Section 5)

$$B(x, y) \approx \sum_{k=0}^{N} \hat{N}_k(x)\hat{N}_k(y)\hat{\lambda}_k \tag{6.1}$$

where it is understood that the approximate eigenfunctions $\hat{N}_k(x)$ and eigenvalues $\hat{\lambda}_k$ are assumed to have been computed by one of the various methods to be described. (A continuum term could be added on the right-hand side if desired.) Once obtained, the above representation can be entered into solutions for the initial-value problem and its effectiveness judged accordingly. A good spectral resolution will be one which appears to solve the initial-value problem well over a specified time range; a bad one will falsify the evolution of the system, at least for some range of time. By constructing increasingly detailed resolutions and observing the convergence in their effect in predicting a time-dependent distribution function, a good idea of their value can be obtained. The point is that all this can be done without special consideration of whether the $\hat{\lambda}_k$ involved are good approximations to some particular *true* eigenvalues or not. Because of the peculiar nature of the spectra in the present type of relaxation problem, the futility of approaching the initial-value problem solely through a search for the "true" eigenvalues and functions can hardly be overemphasized.

A. Direct Methods

1. *Discretization*

An obvious possibility for the solution of the scalar Master Equation is to "discretize" both time and the energy space in as fine a mesh as possible and follow the evolution of the system by a computer "book-keeping" procedure at time intervals Δt. In this way the unsymmetric transport equation (2.6) is replaced by the difference scheme

$$P_i(t + \Delta t) = P_i(t) + \Delta t \left\{ \sum_{j=0}^{m} \tilde{K}_{ij} P_j(t) - \tilde{Z}_i P_i(t) \right\} \qquad (i, j = 0, 1, 2, \ldots, M) \tag{6.2}$$

where the kernel, collision number, and distribution functions have been replaced by vectors and matrices in an obvious way. Though it goes somewhat against the grain to admit it, procedures of this kind can probably give as great an accuracy as the more sophisticated methods which follow, but only of course for a single specified initial condition at a time. It is worth mentioning that, if the difference scheme is truncated at some level i^* and the approximate kernel is not "renormalized" to remove transitions above this, then the effect is statistically that of an absorbing barrier and the degree to which probability is conserved will be a good measure of the reliability of the truncation. The method will be greatly complicated and of much more limited value for the three-dimensional problem (for this, see Ref. 6).

To obtain results that are not tied to a particular initial distribution it is more sensible to discretize the energy-dependent eigenvalue equation after separation of the variables. Working in the symmetric form one arrives at a matrix eigenvalue problem (we shall use the tilde for discretized approximations and the caret for variational quantities):

$$\tilde{N}_{ik}[\tilde{Z}_i - \lambda] = \sum \tilde{G}_{ij} N_{jk} \tag{6.3}$$

$$\tilde{N}^{(k)}[\tilde{\mathbf{Z}} - \lambda \mathbf{I}] = \mathbf{G}\tilde{N}^{(k)} \tag{6.4}$$

where the correspondence with the continuous equation (4.3) is obvious. One may then reconstruct the initial-value solutions in the following form:

$$P(x, t) = \sum_{i<i^*} A_i(x) e^{-\tilde{\lambda}_i t} + \sum_{i>i^*} B_i(x) e^{-\tilde{\lambda}_i t} \tag{6.5}$$

Here i^* is the last mesh point before the continuum and the functions $A_i(x)$ and $B_i(x)$ are found from the vectors and the initial conditions. One may reasonably expect that, for a fine enough mesh, the values of the smaller $\tilde{\lambda}_i$ will approximate to the true eigenvalues λ_i, though the ones in the second summation can clearly have no physical reality. Instead, they move as the fineness of the mesh is increased, the lower ones approaching the continuum threshold and eventually emerging into the discretum. Only the very lowest can be said to "converge" to the true discrete eigenvalues. The pattern is thus as indicated in Figure 5, where three types of $\tilde{\lambda}_k$ can be distinguished:

Type A ($\tilde{\lambda}_k < \lambda^*$ and converged.) These are good estimates of the true eigenvalues λ_k.

Type B ($\tilde{\lambda}_k < \lambda^*$ but unconverged.) These are an attempt to "represent" the infinite set near the accumulation point λ^*.

Type C ($\tilde{\lambda}_k > \lambda^*$). These would appear to give an account of the continuum modes in whatever detail is possible. We shall refer to them as *pseudo-eigenvalues*.

Presumably as the fineness of the mesh is increased they will cover the continuum region ever more densely, but the precise sense in which the whole spread of approximate eigenvalues represents the original operator is an open and interesting mathematical question. Finally, we may note that the algebraic equations (6.4) do contain some reflection of the singularity of the eigenfunctions for $\lambda > \lambda^*$ since, as λ falls near to some $\tilde{Z}_i$ at a mesh point, this must be compensated by an increasing sharpness in the $\tilde{N}_{ki}$ components.

An account of this method is given by Wood,[49] who has obtained very accurate results for the lowest eigenvalues of the hard-sphere kernel at unit mass ratio (see Section VI). The three types of numerical eigenvalues

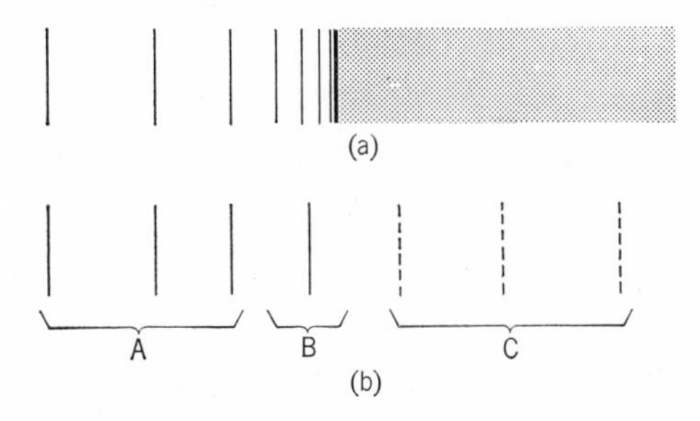

Fig. 5. Actual spectrum for a "hard" system (a) compared with an approximate variational spectrum (b). The lines (A) are converged estimates of λ_1 to λ_3; (B) is an unconverged representation of the higher discrete lines, (C) indicates "pseudo-eigenvalues" representing the continuum.

and the general problem of interpreting discrete representations of the continuum will recur with several of the other numerical methods to be described.

2. *Short-Time Expansions*

There is some compensation for the difficulty of computing continuum eigenfunctions in that the behavior at very short times, which for "hard" systems must reflect the far-continuum modes, can be obtained by direct application of the collision operator. To do this we write the formal solution of the Master Equation (2.6) for initial conditions $P(x, 0)$ as

$$P(x, t) = \exp(\mathscr{A}t).\, P(x, 0) \tag{6.6}$$

$$= \{1 + \mathscr{A}t + \tfrac{1}{2}\mathscr{A}^{(2)}t^2 + \cdots\}P(x, 0) \tag{6.7}$$

where formally the powers correspond to repeated action of $\mathscr{A}$, that is,

$$\mathscr{A}^{(n)} \equiv \{\mathscr{K} + [Z(x)]\}^n \tag{6.8}$$

$$\mathrm{A}^{(n)}(x, y) \equiv [K(x, y) + Z(x)\delta(x - y)]^n \tag{6.9}$$

Since $\mathscr{A}^{(n)}$ contains the multiplicative operator $[Z(x)]$ it requires rather special interpretation. We must notice that $\mathscr{K}$ and $[Z(x)]$ do not commute and then that powers $[Z(x)]^n$ are *not products but convolutions* of the delta function. Thus, using

$$\int \delta(x - w)\delta(w - y)\, dw = \delta(x - y) \tag{6.10}$$

we have that

$$A^{(2)}(x, y) = K^{(2)}(x, y) - (Z(x) + Z(y))K(x, y) + (Z(x))^2\delta(x - y) \tag{6.11}$$

with $K^{(2)}$ the second iterate kernel of K, in other words the *two-step* transition probability for $(x \to y)$. In general, defining the operation of kernel iteration as

$$K^{(n+m)}(x, y) = \int K^{(n)}(x, w)K^{(m)}(w, y)\, dw$$

$$K^{(1)}(x, y) = K(x, y) \tag{6.12}$$

we find that $A^{(n)}(x, y)$ contains a term $K^{(n)}(x, y)$, a term $(Z(x))^n \delta(x-y)$, and a whole variety of "diagrams" of the form

$$(Z(x))^a \cdot (Z(y))^b \cdot K^{n_1} \cdot Z^{(m_1)} \cdot K^{(n_2)} \cdot Z^{m_2} \cdots K^{n_k} \tag{6.13}$$

where the dots denote iteration by the relation (6.12) and the indexes sum to n.

Evidently the use of Equation (6.6) could be made into quite a long story; here we shall simply emphasize that the expansion is not purely formal, that it can probably be carried out in simple cases to order three or four, and that the results might well give insight into the puzzling question of the physical significance of the singular eigenfunctions in energy space.

The limitation of these equations to some three or four mean collision times does not seem unduly restrictive in the range $\gamma \approx 1$. Later results show that it is precisely in this time regime that the hard-sphere continuum eigenfunctions have most of their effect. Moreover, Monte Carlo calculations for the same system at unit mass ratio are well known to show an effective relaxation time of some *four* collisions.[50]

B. Rayleigh-Ritz Methods

As usual in variational eigenvalue calculations, the advantage of the Rayleigh-Ritz method is that it enables approximations to be improved by stages without the need to repeat much of the computation on taking a more detailed trial function. Polynomial expansions in one form or another have appeared for some time in both neutron transport studies and conventional kinetic theory, but their use to map out the detailed spectra of different scattering operators is comparatively recent. A theoretical prescription for obtaining detailed spectra was given by Koppel in 1962,[34] but the main source of numerical results remains the work of Shapiro and Corngold (1965),[24] who systemized several methods and investigated crystalline moderators as well as the Wigner-Wilkins kernel. A detailed investigation of the relaxation spectrum of the hard-sphere gas has also been carried out by Hoare and Kaplinsky[25] with special attention to the Rayleigh regime and the approach to the Fokker-Planck limit. Since the application of the Rayleigh-Ritz method to singular integral operators is not a familiar topic, we shall defer discussion of the results until we have outlined this and some other practical alternatives.

The most straightforward procedure (though not, as we have seen, the most accurate) is to expand the transition kernel $G(x, y)$ by a double series of orthogonal functions as in Equation (3.33). As before, we shall assume a complete, orthonormal set $\{\phi_0(x), \phi_1(x) \ldots \phi_N(x)\}$ such that $\langle \phi_i | \phi_j \rangle = \delta_{ij}$ and $G_{ij} = \langle \phi_i | \mathscr{G} | \phi_j \rangle$. If we then expand the eigenfunctions $N(x)$ in the same basis, the eigenvalue equation (4.3) becomes

$$N(x)[Z(x) - \lambda] = \sum_{i=0}^{N} \sum_{j=0}^{N} \alpha_k G_{ki} \phi_i(x) \tag{6.14}$$

where

$$N(x) = \sum_{i=0}^{N} \alpha_i \phi_i(x) \tag{6.15}$$

We should note that, although the set $\{\phi_i\}$ cannot properly represent a singular (non L_2) function themselves, the singular character of the problem is still retained in the factor $[Z(x) - \lambda]$ of the above equation. We have also tacitly assumed that the kernel $G(x, y)$ is square integrable, so that its orthogonal expansion converges in the mean.

As we remarked earlier, there is no *physical* necessity for this to be the case, though it is known to hold for the Wigner-Wilkins kernel. Lack of L_2 character in a kernel can arise in two ways—by unboundedness at some point such as the origin, or by insufficiently fast convergence in some direction, most likely for $[G(x, x)]^2$ along the diagonal. In the first case the implications for practical techniques are likely to be serious, but in the second much less so. To see this we may truncate the kernel at some point x', y' such that the new kernel is square integrable and the remainder can be treated as a perturbation δG. The whole spectrum will be shifted slightly; in particular the equilibrium eigenvalue will become

$$\lambda_0 = \langle \phi_0 | \delta G | \phi_0 \rangle$$

Since this is nonzero, no true equilibrium state will be reached and probability will not be conserved due to leakage of systems at the absorbing barrier x'. However, if the truncation is at high enough energy, the transient $\exp -\langle \phi_0 | \delta G | \phi_0 \rangle t$ can be made as small as desired and certainly slower than any other transients of interest in the initial-value problem. Replacement of $G(x, y)$ by the truncated kernel will then give acceptable results.

The next step in reducing Equation (6.14) to an algebraic eigenvalue problem can be taken in two quite different directions. We may either form the scalar product of each side with some particular $\phi_i(x)$ as the equation stands, or we may take the term $[Z(x) - \lambda]$ into the denominator on the right before doing so. The first method is simpler, but theoretically unpleasant in that it amounts to attempting a resolution of the delta function in terms of the polynomials; the second leads to singularities in the integral and a more difficult algebraic eigenvalue problem. Following Shapiro and Corngold[24] we shall call these alternatives the "explicit" and "implicit" methods. These will now be considered in further detail.

1. *The Explicit Method*

The explicit method leads immediately to the set of algebraic equations

$$[\mathbf{G} - \mathbf{Z}]\boldsymbol{\alpha} = -\lambda\boldsymbol{\alpha} \tag{6.16}$$

where $\boldsymbol{\alpha}$ is the vector of coefficients in Equation (6.15) and $\mathbf{Z}$ is constructed from the "matrix elements"

$$\begin{aligned} Z_{ij} &= \langle \phi_i(x) \,|\, Z(x) \,|\, \phi_j(x) \rangle \\ &= \int_0^\infty \phi_i(x)\phi_j(x) Z(x)\, dx \end{aligned} \tag{6.17}$$

Thus the eigenvalues of the integral operator are approximated by the roots of the secular equation

$$|\,\mathbf{G} - \mathbf{Z} - \lambda \mathbf{I}\,| = 0 \tag{6.18}$$

There will normally be N finite values and, with sufficient basis functions, the situation depicted in Figure 5 will again occur with the approximate $\hat{\lambda}_k$ distributed between "true," "unconverged," and "pseudo-eigenvalue" types. When expressed in the ϕ_n representation the equilibrium condition (2.7) is found to be simply $G_{i0} = Z_{i0}$ for all i so that, whatever the size of the matrices, the $\lambda_0 = 0$ eigenvalue emerges automatically.

On solving for the eigenvectors $\boldsymbol{\alpha}^{(k)}$ and eigenvalues $\hat{\lambda}_k$, the approximate eigenfunctions may be reconstructed using Equation (6.14).

2. *The Implicit Method*

The disadvantage of the explicit method just described is that, with polynomials of only moderate order, it must be altogether impossible to represent the near-singular behavior of the true eigenfunctions in the region of accumulation $\lambda \to \lambda^*$. This drawback is partly overcome in the "implicit" method which, with slight modifications, can also be used as a continuum variation method to determine the singular eigenfunctions.

Assuming for the present that λ is restricted to the discrete range $\lambda < Z(0)$, the factor $[z(x) - \lambda]$ in Equation (4.3) is taken into the denominator and scalar products are formed on each side. The result is the much more complicated eigenvalue problem

$$\boldsymbol{\alpha} = \mathbf{G}\mathbf{H}(\lambda)\boldsymbol{\alpha} \tag{6.19}$$

where $\boldsymbol{\alpha}$ is the vector of expansion coefficients as before, and the matrix $\mathbf{H}$ is the following function of λ:

$$H_{ij}(\lambda) = \int_0^\infty \frac{\phi_i(x)\phi_j(x)}{Z(x) - \lambda}\, dx \qquad (\lambda < Z(0)) \tag{6.20}$$

The secular equation to be solved is now

$$|\,\mathbf{GH}(\lambda) - \mathbf{I}\,| \tag{6.21}$$

It can be proved that this can have at most $N + 1$ roots under the condition $\lambda < Z(0)$; in practice, fewer than this number are likely to be found.[34]

The eigenfunctions are reconstructed as before from Equation (6.14) but their near-singular character in the region $\lambda \approx Z(0)$ is better displayed on writing the particular expression

$$\hat{N}_k(x) = \frac{1}{Z(x) - \hat{\lambda}_k} \sum_{j=0}^{N} \alpha_j^{(k)} \sum_{i=0}^{N} G_{ij}\, \phi_i(x) \tag{6.22}$$

This prescription was first suggested by Koppel[34] but does not seem to have been carried out in any detailed series of calculations. Its main interest is in connection with the possible computation of singular eigenfunctions, which we shall describe in a later section.

3. *Choice of Basis Functions*

There is some scope for choice in selecting suitable basis functions $\phi_i(x)$. From a theoretical standpoint it is natural to adopt the classical polynomials orthogonal on $(0, \infty)$ with weight function equal to the Boltzmann distribution for the equilibrium system. This leads to different orders of Laguerre polynomials for the possible independent variables: $L_k^{1/2}(x)$ for engery, $L_k^{\,1}(v^2)$ for speed, and $L_k^{\,1}(x)$ when the solution is required in energy flux. We shall quote only the simple energy case. Passing to the symmetric form required above, the functions are

$$\phi_n(x) = l_n^{\,1/2}(x) = [n!/\Gamma(n + \tfrac{3}{2})]^{1/2} x^{1/4} e^{-x/2} L_n^{\,1/2}(x) \tag{6.23}$$

with

$$L_n^{\,1/2}(x) = (n!)^{-1} x^{-1/2} e^{x} (d^n/dx^n) \cdot (x^{n+1/2} e^{-x}) \tag{6.24}$$

It will be shown below that these are just the orthogonal forms of the eigenfunctions of the hard-sphere Rayleigh scattering operator for the limiting case $\gamma \to 0$. An immediate consequence of this is that the expansion matrix G_{ij} will become increasingly diagonal in the regime $\gamma \ll 1$ with its off-diagonal terms reflecting the degree of "non-RFP" character in the relaxation process.

It emerges in a similar way that the most natural treatment at the Lorentz limit is to use *energy flux* as the dependent variable so that the orthogonal set based on $L_k^1(x)$ become eigenvalues of the Lorentz operator for flux relaxation. But, in contrast to the situation at the Rayleigh limit, this property does not provide solutions to the energy Master Equation and a variational calculation is still necessary to obtain the true Lorentz relaxation times (see Section VIIIC).

In view of these properties, it is somewhat surprising to find, according to the detailed studies of Shapiro and Corngold[24] that the above sets are by no means ideal from the point of view of obtaining fast convergence in the Rayleigh-Ritz calculations. It is now widely accepted in neutron transport theory that basis functions constructed in powers of $x^{1/2}$ (velocity polynomials) give a better fit to the eigenfunctions and hence faster convergence, at least in the region of moderate mass ratio.

4. *Approximations to* λ_1

So long as the continuum threshold $Z(0)$ remains distinct from the equilibrium eigenvalue $\lambda_0 = 0$, the first discrete eigenvalue λ_1 will always govern the final stages of approach to equilibrium, when the distribution function is of the form

$$P(x, t) = M(x) + a_1 N_1(x) e^{-\lambda_1 t} \tag{6.25}$$

In the earlier days of neutron relaxation theory before polynomial and degenerate-kernel methods were practicable, a number of approximate estimates of λ_1 were worked out. Whether these were written as a "two-term" polynomial expansion or through an ansatz postulating a single relaxation time with closeness to equilibrium, or by the variational method that follows, the essential content is the same.

The simplest method is to write the Rayleigh variational expression

$$\lambda_1 \leqslant \frac{\int_0^\infty f_1(x) \mathscr{B} f_1(x)\, dx}{\int_0^\infty [f_1(x)]^2\, dx} \tag{6.26}$$

where, in order to get λ_1 the trial function $f_1(x)$ must be made orthogonal to the equilibrium eigenfunction $N_0(x)$. The lowest-order function satisfying this condition is $f_1(x) = N_0(x)(x - 3/2)$. The denominator in the Rayleigh quotient is then just the equilibrium variance $\mathrm{Var}\,(x) = 3/2$ and the numerator can be simplified using the fact that $\mathscr{B} N_0 = 0$. The remaining integral is similar to that appearing in the "matrix elements"

S_{11} (Eq. 3.32). It can be evaluated explicitly for the hard-Sphere kernel with the result

$$\lambda_1 \leqslant \frac{16}{3} A \cdot \frac{\gamma^{1/2}}{(1+\gamma)^{3/2}} \tag{6.27}$$

This is interesting in that it predicts a functional dependence on mass ratio such that, on a mean-collisional time scale, λ_1 tends to zero at the extremes of mass ratio and has a broad maximum at $\gamma = 1/2$. Although the absolute accuracy of the above prediction cannot be expected to be high, the functional behavior should be approximately correct. Actually, since on the same scale $\lambda^* = Z(0) = 2A\gamma^{-1/2}$ we can see that, over a wide range of γ, the value predicted by Equation (6.27) is well into the continuum. Nevertheless the limiting behavior $\lambda_1 \to (16/3)\, A\gamma^{1/2}$ for $\gamma \to 0$ is correct, as we shall see below.

C. Degenerate Kernel Methods

Although the expansion of the kernel in terms of its eigenfunctions for the limit $\gamma \to 0$ is a neat method, well suited to describe the approach to this region, and probably gives as good an overall *spectral resolution* as any other, it is by no means the best for a determination of the *exact* lower discrete eigenfunctions in the region $\gamma \approx 1$. The reason for this is that successive terms in the expansion do not embody any salient feature of the kernel and, in general, approximate it rather badly and with marked oscillations. Moreover, an expansion of the form (3.33) does not preserve the positiveness property $K(x, y) > 0$.

The degenerate kernel approximation discussed in Section IIIC is a powerful alternative which, although computationally more difficult, is favored among neutron transport researchers for the calculation of the lower discrete eigenvalues up to about λ_4 or λ_5. Its great advantage is that it preserves the positiveness of the kernel and that successive terms in the expansion represent *exactly* the contributions of corresponding moments of the true kernel. Like the polynomial expansion method, it can be carried out in both explicit and implicit forms; since we have already described the construction of the approximate kernels in some detail it will be sufficient here to quote the algebraic eigenvalue problem obtained with the formulas for reconstruction of the eigenfunctions from it. The essentials of the method were first described by Shapiro and Corngold,[24] who worked in terms of the unsymmetric kernel. Here we shall give the equivalent symmetric forms which should be more convenient in practical computations.

1. *The Implicit Degenerate-Kernel Method*

Expanding the kernel by Equation (3.39), we obtain

$$N(x) = [Z(x) - \lambda] \sum_{j=1}^{N} f_j(x)\psi_j \tag{6.28}$$

where

$$\psi_j = \int_0^\infty g_j(y)N(y)\,dy \tag{6.29}$$

This leads to the eigenvalue problem

$$\mathbf{\Psi} = \mathbf{F}(\lambda)\mathbf{\Psi} \tag{6.30}$$

where

$$F_{ji}(\lambda) = \int_0^\infty \frac{g_i(x)f_j(x)\,dx}{[Z(x) - \lambda]} \qquad (\lambda < Z(0)) \tag{6.31}$$

The secular equation is thus

$$|\mathbf{F}(\lambda) - \mathbf{I}| = 0 \tag{6.32}$$

On determining a set of $\mathbf{\Psi}^{(k)}$ and λ_k, the approximate eigenfunctions are constructed through Equation (6.28).

2. *The Explicit Degenerate-Kernel Method*

In this method one attempts to represent the whole operator $\mathscr{B}$ including the singular term in degenerate kernels. The result is an approximate kernel of the form

$$G(x, y) - Z(x)\delta(x - y) \cong \sum_{i=0}^{N} \sum_{j=0}^{N} [\mathbf{S} - \mathbf{Z}]_{ij}^{-1} b_i(x)b_j(y) \tag{6.33}$$

where now

$$b_i(x) = g_i(x) - N_0(x)Z(x)x^i \tag{6.34}$$

$$(\mathbf{Z})_{ij} = \int_0^\infty M(x)Z(x)x^{i+j}\,dx \tag{6.35}$$

The following matrix eigenvalue problem is then obtained

$$\xi_i = \int_0^\infty b_i N(x)\,dx$$

$$[\mathbf{S} - \mathbf{Z}]^{-1}\mathbf{Q}\xi = \lambda\xi \tag{6.36}$$

with

$$(\mathbf{Q})_{ij} = \int_0^\infty b_i(x) b_j(x)\, dx \tag{6.37}$$

The secular equation can be written without the inverse matrix, although this will be needed in reconstructing the eigenfunctions. We have

$$|\mathbf{Q} - (\mathbf{S} - \mathbf{Z})\lambda| = 0 \tag{6.38}$$

$$\hat{N}_k(x) = \hat{\lambda}_k^{-1} \sum_{i=1}^{N} \sum_{j=1}^{N} [\mathbf{S} - \mathbf{Z}]_{ij}^{-1} b_i(x) \xi_i^{(k)} \tag{6.39}$$

Calculations with both implicit and explicit methods have been made by Shapiro and Corngold.[24] Their results will be evaluated below. A further modification which they have used is to expand using velocity moments instead of the energy moments defined in Equation (3.26). The equations are very similar and, as in the polynomial method, faster convergence is obtained.

D. WKB Methods: Separable Kernels

1. *The Uniform Gas* ($\gamma = 1$)

In discussing the hard-sphere kernel it was shown that a particularly simple form was taken when the mass ratio became unity. If we rewrite Equation (3.15) slightly, this can be expressed as follows:

$$\begin{aligned} M(x)K(x, y) &= U(x)V(y);\ x > y \\ &= U(y)V(x);\ y > x \end{aligned} \tag{6.40}$$

where

$$\begin{aligned} U(x) &= 2Ae^{-x} \\ V(x) &= \operatorname{erf}(x^{1/2}) \end{aligned} \tag{6.41}$$

$$M(x) = 2\pi^{-1/2} x^{1/2} e^{-x} \tag{6.42}$$

In this form its symmetry can be recognized as that of a Green function for a differential equation and the equation itself can be recovered on differentiating the eigenvalue expression as

$$e^{-x} \int_0^x \operatorname{erf}(y^{1/2}) \psi(y)\, dy + \operatorname{erf}(x^{1/2}) \int_x^\infty e^{-y} \psi(y)\, dy$$

$$= \frac{1}{2\pi^{1/2} A} [Z(x) - \lambda] M(x) \psi(x) \tag{6.43}$$

$\psi(x)$ is now the eigenfunction for the unsymmetric Master Equation; that is, it corresponds to particular solutions of the form $P(x,t)=M(x)\psi_\lambda(x)\exp(-\lambda t)$. Operating on both sides of (6.43) with the operator $(d^2/dx^2)+(d/dx)$, we obtain the rather complicated form

$$\frac{d}{dx}\left\{M(x)[x^{1/2}(d/dx)(x^{1/2}Z(x)]^{-1}\frac{d}{dx}\left[\frac{1}{\lambda}-\frac{d}{dx}[M(x)Z(x)]^{-1}\cdot\psi_\lambda(x)\right]\right\} + M(x)\psi_\lambda(x)=0 \quad (6.44)$$

This is essentially the equation derived in Wigner and Wilkins' original paper.* It is not amenable to simple solution, though Wigner and Wilkins were able to transform it into a Ricätti equation for which certain approximate solutions can be written.[12]

However, interest in separable kernels of the above type is by no means restricted to the uniform gas problem; in fact, a considerable branch of transport theory exists in which a given kernel is "modeled" to an approximation of type (6.40) even though it may not satisfy it exactly. This proves to be a powerful method of studying the eigenvalue problem, and we shall describe it in more general terms than would be necessary for treating the uniform gas alone.

2. *The "Secondary Model"*

The so-called "secondary model," introduced into neutron studies by Cadilhac[52] and extended by Williams,[45] Schaeffer and Allsop,[53] and others, depends on an ingenious reduction of the "collision integral" in the Master Equation to a differential form. Since symmetry of the transport equations is of no particular advantage here we shall revert to the simple Master Equation (2.6) with the natural collision operator $\mathscr{A}$ and distribution function $P(x, t)$.

One seeks to replace the term $\mathscr{A}P(x, t)$ with a differential term $\partial Q/\partial x$ such that

$$\partial P(x, t)/\partial t = \partial Q(x, t)/\partial x \quad (6.45)$$

The form of Q is to be determined. Separation of the time variable shows that two particular solutions will be

$$P(x, t) = X_\lambda(x)\exp(-\lambda t) \quad (6.46)$$

$$Q(x, t) = q_\lambda(x)\exp(-\lambda t) \quad (6.47)$$

* This equation was of importance in computing steady-state neutron fluxes long before the eigenvalue problem became of interest. In this λ is replaced by an absorption term and $\psi(x)$ is sought with suitable boundary conditions.

where the quantity $q_\lambda(x)$ satisfies the integro-differential relations

$$\partial q_\lambda(x)/\partial x = \mathscr{A} X_\lambda(x) = -\lambda X_\lambda(x) \tag{6.48}$$

The key step is the recognition that $Q(x, t)$ can be written in the following explicit form:

$$\int_0^x \int_x^\infty K(\alpha, \beta)P(\alpha, t)\, d\alpha\, d\beta - \int_x^\infty \int_0^x K(\alpha, \beta)P(\alpha, t)\, d\alpha\, d\beta \tag{6.49}$$

This simply states the physical necessity that Q represents the net probability flux through a state x per unit time.* Now, assuming the transition kernel to be of the form (6.40) and writing a similar equation for $q_\lambda(x)$ we have

$$q_\lambda(x) = h(x) \int_x^\infty U(\alpha)\psi_\lambda(\alpha)\, d\alpha - g(x) \int_0^x V(\alpha)\psi_\lambda(\alpha)\, d\alpha = \mathscr{Q}\psi \tag{6.50}$$

where

$$h(x) = \int_0^x V(x)\psi(x)\, dx; \qquad g(x) = \int_x^\infty U(x)\psi(x)\, dx \tag{6.51}$$

To obtain a useful differential form we must find a differential operator $\mathscr{J}$ which applied to the integral operator $\mathscr{Q}$ leads to an explicit relation between $q_\lambda(x)$ and $\psi(x)$.

Assuming $\mathscr{J}$ to be second order and of the form

$$\mathscr{J} \equiv j(x) - \frac{d}{dx} k(x) \frac{d}{dx} \tag{6.52}$$

the following relationships result:

$$k(x) = 1/M(x)Z(x) \tag{6.53}$$

$$j(x) = \frac{d}{dx}\left[\frac{V(x)}{M(x)Z(x)}\right] \cdot \int_0^x V(\alpha)\, d\alpha \tag{6.54}$$

Performing the last operation with the explicit form of $V(x)$ and using (6.45) and (6.50) $q_\lambda(x)$ itself may be eliminated to give the differential equation

$$\frac{d}{dx}\left\{\frac{1}{j(x)} \cdot \frac{d}{dx}\left[\left(\frac{1}{\lambda} - k(x)\right)\psi_\lambda(x)\right]\right\} + M(x)\psi_\lambda(x) = 0 \tag{6.55}$$

* The "slowing-down density" in neutron transport theory.

This is now the eigenvalue condition for λ and ψ_λ. The boundary conditions must be reconstructed by requiring regularity for the term $k(x)\psi_\lambda(x)$ at zero and exponential order for the last term at infinity.

For the uniform hard-sphere gas ($\gamma = 1$) we know the resolution of the kernel to be exact; substitution of the functions (6.40–6.42) leads to the following coefficients in the equation

$$j(x) = \frac{x^{1/2}}{M(x)} \frac{d}{dx} \left[\frac{1}{x^{1/2} Z(x)} \right] \tag{6.56}$$

$$\frac{1}{k(x)} = M(x) \left[(2x^{1/2} + x^{-1/2}) \frac{\pi^{1/2}}{2} \operatorname{erf}(x^{1/2}) + e^{-x} \right] \tag{6.57}$$

(Note that λ has been scaled in units of the collision frequency A.) This leads back to the Wigner-Wilkins equation (6.44).

When the kernel is not of precisely separable form, certain prescriptions can be given for "modeling" it to a good approximation. It turns out that, although the condition given for $k(x)$ is always required to obtain the correct limit-point behavior as $\lambda \to (0)$, there is some arbitrariness in the choice of the function $j(x)$. One may choose to represent a particular moment $k_n(x)$ of the kernel correctly, though Williams has shown that it is the logarithmic moment which is appropriate if the low-energy behavior is to be correctly interpreted. This leads to the condition

$$j(x) = \frac{d}{dx} \left[\frac{k_n(x)}{Z(x)} \right] \left[\int_0^x M(y)[k_n(y) - y^n Z(y)]\, dy \right]^{-1} \tag{6.58}$$

where $k^n(x)$ are the moments of the kernel defined by Equation (3.26).

Since, unfortunately, there is no simple method of estimating the error involved in the separability approximation,* the usefulness of the secondary method for pure eigenvalue determinations is somewhat limited and most quantitative calculations have been restricted to the uniform gas condition. Nevertheless, as Williams has shown, the above equation can be made to yield useful qualitative information about the general case—in fact, it forms a more constructive alternative to the Kuščer-Corngold theory of the spectrum outlined above.

* A possible approach would be to represent the given kernel as a large separable part, K, plus a perturbation δK. It would be relatively easy to compute the effect of δK on eigenvalues obtained in the secondary method.

3. *WKB Analysis*

Since Equation (6.55) is linear and of second order, a transformation can be found which converts it into "Schrödinger-form"

$$Y'' + WY = 0 \tag{6.59}$$

The transformation is

$$Y(x, \lambda) = \left[\frac{1}{j(x)}\right]^{1/2}\left[1 - \frac{\lambda}{Z(x)}\right]\psi_\lambda(x) \tag{6.60}$$

and the "potential-term" W becomes

$$W(x, \lambda) = \left[\frac{1}{c}\left(\frac{Z(x)}{Z(x) - \lambda}\right) - 1\right]\frac{M(x)Z(x)j(x)}{(2x)^{1/2}} - [j(x)]^{1/2}\frac{d^2}{dx^2}[j(x)]^{-1/2} \tag{6.61}$$

Here the parameter c of the conjugate eigenvalue problem has been inserted to provide a qualitative analysis of the spectrum. In numerical calculations c will become 1.

In this form the singularity of the solutions for $\lambda \to Z(0)$ is clearly apparent. Although the appearance of λ as an implicit parameter in the eigenfunctions makes the problem rather different from the usual quantum-mechanical one, a quite conventional application of the WKB method may be used to obtain at least the higher levels of the discrete spectrum. The only complication is that, for small x (putting small-velocity approximations into $Z(x)$ and $j(x)$ (Eqs. (3.18) and (3.21)), $W(x)$ is found to have a singularity of the form $1/x^2$. This, according to standard WKB theory,[54] requires the use of an "effective potential" $W_{\text{eff}} = W - 1/(4x^2)$, in terms of which the eigenvalue condition reads

$$I_\lambda(x_1, x_2) = \int_{x_1}^{x_2} [W_{\text{eff}}(x, \lambda)]^{1/2}\, dx = (n + \tfrac{1}{2})\pi \tag{6.62}$$

Since there is no hope of obtaining explicit forms from this, a numerical search must be carried out to obtain the eigenvalue λ_n for each n. It is to be expected that the accuracy will improve rapidly with n and Williams' results for the uniform gas confirm this. In fact only λ_1 seems seriously in error.[46]

As we have seen, it is virtually impossible, using polynomial or degenerate-kernel methods, to resolve more than two eigenvalues below the continuum for the case $\gamma = 1$. There is no such difficulty with the WKB method, as can be seen in Table I. This brings out the remarkably fast accumulation of eigenvalues as $\lambda \to \lambda^*$. Williams was also able to prove that the Kuščer-Corngold limiting law (5.17) is recovered on putting small-velocity approximations into the eigenvalue equation. A test of this law is also shown in Table I.

It is also a relatively simple matter to obtain the WKB eigenfunctions, which take the familiar trigonometric or exponential form depending on the region of x. To within a normalizing factor they can be written as

$$\psi_n(x) \underset{\text{WKB}}{\approx} \lambda_n Z(x)\, Y_n(x)[j(x)]^{1/2}/[Z(x) - \lambda] \tag{6.63}$$

where

$$\begin{aligned} Y_n(x) &= (-)^n W_{\text{eff}}^{-1/4} \exp\{-|I_\lambda(x, x_1)|\}; \qquad x < x_1 \\ &= W_{\text{eff}}^{-1/4} \cos\left\{I_\lambda(x_1, x) - \frac{\pi}{4}\right\}; \qquad x_1 < x < x_2 \\ &= (-)^n W_{\text{eff}}^{-1/4} \exp\{-|(n + \tfrac{1}{2})\pi + I_\lambda(x_2, x)|\} \end{aligned} \tag{6.64}$$

In the near-continuum region they may be further approximated. We shall only quote the special results for the uniform gas. These are

$$\begin{aligned} Y_n(x) &\approx (-)^n W_{\text{eff}}^{-1/4} \exp\left\{-\frac{1}{4}\sqrt{23}\left|\log\left(\frac{8x}{\lambda^* - \lambda}\right)\right|\right\}; \qquad x < x_1 \\ &\approx W_{\text{eff}}^{-1/4} \cos\left\{\frac{1}{4}\sqrt{23}\log\left(\frac{8x}{\lambda^* - \lambda}\right) - \frac{\pi}{4}\right\}; \qquad x_1 < x < x_2 \\ &\approx (-)^n W_{\text{eff}}^{-1/4} \exp[-(n + \tfrac{1}{2})\pi]\left(\frac{x_2}{x}\right)^{1/4\sqrt{23}}; \qquad x_2 < x \end{aligned} \tag{6.65}$$

These appear to be the only explicit forms which have been derived for the hard-sphere eigenfunctions. One can see very clearly how the logarithmic singularity causes them to oscillate more or less violently as $\lambda \to \lambda^*$. This behavior is clear enough indication of why the eigenfunctions for $\gamma = 1$ cannot possibly be approximated to any accuracy by low-order polynomials and why the true eigenfunctions are so far from being a complete set in this case. For this reason it also seems unlikely that they could form part of a "good" spectral resolution of the operator $\mathscr{B}$ which would be of any use in practical calculations of the initial-value problem.

E. The Hard-Sphere Spectrum

It is hardly feasible here to give complete numerical tables of the hard-sphere eigenvalues obtained by the foregoing methods, so we shall concentrate on discussing their qualitative behavior as a function of mass ratio. Taking the spectral theory of Section V in conjunction with variational calculations, a reasonably clear picture of the behavior of the discretum over a wide range of mass ratios can now be given.

The first detailed calculations of the discrete spectrum of the hard-sphere gas are to be found in the neutron study by Shapiro and Corngold.[24] These authors used and compared several of the procedures described earlier and concluded that the implicit degenerate-kernel one is superior. The accuracy of their values for λ_1 to λ_4 (Fig. 6) has not been bettered. Although some further detail is given by Shapiro,[51] their results for the scalar eigenvalues are only part of a broad study including also solid moderators and aspects of the spatially dependent problem for neutrons; they could be expanded considerably in a more general kinetic-theory context. The lower eigenvalues obtained can be seen to show the characteristic bell-shaped behavior predicted by the "two-by-two" approximation with the correct tendency to emerge tangentially to the continuum and then "peel-off" and tend to zero at the two extremes of mass ratio. The maximum in λ_1 at $\gamma = 1/2$ is clearly shown, confirming the "two-by-two" variational result. These authors do not quote unconverged "pseudo-eigenvalues" lying in the continuum, though these would be of use in initial-value calculations.

A similar study was carried out by Hoare and Kaplinsky and completed largely in ignorance of the results and techniques of neutron transport

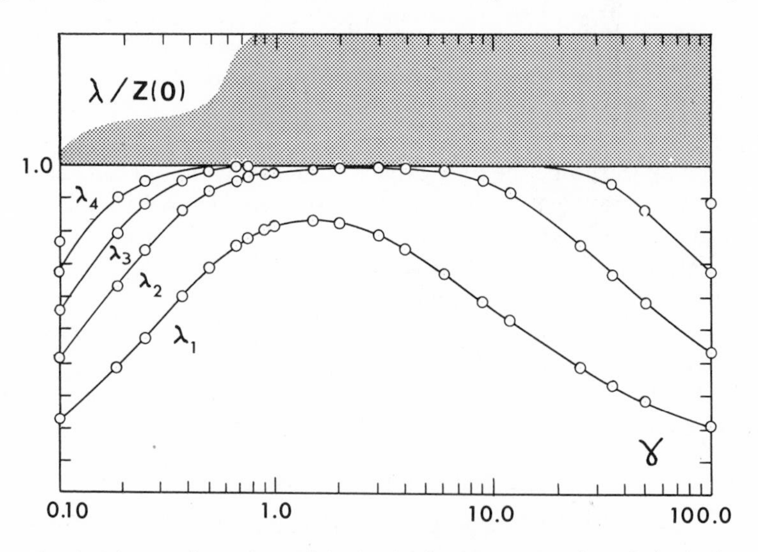

Fig. 6. Discrete eigenvalues of the hard-sphere gas energy kernel obtained by the degenerate-kernel method. From Shapiro and Corngold.[24]

theory. Using an energy-polynomial Rayleigh-Ritz method, it was not possible to achieve the accuracy of Shapiro and Corngold in the near-continuum region; on the other hand a larger (21 × 21) spectral resolution was obtained and attention was concentrated on the Rayleigh regime ($\gamma < 1$) which is of no relevance in neutron transport. The results of this are shown in Figures 7 and 8. Some sample figures for converged eigenvalues are shown in Table II.

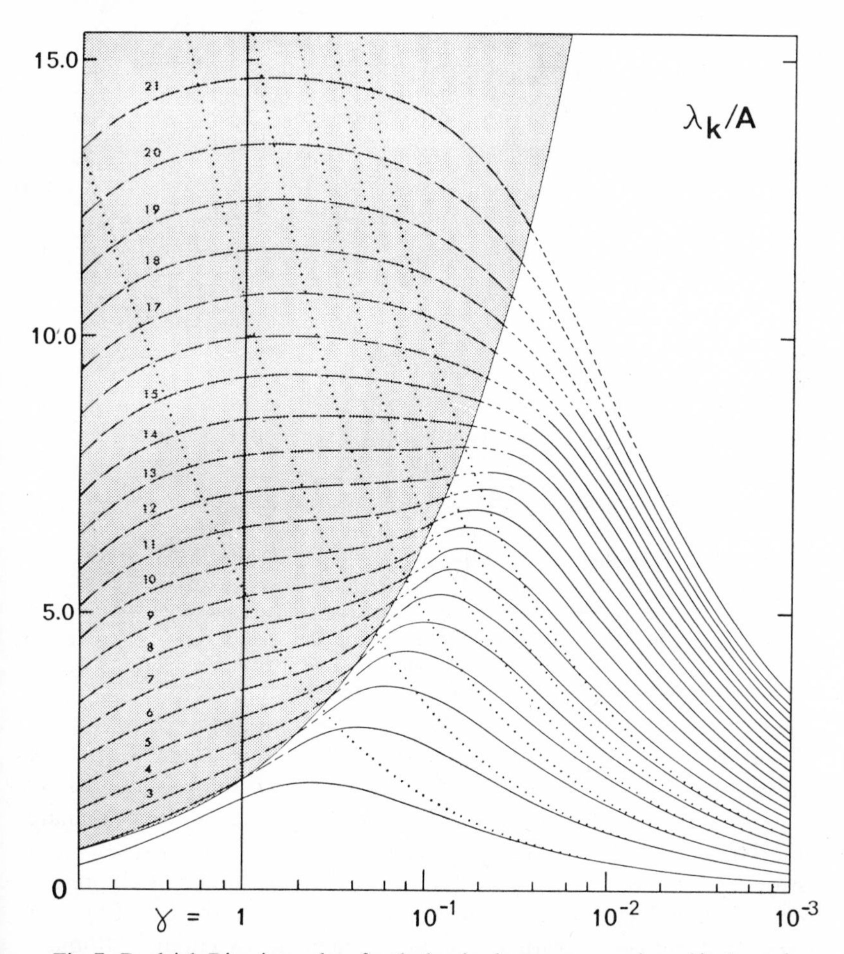

Fig. 7. Rayleigh-Ritz eigenvalues for the hard-sphere gas energy kernel in the regime $\gamma < 1$. Converged (Type A) eigenvalues are given by the solid lines, Types (B) and (C) by the broken lines outside and inside the continuum, respectively. The ascending dotted lines show the first six eigenvalues of the energy Fokker-Planck equation on the same scale. From Hoare and Kaplinsky.[25]

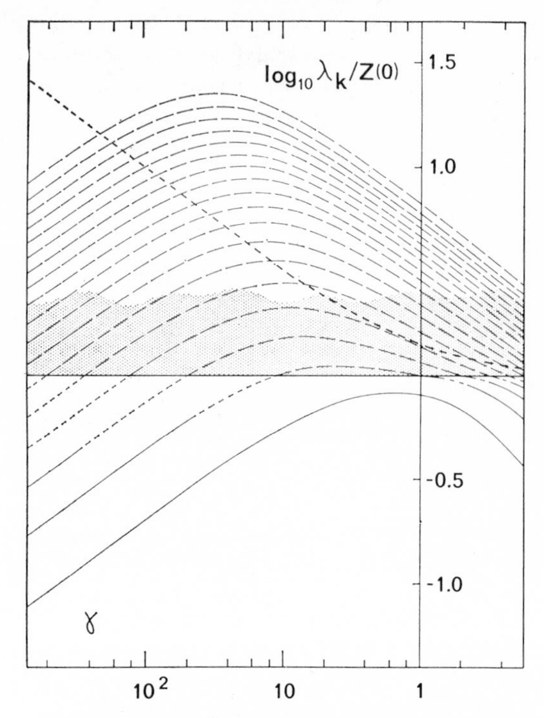

Fig. 8. Rayleigh-Ritz eigenvalues for the hard-sphere gas energy kernel in the Lorentz regime $\gamma > 1$. The eigenvalues are plotted logarithmically and scaled in terms of the mean equilibrium collision number $Z(0)$. The dashed line ascending to the left is the mean equilibrium collision number Z_{00} on the same scale. The discrete and pseudo-eigenvalues are indicated as in Figure 7. From Hoare and Kaplinsky.[25]

Where the eigenvalues are converged the first four agree with those by Shapiro and Corngold but, as is to be expected, the resolution near the continuum is less satisfactory. However the tendency to accumulate at the boundary is just detectable and the eigenvalues seem to move in this way only to fan out just before the point λ^* in an attempt to represent the continuum. The general picture shows a distinct sparseness of lines in the region around $\gamma = 1$ with an accumulation at the two limits. Although it would appear that the discretum is filled ever more densely at the two limits, accumulation seems to be more pronounced at the Rayleigh end. An interesting feature is the distinct flatness of the higher pseudo-eigenvalues as a function of mass ratio in the region $\gamma \approx 5$ to $\gamma \approx 0.1$. If these

TABLE I

Eigenvalues for the Hard-Sphere Energy Kernel $(\lambda_k/Z(0))(\gamma = 1)$

	Discretization[a]	Degenerate-kernel method[b]	V-polynomial method[c]	E-polynomial method[d]	J-WKB method[e]	$\left(\dfrac{\lambda^* - \lambda_k}{\lambda^* - \lambda_{k+1}}\right)$
$\lambda_1 =$	0.8190	0.8190	0.8190	0.8190	0.821	5.6
$\lambda_2 =$	0.9797	0.9802	0.9802	0.998	0.9804	9.1
$\lambda_3 =$	0.9983				0.99846	12.7
$\lambda_4 =$					0.99989	13.7

[a] Wood.[49]
[b] Shapiro[51] ($N = 5$).
[c] Shapiro[51] ($N = 8$).
[d] Hoare and Kaplinsky[25] ($N = 21$).
[e] Williams.[46]

N.B. $\exp(4\pi/\sqrt{23}) = 13.7$.

TABLE II

Discrete Eigenvalues for the Hard-Sphere Energy Kernel as a Function of Mass Ratio $\gamma = m_2/m_1$ $(\lambda_k/Z(0))$. From Hoare and Kaplinsky[25]

$\gamma^{-1} =$	1	2	4	8	16	32	64	128	256
	0.8190	0.6891	0.4707	0.2784	0.1520	0.0796	0.0407	0.0206	0.0104
		0.912	0.7411	0.4882	0.2836	0.1535	0.0799	0.0408	0.0206
			0.8864	0.6465	0.3976	0.2222	0.1177	0.0606	0.0308
			0.957	0.7643	0.4957	0.2861	0.0542	0.0801	0.0409
				0.8490	0.5824	0.3456	0.1829	0.0923	0.0508
				0.907	0.6566	0.4010	0.2232	0.1180	0.0607
				0.952	0.7206	0.4526	0.2258	0.1364	0.0705
					0.7752	0.5006	0.2873	0.1545	0.0802
						0.5453	0.3177	0.1723	0.0898
						0.5870	0.3471	0.1897	0.0993
							0.3754	0.2068	0.1088
							0.4027	0.2237	0.1181
							0.4290	0.2402	0.1274
								0.2564	0.1366
								0.2723	0.1457
$\gamma =$	2	4	8	16	32	64	128	256	512
	0.826	0.746	0.615	0.474	0.351	0.254	0.182	0.129	0.092
						0.545	0.397	0.285	0.203
							0.67	0.49	0.347

are a correct indication of short-time relaxation behavior in this region, then this might be due to an effective compensation between increasing collision number and decreasing efficiency of energy transfer. On the other hand, it could be a peculiarity of the method that the higher λ_k are forced into a plateau region before accumulating in an attempt to represent the far-Lorentz region. Only further understanding of the true nature of the continuum will decide this.

Although these results could certainly be improved somewhat they are probably close to the limits of what can be achieved at present within the restrictions imposed by computing time and rounding error. There does not seem to be much future in simply increasing the number of linear trial functions, since several hundred would probably be required to obtain very significant improvement of the curves shown in Figure 7. In fact, using only the fully converged values shown by the solid lines, a very good guess at the true behavior in the near-continuum region can be made simply by producing the curves to be tangential to the continuum threshold.

Some specimen calculations of relaxation behavior at different mass ratios are given in Reference 25. In computing these it is very clear that a spectral resolution of order about 20×20 is still very limited when it is desired to represent the evolution from initial conditions that are at all sharply peaked or shifted far from the thermal region. It is very easy indeed to find initial distributions $P(x, 0)$ of practical interest which have strong Fourier components outside the range of a reasonable basis set, and in the extreme case of thermalization from very high energies the methods described here are of little relevance. A considerable body of technique exists to deal with this case and can be found in standard works on neutron transport under the heading of "slowing-down theory." (See Ref. 4, Chap. VIII *et seq.*) We shall not expand on this aspect of the problem here.

VII. THE SINGULAR EIGENFUNCTIONS

As has been emphasized, the possession of a "true" set of the singular eigenfunctions of the operator $\mathscr{A}$ is not necessary for the construction of practical solutions to the homogeneous initial-value problem. Nevertheless the theoretical interest of these functions is considerable and, although their exact role is still somewhat obscure, it seems inescapable that they represent a type of solution fundamentally different from any that can be constructed from regular, L_2 functions.

Although computational methods are difficult and have been little used, a definite prescription can be given for the calculation of approximate singular eigenfunctions $\hat{N}(x, \lambda)$ and the continuum weight function $\omega(\lambda)$.

This was first described by Koppel[34] and extended somewhat for the case of the linear Boltzmann equation by Ferzinger.[35] Both treatments derive from earlier work by Case,[27] Mika,[28] Zelazny and Kurszell,[29] and Van Kampen.[30] Here we shall explain only the essential algorithm for obtaining distributions $\hat{N}(x\ \lambda)$ and constructing singular initial-value solutions. In addition to the proofs of completeness, the papers by Koppel and Ferzinger contain a wealth of further properties relating these to the Hilbert transform and other topics in the theory of singular integral equations.*

A. The Koppel Method

This begins by putting the orthogonal-function representation of the kernel (3.33) into the singular eigenvalue condition (5.1). The result is

$$N(x, \lambda) = \mathfrak{P} \frac{1}{[Z(x) - \lambda]} \sum_{i=1}^{N} \sum_{j=1}^{N} G_{ij} \alpha_j(\lambda) \phi_i(x) + \omega(\lambda) \delta(x - x_\lambda) \tag{7.1}$$

where now

$$\alpha_j(\lambda) = \int_0^\infty N(x, \lambda) \phi_j(x)\, dx \tag{7.2}$$

and $Z(x_\lambda) = \lambda$ as before.

If scalar products are taken $\phi_j(x)$ on each side, this leads to the algebraic equations

$$\alpha_j(\lambda) = \sum_{i=0}^{N} \sum_{j=0}^{N} G_{ij} L_{ij}(\lambda) \alpha_j(\lambda) + \omega(\lambda) \phi_j(x_\lambda) \tag{7.3}$$

with the matrix $\mathbf{L}$ defined through the singular integrals

$$L_{ij}(\lambda) = \mathfrak{P} \int_0^\infty \frac{\phi_i(x) \phi_j(x)}{[Z(x) - \lambda]}\, dx \tag{7.4}$$

With λ a given value in the continuum region, the matrix equations $\boldsymbol{\alpha} = \mathbf{GL}(\lambda)\boldsymbol{\alpha}$ can then be solved for the $N + 1$ quantities $\alpha_0(\lambda), \alpha_1(\lambda) \ldots \alpha_N(\lambda)$ provided that an additional condition is supplied to determine the unknown function $\omega(\lambda)$. This condition can be either the normalization

$$\sum_{i=1}^{N} \alpha_i(\lambda) = 1 \tag{7.5}$$

or, alternatively,

$$\int_0^\infty N(x, \lambda)\, dx = 1 \tag{7.6}$$

* The standard source for these deeper aspects of the problem is Muskhelishvili.[55]

the latter choice giving

$$\omega(\lambda) = 1 - \mathfrak{P} \int_0^\infty \frac{\phi_i(x)\, dx}{[Z(x) - \lambda]} \sum_{i=0}^{N} \sum_{j=0}^{N} G_{ij}\eta_j(\lambda) \tag{7.7}$$

In either case the $N + 2$ unknowns for given λ are determined and the singular eigenfunctions reconstructed by substitution back into Equation (7.1).

We can now contemplate constructing singular initial-value solutions for some given distribution $h(x, 0)$ using a more explicit form of Equation (4.8):

$$h(x, t) = N_0(x) + \sum_{k=1}^{N} a_k N_k(x) e^{-\lambda_k t} + \int_{\lambda \in c} A(\lambda) \hat{N}(x, \lambda) e^{-\lambda t}\, d\lambda \tag{7.8}$$

To do this we must be satisfied that the combined set $\{N_k(x), N(x, \lambda)\}$ is complete for arbitrary "reasonable" functions, and then find a means of computing the expansion coefficients a_k and $A(\lambda)$.

The orthogonality properties of the different eigenfunctions are clearly crucial. Since $G(x, y)$ is symmetric the problem is self-adjoint and all distinct eigenfunctions are orthongonal; we can check this by cross-multiplying and subtracting two versions of the eigenvalue equation (4.3) to obtain

$$\int_0^\infty N_i(x) N_j(x)\, dx = R_i \delta_{ij} \tag{7.9}$$

$$\int_0^\infty N(x, \lambda) N(x, \mu)\, dx = R(\lambda)\delta(\lambda - \mu) \tag{7.10}$$

$$\int_0^\infty N_i(x) N(x, \lambda)\, dx = 0; \quad \lambda \varepsilon c \tag{7.11}$$

The quantities R_k and $R(\lambda)$ are normalization integrals, which are not arbitrary since we have already used a normalization in determining $\omega(\lambda)$.

Proof of completeness depends on showing that an arbitrary function $f(x)$ orthogonal to the whole discrete set $N_k(x)$ can always be expanded in the singular eigenfunctions as

$$f(x) = \int_{\lambda \varepsilon c} A(\lambda) N(\lambda, x)\, d\lambda \tag{7.12}$$

This is achieved by substituting the formal expression (7.1) for $N(x, \lambda)$ into this integral and treating the result as a singular integral equation for $A(\lambda)$. Proof that this always has a solution is then proof of completeness.[34,35]

With completeness assumed it only remains to determine the expansion coefficients, which depend in turn on the normalization integrals R_k and $R(\lambda)$.

$$A_k = \frac{1}{R_k} \int_0^\infty h(x, 0) N_k(x)\, dx \tag{7.13}$$

$$A(\lambda) = \frac{1}{R(\lambda)} \int_0^\infty h(x, 0) N(x, \lambda)\, dx \tag{7.14}$$

A series expression for R_k follows on forming the square integral of the right-hand side of Equation (7.1). Thus

$$R_k = \int_0^\infty \left[\frac{\sum_{j=0}^{j=N} a_j^{(k)} \sum_{i=0}^{i=N} G_{ij}\, \phi_i(x)}{Z(x) - \lambda} \right]^2 dx \qquad (\lambda < Z(0)) \tag{7.15}$$

The determination of the continuum normalization function $R(\lambda)$ is less elementary. However the square integral of the expression in Equation (7.1) can be correctly interpreted provided one uses the Poincaré-Bertrand theorem for double principal-value integrals.[22,55]. The result is as follows:

$$R(\lambda) = [\omega(\lambda)]^2 [dZ(x)/dx]_{x=\lambda'} + \frac{\pi^2}{[dZ(x)/dx]_{x=\lambda'}} \left\{ \sum_{i=0}^{N} \sum_{j=0}^{N} G_{ij}\, \phi_i(\lambda') \alpha_j(\lambda') \right\}^2 \tag{7.16}$$

where $\lambda' = \lambda - Z(0)$.

Unfortunately no results appear to have been published for the simple nonabsorbtive, infinite medium hard-sphere gas problem. Implementation of the method seems likely to be difficult, though worth attempting at least with a small value of N.

B. Degenerate Kernel Methods

A method very similar to Koppel's can be carried out using a degenerate-kernel representation instead of one in orthogonal functions. To apply this we must solve the matrix eigenvalue problem $\mathbf{\Psi} = \mathbf{F}(\lambda)\mathbf{\Psi}$ (Eq. (6.30)) with $\mathbf{F}(\lambda)$ now defined in terms of the principal-value integrals:

$$F_{ji}(\lambda) = \mathfrak{P} \int_0^\infty \frac{f_j(x) g_i(x)}{Z(x) - \lambda}\, dx \qquad (\lambda > Z(0)) \tag{7.17}$$

On obtaining the eigenvectors $\boldsymbol{\Psi}(\lambda)$, which are now functions of the continuous parameter λ, the singular eigenfunctions may be constructed by the following prescription similar to (6.28):

$$N(x, \lambda) = \frac{\mathfrak{P}}{[Z(x) - \lambda]} \sum_{i=0}^{N} f_i(x)\psi_i + \omega(\lambda)\delta[Z(x) - \lambda] \qquad (7.18)$$

The continuum weighting function $\omega(\lambda)$ can then be found by defining a normalization as before. The use of the singular eigenfunctions in the initial-value problem is then as described in the previous section.

Solutions for the "Amnesia" Kernel. Since the "amnesia" kernel (Eq. (3.43)) is simply a "one-by-one" version of the previous approximation, we might expect an interesting outcome in this case. In fact it gives an almost unique example of a model for which the singular eigenfunctions can be obtained in explicit form.[4]

We need only make the following identifications with the terms in the previous section:

$$G_0(x, y) = Z_{00}^{-1} N_0(x) N_0(y) Z(x) Z(y)$$

so that

$$f_0(x) = Z_{00}^{-1} N_0(x) Z(x) \qquad (7.19)$$

$$g_0(x) = N_0(y) Z(y) \qquad (7.20)$$

$$F_{00}(\lambda) = \frac{\mathfrak{P}}{Z_{00}} \int_0^\infty \frac{M(x)(Z(x))^2}{Z(x) - \lambda}\, dx \qquad (7.21)$$

and $F_{ij} = 0$ for $i, j \neq 0$.

We first consider possible solutions for the discretum region. Equation (6.30) reduces to the trivial form

$$\psi_0{}^0 = F_{00}(\lambda)\psi_0{}^0 \qquad (7.22)$$

which means that the eigenvalue condition is simply

$$\frac{1}{Z_{00}} \int_0^\infty \frac{M(x)[Z(x)]^2}{Z(x) - \lambda}\, dx = 1 \qquad (\lambda < Z(0)) \qquad (7.23)$$

with $\psi_0{}^0 =$ constant.

This clearly admits only the equilibrium solution $\lambda_0 = 0$, and the rest of the discretum is therefore empty. Since the corresponding eigenfunction must be $N_0(x)$, it follows from Equations (6.29) and (7.20) that this normalization implies $\psi_0{}^0 = Z_{00}$.

Turning now to the singular eigenfunctions, we can see that the correct prescription for these (Eq. (7.1)) is

$$N(x, \lambda) = \mathfrak{P} \frac{N_0(x)Z(x)}{Z(x) - \lambda} + \omega(\lambda)\delta(x - x_\lambda) \qquad (Z(x_\lambda) = \lambda) \quad (7.24)$$

The continuum weighting function $\omega(\lambda)$ must then be found by fixing a suitable normalization. A convenient choice in this case is to multiply both sides of the above by $g_0(x)$ ($= N_0(x)Z(x)$) and put the left-hand result equal to unity. We then have

$$1 = F_{00}(\lambda) + \omega(\lambda) \int_0^\infty N_0(x)Z(x)\delta(\lambda - Z(x))\, dx \quad (7.25)$$

The integral can be evaluated more explicitly and the final result is

$$\omega(\lambda) = \frac{Z'(x_\lambda)}{\lambda N_0(x_\lambda)} \{1 - F_{00}(\lambda)\} \qquad (\lambda > Z(0)) \quad (7.26)$$

where, as usual, $(Z(x_\lambda) = \lambda)$.

The completeness of the "amnesia" continuum eigenvalues, supplemented by the single function $N_0(x)$, is guaranteed by the Koppel-Ferzinger analysis, but may also be demonstrated directly as a special case.[4] This implies that any "reasonable" function $f(x)$ may be written in the form

$$f(x) = a_0 N_0(x) + \int_{Z(0)}^\infty A(\lambda)N(x, \lambda)\, d\lambda \quad (7.27)$$

with the continuum eigenfunctions $N(x, \lambda)$ obtained as above.

C. The Rahman-Sundaresan Formulation

Although Koppel's algorithm for obtaining the singular eigenfunctions is, however difficult, a practical proposition, it has a certain awkwardness from the theoretical point of view and raises the question of whether there might exist some alternative which accepts the singularness of the solutions at the outset in a more natural way. A possible advance in this direction has been suggested by Rahman and Sundaresan,[56,57] though their approach so far remains very formal and can hardly be presented here under the heading of a practical method.

The main achievement of these authors is to reformulate the singular eigenvalue problem in terms more suited to the theory of singular integral equations as developed in other fields.[55] The first step is to transform the energy part of the Master Equation into an integral equation with a Cauchy-type kernel, that is, one with a simple logarithmic singularity instead of the more complicated one in $[Z(x) - \lambda]^{-1}$. We shall outline this process as they present it with some minor changes of notation and scale.

The method proceeds from the ansatz that the singular eigenfunctions $N(x, \lambda)$ can be represented as distributions of the form

$$N(x, \lambda) = \frac{\mathfrak{P} f(x)}{Z(x) - \lambda} + \omega(\lambda)\delta(Z(x) - \lambda) \tag{7.28}$$

This, looked at in the light of the distribution theory identity (cf. Ref. 32, Eq. 2.30),

$$\frac{d}{dx} \ln x = -i\pi\delta(x) + \mathfrak{P}(1/x) \tag{7.29}$$

can be seen to imply that the singularity for $Z(x) - \lambda$ has a logarithmic character. One therefore expects a Cauchy-type kernel to occur. Introducing the ansatz into the eigenvalue equation we obtain, with careful interpretation of the effect of the delta-function,

$$f(x, x_\lambda) = \frac{\omega(\lambda)G(x, x_\lambda)}{Z'(x_\lambda)} + \mathfrak{P} \int_0^\infty \frac{f(y, x_\lambda)G(y, x)}{Z(y) - \lambda} \, dy \tag{7.30}$$

where $\omega(\lambda)$ is a function to be determined, $Z'(x)$ is the derivative of the collision number, and x_λ is defined as before by $Z(x_\lambda) \sim \lambda$. As the parameter x_λ varies from 0 to ∞, so λ spans the continuum range $Z(0)$ to infinity. The equation can then be reduced to Cauchy form on defining a new kernel

$$L(x, y) = G(x, y)(y - x_\lambda)/[Z(y) - Z(x_\lambda)] \tag{7.31}$$

In terms of this,

$$f(x, x_\lambda) = \frac{\omega(\lambda)G(x, x_\lambda)}{Z'(x_\lambda)} + \mathfrak{P} \int_0^\infty \frac{f(y, x_\lambda)L(y, x)}{y - x_\lambda} \tag{7.32}$$

This is still not in standard form since the kernel contains three parameters x, y, x_λ instead of the normal two. A final transformation takes care of this. New variables are introduced

$$h(x, x_\lambda) = f(x, x_\lambda) - f(x_\lambda, x_\lambda) \tag{7.33}$$

$$r(x, x_\lambda) = \rho(x, x_\lambda) - \rho(x_\lambda, x_\lambda) \tag{7.34}$$

with $\rho(x, x_\lambda)$ standing for the first term on the right of Equation (7.30). With these substitutions the integral equation reads

$$\begin{aligned} h(x, x_\lambda) + f(x_\lambda, x_\lambda) = {} & r(x, x_\lambda) + \rho(x, x_\lambda) \\ & - f(x, x_\lambda) \, \mathfrak{P} \int_0^\infty \frac{L(y, x)}{y - x_\lambda} \, dy \\ & - \mathfrak{P} \int_0^\infty \frac{L(y, x)h(y, x_\lambda)}{y - x_\lambda} \, dy \end{aligned} \tag{7.35}$$

Putting $x = x_\lambda$ in this, a second equation is obtained:

$$\mathfrak{P}\int_0^\infty \frac{L(y, x)h(y, x_\lambda)}{y - x_\lambda}\,dy = \rho(x, x_\lambda) - f(x_\lambda, x_\lambda)\left\{1 + \mathfrak{P}\int_0^\infty \frac{L(y, x_\lambda)}{y - x_\lambda}\,dy\right\} \tag{7.36}$$

and making the same substitution in the original (7.30), a third equation results:

$$f(x_\lambda, x_\lambda) = \rho(x, x_\lambda) - \mathfrak{P}\int_0^\infty \frac{L(y, x_\lambda)}{y - x_\lambda} f(y, x_\lambda)\,dy \tag{7.37}$$

The various integrands can be shown to be sufficiently well-behaved so that, in principle, the equations are amenable to solution by the methods developed for Cauchy-type kernels. A possible procedure is suggested based on the method of Carleman and Vekua,[55] and this is shown to lead to two singular Fredholm equations, one of each kind. Although these are simple enough that numerical solution can be contemplated, none seems to have been attempted up to this time.

The complete algorithm for obtaining the singular eigenfunctions can then be summarized as follows:

(i) Selecting a particular λ, solve Equation (7.37) for the function $f(x_\lambda, x_\lambda)$.
(ii) Solve Equation (7.36) for $h(x, x_\lambda)$ in terms of $f(x_\lambda, x_\lambda)$, obtaining $f(x, x_\lambda)$ from Equation (7.33).
(iii) Construct the singular eigenfunctions $N(x, \lambda)$ using the ansatz (7.28).
(iv) Obtain the continuum weight function $\omega(\lambda)$ by defining a normalization as in Koppel's method.

A number of other insights into the singular eigenvalue problem have been obtained by Rahman. In a very recent paper[58] he obtains expressions for the form of the singular solutions at $\gamma = 1$ as they appear in an expansion about a point x_λ. The Wigner-Wilkins differential equation is reduced to a simpler, hypergeometric form by expanding the coefficients to second order about the given point x_λ. Explicit solutions can then be written which contain logarithmic singularities. When the limit $x_\lambda \to 0$ ($\lambda \to Z(0)^+$) is taken, the solutions agree with those obtained by Kuščer and Corngold approaching the limit from the discrete side $\lambda \to Z(0)^-$.

Although it is not clear exactly what part these results might play in actual initial-value calculations, the approaches outlined in this section seem to hold out more hope of a full understanding of the continuum region than any others.

VIII. FOKKER-PLANCK EQUATIONS

A. General Aspects

The part of linear transport theory dealing with the extremes of mass ratio $\gamma \to 0$ and $\gamma \to \infty$ has become a field in its own right, having developed almost in isolation from the general theory discussed here. As a result the theory of Brownian motion has reached such a dominant position in statistical physics that a large part of the interest in solving the general relaxation problem is that, for the first time, it has been possible to check the range of validity of the Brownian motion assumptions against the "exact" (i.e., integral-equation) solutions for a model such as the hard-sphere gas.

As Rayleigh demonstrated, it is possible to bypass the integral equation for his model when $\gamma \ll 1$ on the assumption that the test-particle velocities are always narrowly distributed and change only infinitesimally on each impact. This diffusionlike behavior leads to the following second-order partial differential equation:

$$\frac{\partial P(V, t)}{\partial t} = 4N_2\gamma(2kT/\pi m_2)^{1/2} \frac{\partial}{\partial V}\left[VP(V, t) + \frac{kT}{m_1}\frac{\partial P(V, t)}{\partial V}\right] \tag{8.1}$$

Rayleigh was able to solve this by a substitution which avoids consideration of the eigenvalue problem. The "fundamental solution," that is, the evolution of an initial delta distribution $P(V, 0) = \delta(V - V_0)$, was found to be

$$P(V, \tau \mid V_0) = \left[\frac{m_1}{2\pi kT(1 - e^{-2\tau})}\right]^{1/2} \exp\left[-\frac{m_1(V - V_0e^{-\tau})^2}{2kT(1 - e^{-2\tau})}\right]$$

$$(\tau = 4N_2\gamma(2kT/\pi m_2)^{1/2}t) \quad (8.2.)$$

Solutions for other initial conditions can be found by superposition (Ref. 73, Section 15.12).

The three-dimensional analog of Equation (8.1) was obtained sixty years later by Green,[59] who successfully adapted Rayleigh's approach to the hard-sphere gas. A fairly complicated derivation leads to the equation

$$\frac{\partial}{\partial \tau} P(\mathbf{V}, \tau) = \nabla[\mathbf{V} + kT\nabla]P(\mathbf{V}, \tau) \qquad (\tau = \tfrac{8}{3}m_1N_2\sigma^2(2\pi\gamma kT/m_1)^{1/2}t) \tag{8.3}$$

No explicit solution of this equation in vector form appears to be possible, though a scalar solution in energy or speed can be obtained, as we shall see.

Both the above examples can be seen as special realizations of the Fokker-Planck equation:

$$\frac{\partial P(x, t)}{\partial t} = -\frac{\partial}{\partial x}\alpha_1(x)P(x, t) + \tfrac{1}{2}\frac{\partial^2}{\partial x^2}\alpha_2(x)P(x, t) \tag{8.4}$$

(The variables and operators must be interpreted as vector or scalar, as appropriate.)

The question of the precise relationship between the Fokker-Planck equation and its corresponding integral transport-equation is a long-standing one of considerable difficulty. There can be no question of an exact correspondence between an integral operator such as $\mathscr{A}$ and a second-order differential one, except in the special case where the kernel is of Green function type, and even then the domain of the operators will be somewhat different. In particular, only the integral operator will correctly represent the persistence of a delta-function component in the solution for finite times.

The points at issue are, first, whether any systematic correspondence can be made; second, how is the differential form related to the kernel of the integral operator; and third, what is the range of validity of the approximation in terms of parameters in the kernel and possible initial conditions $P(x, 0)$?

A partial answer to these questions is suggested by early derivations from the Smoluchowski equation of Brownian motion theory[60–62] and has been formalized in the work of Kramers,[63] Moyal,[64] and Keilson and Storer.[65] By somewhat unrigorous manipulations it may be shown that a linear integral operator with kernel $A(x, y)$ may be represented formally by the following differential operator of *infinite order*:

$$\mathscr{A} \equiv \sum_{k=1}^{\infty} \frac{1}{k!}\left(-\frac{\partial}{\partial x}\right)^k a_k(x) \tag{8.5}$$

Here the functions $a_k(x)$ are the transfer moments of the kernel introduced previously (Eq. (3.27)). Clearly any practical application of the right-hand side will depend on its truncation after a small number of terms, two in the case of the Fokker-Planck approximation.

As all the above-mentioned authors have admitted, it seems impossible to justify rigorously either the expansion as a whole or the truncation process, even for limited classes of functions. For this it would be necessary to demonstrate that the transfer moments decrease rapidly with the terms of the series and also that legitimate solutions are of such "smoothness" that the higher derivatives may be safely neglected. A detailed analysis by Van Kampen[66] shows that the first condition follows when the expansion is effectively in powers of a small parameter, for instance γ or γ^{-1}. One can then construct a modified Kramers-Moyal expansion in which the coefficients are not the exact transfer moments $\alpha_n(x)$ but these quantities corrected to appropriate order in the small parameter. However, although some attempts have been made to evaluate the higher terms in these expansions, it is now widely accepted that only the two-term version is useful. This is not simply a matter of practical difficulty; a proof by Pawla[67] shows that there is a fundamental inconsistency in retaining any number of terms other than two or infinity, and, apart from this, it is known that attempts to use the third and higher terms lead to loss of the positive-definite character of the operator, that is, nonconservation of probability.

B. The Rayleigh Gas

The limiting behavior of the equation $\partial P/\partial t = \mathscr{A}P$ for $\gamma \to 0$ has long formed a part of Brownian motion theory, but it was only comparatively recently that a systematic derivation of the Fokker-Planck operator from the Wigner-Wilkins kernel was given. This was carried out by Andersen and Shuler[13] working with the scalar equation in energy.

Their method was to evaluate the exact transfer moments $a_1(x)$, $a_2(x)$, $a_3(x)$, examine their dependence on γ, and reduce them to the lowest order in this parameter. The following results were obtained

$$a_1(x) = \tfrac{8}{3}\gamma(\tfrac{3}{2} - x)Z(x) + O(\gamma x) \tag{8.6}$$

$$a_2(x) = \tfrac{16}{3}\gamma x Z(x) + O(\gamma x) \tag{8.7}$$

$$a_3(x) = 128\gamma(\tfrac{2}{5}x - 1)Z(x) + O(\gamma x) \tag{8.8}$$

$$Z(x) = 2A \exp(-\gamma x)[1 + O(\gamma x)] \tag{8.9}$$

Evidently $a_3(x)/a_1(x) = O(\gamma x)$ and $a_3/a_2 = O(\gamma x)$. Dropping $a_3(x)$ then leads to the following Fokker-Planck equation:

$$\frac{\partial P(x, \tau_R)}{\partial \tau_R} = \frac{\partial}{\partial x}\left[(x - \tfrac{3}{2})P(x, \tau_R) + \frac{\partial}{\partial x}(xP(x, \tau_R)\right]$$

$$= \mathscr{F}_R P(x, \tau_R) \tag{8.10}$$

Here τ_R is the scaled time $\tau_R = \frac{16}{3}\gamma^{1/2}At$.

The solution of this equation follows straightforwardly after separation of variables; the eigenfunctions of the operator $\mathscr{F}_R$ are the Laguerre functions of order one-half and the eigenvalues are simply the integers $v = 1, 2, 3 \ldots$ in the time scale used. Explicitly,

$$P(x, \tau_R) = x^{1/2}e^{-x} \sum_{\nu=0}^{\infty} c_\nu L_\nu^{1/2}(x) \exp(-\nu\tau_R) \tag{8.11}$$

that is, $\lambda_n = (16/3)\gamma^{1/2}An$ in the real time scale.

The coefficients would be obtained using the relation

$$c_\nu = \Gamma(\nu + 1)/\Gamma(\nu + \tfrac{3}{2}) \int_0^\infty P(x, 0)L_\nu^{1/2}(x)\, dx \tag{8.12}$$

If the equation had been written in self-adjoint form the eigenfunctions would have been just the set $\phi_v(x)$ defined in the discussion of the Rayleigh-Ritz method (Eq. (6.23)). On introducing the "fundamental" initial condition $P(x, 0) = \delta(x - x_0)$, the solution becomes

$$P(x, \tau_R) = x^{1/2}e^{-x} \sum_{\nu=0}^{\infty} \frac{\Gamma(\nu + 1)}{\Gamma(\nu + \frac{3}{2})} L_\nu^{1/2}(x_0)L_\nu^{1/2}(x)e^{-\nu\tau_R} \tag{8.13}$$

The summation can be carried out, with the result

$$P(x, \tau_R) = \frac{e^{1/2\tau_R}}{2[\pi x_0(1 - e^{-\tau_R})]^{1/2}} \times \left\{\exp\left[-\frac{(x^{1/2} - (x_0 e^{-\tau_R})^{1/2})^2}{1 - e^{-\tau_R}}\right] - \exp\left[-\frac{(x^{1/2} + (x_0 e^{-\tau_R})^{1/2})^2}{1 - e^{-\tau_R}}\right]\right\} \tag{8.14}$$

A number of interesting properties follow from this solution, which has obvious similarity to Rayleigh's one-dimensional case. First, it can be shown that a Maxwellian distribution at some temperature different from that of the heat bath preserves its shape during relaxation, evolving with a well-defined, time-dependent temperature. This very special behavior has been called "canonical invariance."[68] A similar simplicity is found in the relaxation behavior of the mean energy $\langle x(t)\rangle$. Either by averaging the original Fokker-Planck equation or by working from the solution (8.14), it can be shown that the time-dependence is exponential with a single transient $\exp(-\lambda_1 t)$. In terms of the initial and equilibrium energies, this can be written

$$[(\langle x(t)\rangle - \langle x(\infty)\rangle)/(x\langle(0)\rangle - \langle x(\infty)\rangle)] = e^{-\lambda_1 t} \tag{8.15}$$

where $\lambda_1 = \frac{16}{3}A\gamma^{1/2}$ is the first eigenvalue for the full relaxation problem. We may note that this agrees with the limiting form of the approximate estimate in Equation (6.27).

Another result of great theoretical importance is for the *equilibrium autocorrelation function*. This is defined as

$$S(t) = \langle \eta(0)\eta(t)\rangle_{\text{eq}} \tag{8.16}$$

where $\eta(t) = \langle x(t) \rangle - \langle x(\infty) \rangle$ and the final average is over the equilibrium ensemble. One obtains

$$S(t) \underset{\gamma \to 0}{=} \tfrac{3}{2} e^{-\lambda_1 t} \tag{8.17}$$

with λ_1 the first eigenvalue as before. This is the characteristic form for the Gaussian Markov process. The more general problem of equilibrium fluctuations is taken up in the next section.

One question which remains unanswered in the above development is, what has become of the continuum? Clearly it has been spirited away somewhere in the expansion and truncation process leading to the Fokker-Planck equation, which is not surprising in that any singular component in the solution would hardly be analytic in the expansion variable. Although it is tempting merely to note that since, in real time, $Z(0) \to \infty$ as $\gamma \to 0$ and the continuum disappears (Fig. 3d), this is misleading since the Fokker-Planck equation is a workable approximation for a *finite* γ to which there corresponds a definite continuum threshold $\lambda^* = Z(0) = 2A/\gamma^{1/2}$. This must therefore be regarded as an ultimate limit to the "reality" of the Fokker-Planck eigenvalues since, referring to the formula for λ_n, it will be seen that for $\nu > (3/8\gamma)$ the threshold will be exceeded and the term will be a "pseudo-eigenvalue." Thus with $\gamma = 1/10$ only some four eigenvalues can be "real" ones (i.e., class B or C) while for $\gamma = 1/50$ as many as twenty are.

Since the variational results described earlier give very accurate estimates of the "true" eigenvalues as the Rayleigh limit is approached, we are now in a position to answer some of the questions posed earlier as to the actual accuracy and meaningfulness of the Fokker-Planck solutions in relation to the "true" solutions of the integral equation. In Figure 7 the first six Fokker-Planck eigenvalues are plotted in the appropriate time scale as dotted lines. As deduced above, only a small number fall outside the continuum for the higher mass ratios and these are noticeably inaccurate. However, as the mass ratio tends to zero the approximations improve progressively though always with some deterioration as the index of the eigenvalue is increased. Evidently at no finite mass ratio will more than a certain number of eigenvalues be approximated to a given accuracy; equivalently there will always exist a lower limit to the time scale for accurate representation of an evolving distribution which contains higher Fourier components. Although the accuracy of the hard-sphere Fokker-Planck eigenvalues is quite clear from the diagram it is still difficult to translate this into a definite statement of the smallness of mass ratio required for accuracy of the Fokker-Planck solution to be adequate in a given case. The most that can really be said is that, for a process dominated by λ_1

(which could be either because it started with a strong $\phi_1(x)$ component in the distribution or because it has aged into this condition), the Fokker-Planck description will be adequate for mass ratios of less than about 1/20. Progressively smaller mass ratios will be required if higher transients are to be successfully approximated and, as always, the predicted eigenfunctions will be of somewhat less quality than the corresponding eigenvalue.

C. The Lorentz Gas

The "inverse Brownian motion" process in the Lorentz limit $\gamma \gg 1$ is less well understood than the simple Brownian motion just considered, and the complications introduced by the continuum and the different ways of proceeding to the limit have made this a fertile field for misunderstanding. In fact several parallel treatments of this case have evolved, in "ordinary" kinetic theory,[11] in plasma physics,[69] and above all, in neutron thermalization; and each of these covers very much the same ground though with different language and emphasis. It will not be possible to unravel all the details of these different developments here, but we shall attempt to describe the clearest results and indicate the main connections between the various fields.

As we have seen in Section V, the hard-sphere relaxation problem for $\gamma \gg 1$ is dominated by the fact that, with the mass of the heat-bath atom tending to infinity, we have $Z(0) \approx 2A/\gamma^{1/2} \to 0$ so that, on a real time scale, the continuum threshold appears to collapse down onto the equilibrium eigenvalue $\lambda_0 = 0$, trapping the whole infinity of discrete λ_k next to the origin. Thus it is clear that, on a real-time basis, the continuum modes completely dominate the relaxation process. However, both in spite of and because of this, it is usual to scale time in the relaxation equation, effectively by a factor $\gamma^{1/2}$, which simplifies the mathematical problem by throwing the whole emphasis onto the relaxation transients slower than $\exp(-Z(0)t)$. The justification for this is not usually declared and cannot be taken as obvious; doubtless in some experimental situations the very final stages of thermalization will be of main interest, but this is hardly sufficient reason for casting the whole mathematics into a form quite unsuited to describing the short-time behavior. It seems to the author that the correct attitude in this situation is *either* to declare one's interest limited to the slowest relaxation times $\lambda_1{}^{-1}$, $\lambda_2{}^{-1}$ ... which are sought as "true" values, *or*, in the spirit of previous remarks, to forget about the true qualitative details of the spectrum and simply offer the Lorentz approximation as a particular spectral resolution of the operator $\mathscr{A}$ which takes advantage of the relation $\gamma \gg 1$. Because of the scaling factor this will, in fact, be a very bad resolution at short times, though it might still represent the continuum transients in a useful way.

Anderson and Shuler[13] have studied the hard-sphere Lorentz limit by methods similar to those just described for the Rayleigh case. Since it is now the *speed* and not the velocity which is virtually unchanged on each collision, a Fokker-Planck equation in the energy variable should still be obtainable provided the expansion can be made in powers of γ^{-1}. Treating the transfer moments in this way, these authors find the following relationships:

$$a_1(x) = 2\gamma^{-1}(2 - x)Z(x) + O(1/\gamma x) \tag{8.18}$$

$$a_2(x) = 4\gamma^{-1}xZ(x) + O(1/\gamma x) \tag{8.19}$$

$$a_3(x) = 32\gamma^{-2}(x - 3)xZ(x) + O(1/\gamma x) \tag{8.20}$$

$$Z(x) \approx \tfrac{1}{2}\pi^{1/2}Ax^{1/2} = N_2\pi\sigma^2(2kTx/m_1)^{1/2}; \; x > 0 \tag{8.21}$$

The last follows on taking the limit $\gamma \to \infty$ in Equation (3.16) while requiring that $x > 0$. Since evidently a_3/a_1 and $a_3/a_2 = O(x/\gamma)$, the third-moment term can be dropped and the Fokker-Planck equation becomes

$$\begin{aligned}\frac{\partial P(x, \tau_L)}{\partial \tau_L} &= \frac{\partial}{\partial x}\left[(x^{3/2} - 2x^{1/2})P(x, \tau_L) + \frac{\partial}{\partial x}(x^{3/2}P(x, \tau_L))\right] \\ &= \mathscr{F}_L P(x, \tau_L)\end{aligned} \tag{8.22}$$

with the scaled time

$$\tau_L = 2\gamma^{-1/2}N_2\pi\sigma^2(2kT/m_2)^{1/2} \tag{8.23}$$

Unlike the Rayleigh case, there must be some restrictions here on the behavior of $P(x, 0)$ at the origin. Obviously enough we require $P(x, 0) \neq \delta(x)$; in addition the following should be satisfied: *either* $(\partial P(x, 0)/\partial x) < \infty$ at $x = 0$ *or* $x(\partial P(x, 0)/\partial x) \to 0$ for $x \to 0$.

No simple solution to Equation (8.22) has been found since the eigenfunctions of the operator $\mathscr{F}_L$ do not seem to be expressible in terms of standard functions. However, even with the necessity for numerical methods, this form is a considerable simplification of the original integral equation. Whether this practical advantage compensates for the doubt introduced in going through the Fokker-Planck approximation is still debatable. Provided that similar uncertainties are not introduced in another guise, it would still seem most preferable to work with the integral equation whenever possible.

The presence of fractional powers in Equation (8.22) invites a transformation to the "speed" variable $y = x^{1/2}$, and with a further change of dependent variable,

$$P(y^2, \tau_L) = \tfrac{1}{2} y e^{-y^2} \chi(y, \tau_L) \tag{8.24}$$

Andersen and Shuler arrive at the equation

$$\frac{\partial \chi(y, \tau_L)}{\partial \tau_L} = \frac{1}{4}\left[y \frac{\partial^2 \chi(y, \tau_L)}{\partial y^2} + (3 - 2y^2) \frac{\partial \chi(y, \tau_L)}{\partial y} \right] \tag{8.25}$$

The eigenfunctions of the operator on the right are still not obtainable in known functions, so that numerical integration must be used. Values found in this way are given in Reference 13.

This approach overlooks a simpler transformation which, though still not leading to exact solutions, offers some advantage for numerical methods. If one defines an *energy-flux* variable $\phi(x, \tau_L) = x^{1/2} P(x, \tau_L)$, Equation (8.22) falls immediately into the form

$$\begin{aligned} \frac{\partial \phi(x, \tau_L)}{\partial \tau_L} &= x^{1/2} \frac{\partial}{\partial x}\left\{ (x - 2)\phi(x, \tau_L) + \frac{\partial}{\partial x}[x\phi(x, \tau_L)] \right\} \\ &= x^{1/2} \mathscr{D}_L \phi(x, \tau_L) \end{aligned} \tag{8.26}$$

where $\mathscr{D}_L$ is the operator

$$\mathscr{D}_L \equiv x \frac{\partial^2}{\partial n^2} + x \frac{\partial}{\partial n} + 1 \tag{8.27}$$

A further substitution then shows that the eigenfunctions of $\mathscr{D}_L$ are

$$F_k(x) = x e^{-x} L_k(x) \tag{8.28}$$

with eigenvalues $k = 0, 1, 2, \ldots \infty$.

This still does not solve the equation since what are needed are the eigenfunctions of $x^{1/2} \mathscr{D}_L$ rather than $\mathscr{D}_L$. Nevertheless, using the functions $F_k(x)$ as an expansion set, a very simple algebraic eigenvalue problem is obtained. Separating the variables by $\phi(x, t) = \Theta(t) H(x)$ we have for the energy part

$$\mathscr{D}_L H(x) = -\lambda x^{-1/2} H(x) \tag{8.29}$$

Thus expanding

$$H(x) = \sum_{k=0}^{N} \Gamma_k F_k(x) \tag{8.30}$$

and using the orthogonality of the Laguerre polynomials, we obtain the matrix system

$$\mathbf{W\Gamma} = -\lambda^{-1}\mathbf{\Gamma} \tag{8.31}$$

where

$$(\mathbf{W})_{ij} = \frac{\dfrac{1}{i}\displaystyle\int_0^\infty x^{1/2}e^{-x}L_i^{\,1}(x)L_j^{\,1}(x)\,dx}{\displaystyle\int_0^\infty xe^{-x}L_i^{\,1}(x)L_i^{\,1}(x)\,dx} \tag{8.32}$$

The integral in the denominator is $i(i+1)$, but that in the numerator does not appear to be standard. After solving the algebraic eigenvalue problem, the full solution for $P(x, t)$ is easily reconstructed. This is a standard technique in the "heavy gas" approximation of neutron transport theory (Ref. 4, p. 127). Usually, however, it is applied in a more complicated form with absorption and diffusion terms present. At this time no sets of eigenvalues appear to have been published for the simple, homogeneous, nonabsorptive case, but there is every reason to suppose that they would agree with those obtained by Andersen and Shuler.

The "heavy-gas" approximation was first studied by Wilkins[70] and is introduced into neutron transport theory in various ways, most of which have some relevance to the Lorentz gas problem. One particularly neat method (Ref. 4, p. 45) bypasses the development given here by performing an expansion of the integrand in an integral expression for the hard-sphere transition kernel (3.12) in powers of γ^{-1}. The result is a symbolic kernel formed of integral representations of the derivatives of the delta function. If these are applied, the Lorentz-Planck equation is obtained as before. Since neither method can be said to be rigorous, it is rather a matter of taste whether one prefers the unpleasantness of the Kramers-Moyal expansion (8.5) to the use of the delta functions.

It is also possible to introduce the "heavy-gas" approximation into any of the practical methods described in Section VI simply by forming the various expansion matrix elements and moment expressions to order γ^{-1}. When general expressions are available there would seem to be little advantage in this other than economy of computing time, but with these approximations in force, the polynomial and degenerate-kernel methods become special treatments of the Lorentz gas. Indeed when the expansion set used is based on the Laguerre polynomials $L_k^{\,1}(x)$ the polynomial method will reduce to the form just given in Equation (8.31). Explicit expressions for the heavy-gas approximation to moments of the hard-sphere transition kernel are given by Shapiro and Corngold.[24]

Some interesting results have also been published on the Lorentz-limit relaxation for gases with interaction laws other than the hard-sphere one. This is a long-standing problem which is already described in some standard works (e.g., Ref. 11, Chap. 18). Oser, Shuler, and Weiss[71] obtain an equation similar to (8.25) for the case of scattering by a potential law $V(r) = \text{const}\, r^{-s}$. This takes the form

$$\frac{\partial \chi(y, \tau_L)}{\partial \tau_L} = \frac{1}{4}\, y^{4/(1-s)} \left\{ y \frac{\partial^2 \chi}{\partial \tau_L} + \left[\left(\frac{3s-7}{s-1} \right) - 2y^2 \right] \frac{\partial \chi}{\partial y} \right\} \tag{8.33}$$

where the time τ_L has been scaled in terms of the hard-sphere mean collision time. The equation clearly displays the essential difference between the cases $s > 5$ and $s < 5$ which we noted before when discussing the energy dependence of the collision number $Z(x)$. Evidently as $s \to \infty$ the equation tends to the previous hard-sphere form. The only case which may be solved exactly is that of the Maxwell gas ($s = 5$) where the strange result is obtained that the relaxation behavior is identical with that for the hard-sphere *Rayleigh* Fokker-Planck equation. There seems to be no significance in this beyond the fact that—for entirely different reasons—both models are governed by an effectively constant collision number.

IX. EQUILIBRIUM FLUCTUATIONS

A. General Characteristics

Although the initial-value problem for $P(x, t)$ dominates the practical applications of the Master Equation, an equally interesting topic from the theoretical point of view is that of equilibrium fluctuations and their relationship to the spectral character of the relaxation process. As in other areas, the necessary theory has grown up in such intimate connection with Brownian motion that its extension to the more general linear case is likely to seem unfamiliar. Here we shall indicate briefly how this extension is possible for the case of scalar process and how the calculations of spectra described earlier may be turned into definite results for the autocorrelation and power spectrum of energy fluctuations.

The essential theory of the autocorrelation function in linear systems has been given by Lax,[72] though in a form which does not take into account the possibility of continuum modes. Here we shall simplify his approach to the case of a scalar variable (energy) but at the same time extend it to allow for the continuum and the use of variational estimates of the eigenvalues.

The first step in obtaining the autocorrelation function $S(t)$ (Eq. (8.16)) is to write the relaxation expression for an initial delta distribution and form the first-moment equation from this by multiplication with x and integration over all energy. To simplify the notation we shall first assume that we are operating with an effective spectral resolution $\hat{N}_k(x)$, $\hat{\lambda}_k$ in the sense of Equation (6.1). This process leads to the expression

$$\langle x_0(t)\rangle - \langle x(\infty)\rangle = \frac{1}{N_0(x_0)} \sum_{k=1}^{N} \hat{N}_k(x_0)\hat{E}_{0k}e^{-\hat{\lambda}_k t} \tag{9.1}$$

where E_{0k} is the "matrix element"

$$\hat{E}_{0k} = \langle N_0(x)|\, x \,|\hat{N}_k(x)\rangle \tag{9.2}$$

and

$$\langle x(t)\rangle = \int_0^\infty xP(x, t)\, dx \tag{9.3}$$

Now, since $\langle x_0(0)\rangle - \langle x(\infty)\rangle$ is the same expression without the exponential factor, and since forming the equilibrium ensemble average corresponds to multiplication of the product of the two series by $M(x) = [N_0(x)]^2$ and integrating, we have, using the orthogonality $\langle N_i(x) \mid N_j(x)\rangle = \delta_{ij}$, that

$$S(t) = \sum_{k=1}^{N} \hat{E}_{0k}{}^2 e^{-\hat{\lambda}_k t} \tag{9.4}$$

(cf. Ref. 72, p. 55, Eq. (13.55). We may note, on closer inspection of the above series, that $S(0) = \text{Var}(x) = 3/2$ in the units used, while $x(\infty) = 3/2$ also. In the same way it can be shown that the autocorrelation of higher powers of the energy variable x can be evaluated in terms of the matrix elements $\hat{E}_{0k}^{(n)} = \langle N_0(x)|\, x^n \,|\hat{N}_k(x)\rangle$. But in all cases the spectrum for the decay of fluctuations is the same as that for the initial-value problem in the same variable. (This is not true, e.g., in the case of entropy production where, in the near-equilibrium region, the relaxation times are *twice* those for the simple process.)

The above derivation may be repeated for the *true* spectrum with both discrete and continuous components. The orthogonality properties (7.9) to (7.11) must now be used and this time the normalization integrals R_k and $R(\lambda)$ play a part. By manipulations similar to the above we now obtain

$$S(t) = \sum_{k=1} (E_{0k}{}^2/R_k)e^{-\lambda_k t} + \int_{\lambda \in c} [(E_0(\lambda))^2/R(\lambda)]e^{-\lambda t}\, d\lambda \tag{9.5}$$

where now $E_0(\lambda)$ is the quantity $\langle N_0(x)|\, x \,|N(x, \lambda)\rangle$. The formidable problems of evaluating the terms in this equation, for example, by Koppel's method, will be evident enough. (Among other difficulties $E_0(\lambda)$ will involve principal-value integration.)

Nevertheless, in broad terms the decay of fluctuations will show the same characteristics as the solutions of the initial-value problem. For "hard" systems the continuum modes will dominate at short times, when the decay will not be exponential, while the discrete terms will take over as time tends to infinity; with "soft" systems the situation will be precisely the reverse. With "hard" systems the effect of the continuum will vanish in the Rayleigh limit; in the Lorentz limit the continuum will dominate more or less according to whether interest is in the time scale of collisions or in a scale adjusted to the degree of accomplishment of the relaxation process. In the Rayleigh limit the Gaussian behavior of Equation (8.17) is obtained by virtue of the fact that

$$\begin{aligned} E_{0k} &\xrightarrow[\gamma\to 0]{} 0 \quad \text{for} \quad k \neq 0 \quad \text{or} \quad 1 \\ E_{01} &\longrightarrow \tfrac{3}{2} \quad \text{for} \quad k = 1 \end{aligned} \tag{9.6}$$

Thus we find $S(t) = 3/2 \exp(-\lambda_1 t) = 3/2 \exp(-(16/3)\gamma^{1/2} A)$ as before.

It need hardly be stressed that the presence of the continuum term in Equation (9.5) changes the nature of the autocorrelation function entirely even though, for practical purposes, its replacement by a discrete spectral resolution might be fairly realistic.

This difference is perhaps more evident on considering the *power spectrum* for the process $J(\omega)$ rather than its autocorrelation function $S(t)$. Using the Wiener-Kinchine theorem[72, 73] we have

$$J(\omega) = \frac{1}{\pi} \int_0^\infty S(t) \cos \omega t \, dt \tag{9.7}$$

$$= \frac{1}{\pi} \sum_{k=1} [(E_{0k})^2/R_k][\lambda_k/(\lambda_k^2 + \omega^2)] + \frac{1}{\pi} \int_{\lambda \in c} \frac{\lambda E_0(\lambda)}{R(\lambda)(\lambda^2 + \omega^2)} \, d\lambda \tag{9.8}$$

At least three cases can be distinguished. Under Brownian motion conditions, the function $J(\omega)$ will consist of the single term in $\lambda_1/(\lambda_1^2 + \omega^2)$; when the spectrum is either actually or approximated as discrete there will be a combination of such terms; when the continuum is taken into account everything will depend on the rather indeterminate nature of the last integral.

B. Short-Time Approximations

Just as in the initial-value problem, there is an alternative available for studying the short-time or high-frequency characteristics of the process. We can again obtain some account of this behavior by application of the operator $\exp(\mathscr{A}t)$ without any need to solve the singular eigenvalue problem. Although the method only furnishes values of the lower derivatives of $S(t)$ at $t = 0$ it still seems to be of considerable interest.

The necessary operations can be carried through symbolically. We have, following Equation (6.6),

$$\langle x_0(t)\rangle = \int_0^\infty x \exp(\mathscr{A}t)\delta(x - x_0)\, dx \tag{9.9}$$

where the left-hand side signifies the evolution of the mean energy from an initial delta function at $x = x_0$. From this we find

$$S(t) = \left\langle x_0 \int_0^\infty x[\exp(\mathscr{A}t) - 1]\delta(x - x_0)\, dx \right\rangle_{x_0} \tag{9.10}$$

where the outer brackets indicate an ensemble average over the equilibrium distribution $M(x) = [N_0(x_0)]^2$.

Operating on this expression we can show that

$$\left(\frac{\partial^n S}{\partial t^n}\right)_{t=0} = \left\langle x_0 \int_0^\infty x\mathscr{A}^n\, \delta(x - x_0)\, dx \right\rangle_{x_0} \tag{9.11}$$

To remove any possible ambiguity in this expression we shall interpret it for the first derivative. Thus

$$\left(\frac{\partial S}{\partial t}\right)_{t=0} = \left\langle x_0 \int_0^\infty x\mathscr{A}\, \delta(x - x_0) \right\rangle_{x_0} \tag{9.12}$$

$$= \left\langle x_0 \int_0^\infty x \left[\int_0^\infty K(y, x)\, \delta(y - x_0) - Z(x)\, \delta(y - x)\, \delta(x - x_0)\right] dy\, dx \right\rangle_{x_0} \tag{9.13}$$

$$= \left\langle x_0 \int_0^\infty x\{K(x_0, x) - Z(x)\, \delta(x - x_0)\}\, dx \right\rangle_{x_0} \tag{9.14}$$

$$= \langle x k_1^{(1)}(x) - x^2 Z(x)\rangle_{\text{eq}} \tag{9.15}$$

Here we have written $k_1^{(1)}(x)$ for the first energy moment of the transition kernel defined previously in Equation (3.26). The final average is then a multiplication by $M(x)$ and integration over all energies.

The second derivative follows similarly on interpreting $\mathscr{A}^2$ as in Equation (6.11). The result may be simplified using the detailed-balance property of the kernel. We find

$$\left(\frac{\partial^2 S}{\partial t^2}\right)_{t=0} = \langle x k_1^{(2)}(x) - 2xZ(x)k_1^{(1)}(x) + x^2[Z(x)]^2\rangle_{\text{eq}} \tag{9.16}$$

Here $k_1^{(2)}(x)$ is the first energy moment of the iterated kernel $K^{(2)}(x, y)$. All the integrals involved appear reasonably straightforward, though explicit values for the hard-sphere gas are not available at this time. If pushed, one should be able to carry the method to third or fourth order, and the resulting expansion of $S(t)$ about zero might well account accurately for the whole time range in which the continuum modes are dominant. In fact the crucial range for the decay of fluctuation when $\gamma \sim 1$ is certainly within the first five collision times (see below).

C. Variational Solutions

We may now return to the more practical question of converting the earlier variational results for eigenfunctions $\hat{N}_k(x)$ and eigenvalues $\hat{\lambda}_k$ into definite expressions for the autocorrelation function $S(t)$. Obviously any good set of approximate $\hat{N}_k(x)$ can be used to calculate the matrix elements E_{0k} in the required series expansion, but there is a special advantage in using the approximate set which results from the Rayleigh-Ritz method with the basis functions $\phi_k(x) \sim l_k^{1/2}(x)$ which are the exact set in the Rayleigh limit.

If $\alpha_i^{(k)}$ are the expansion parameters (6.15) determined in the algebraic eigenvalue problem, it follows that

$$E_{0k} = \sum_{i=0}^{N} \alpha_i^{(k)} \langle \phi_0(x) | x | \phi_i(x) \rangle \tag{9.17}$$

But it is easy to show that, in fact, $\langle \phi_0 | x | \phi_i \rangle = (3/2)^{1/2} \delta_{i1}$ by a special property of the Laguerre functions, so that the elements $\hat{E}_{0k}$ take on the very simple form $\hat{E}_{0k} = (3/2)^{1/2} \alpha_1^{(k)}$. Putting this into the expression for the autocorrelation function, we obtain the latter directly in terms of the algebraic eigenvector components. Thus

$$S(t) = \tfrac{3}{2} \sum_{k=1}^{N} [\alpha_1^{(k)}]^2 e^{-\hat{\lambda}_k t} \tag{9.18}$$

Since the eigenvectors $\alpha^{(k)}$ are orthonormal, we can equally well write $\alpha_k^{(1)}$ for $\alpha_1^{(k)}$ and see that the property $S(0) = \text{Var}\,(x(\infty)) = 3/2$ becomes explicit. Furthermore, since in the limit $\gamma \to 0$, $\alpha_1^{(k)} \to \delta_{k1}$, the tendency to a limiting Gaussian process is also exhibited clearly (cf. Eq. (8.17)).

Calculations with the above equation were carried out by Hoare and Kaplinsky using their Rayleigh-Ritz spectral resolutions for the hard-sphere gas. The results are of particular importance in deciding the following questions. (a) How small a mass ratio is actually required for the truly Gaussian behavior of Equation (8.17) to occur? (b) To what extent is the process for other mass ratios "pseudo-Gaussian" in character, that is, dominated by the first eigenvalue transient $\exp(-\lambda_1 t)$ even though not itself a Brownian motion (Ornstein-Uhlenbeck) process? (c) What degree of symmetry is there in the autocorrelation behavior on either side of $\gamma = 1$ and in tending to the Rayleigh and Lorentz limits (i.e., between *normal* and *inverse* Brownian motion)?

So far as the hard-sphere gas is concerned, and excluding the very shortest times $t \to 0$, these questions are answered by the results in Figure 9, which gives plots of $\log_e S(t)$ with time measured in units of the mean collision time $Z_{00}{}^{-1}$. On this basis "pseudo-Gaussian" behavior will show as a straight-line relationship, with deviations from this visible either as curvature or segmentation of the lines, depending on whether the higher eigenvalues responsible are closely distributed or effectively distinct. In fact deviations from straight-line behavior are only just visible, being most pronounced in the near-Lorentz regime $5 < \gamma < 20$. On the whole the destruction of correlations proceeds at a rate in keeping with common-sense predictions, being fastest for $\gamma = 1$, and there requiring of the order of 3–5 collisions. This accords well with the oft-quoted figure of four collisions obtained in Monte Carlo calculations.[50] There is rather little effect on the autocorrelation when the mass ratio is changed by a factor of two on either side of unity and at all mass ratios there is remarkable (though not exact) symmetry between the curves for γ and γ^{-1}. These features are all consistent with the previous observation that, on the time scale used, the curve of λ_1 versus γ is peaked at about $\gamma = \frac{1}{2}$, flat-topped in this region, and quite well separated from λ_2 over the whole range for which Brownian motion conditions do not apply. This is likely to be a special property of the hard-sphere kernel which might not be obtained with other scattering laws.

The onset of Brownian motion behavior is easily demonstrated by plotting $\log_e S(t) = -\lambda_1 t$ from (8.17) in the appropriate time scale (dotted lines.) As to be expected from the eigenvalue curves themselves (Fig. 7), the Brownian motion condition could be said to become effective when the mass ratio is somewhere in the region of $\gamma = 1/20$ to $\gamma = 1/30$—that is, where correlations diminish to a factor e^{-1} in some *ten* collisions.

These results are not entirely insensitive to assumptions about the continuum since the pseudo-eigenvalues appearing in Figure 7 have been used in the calculations. However a simple check shows that this can have very little effect on the results quoted. On the mean-collisional time scale used, the continuum threshold relaxation time $\tau^* = Z(0)^{-1}$ becomes $\tau^* = (\gamma + 1)^{1/2}$. This is close to *one* collision time for most of the Rayleigh regime ($Z_{00} \approx Z(0)$ for $\gamma < 1$) and increases only slowly for $\gamma > 1$. The curves in Figure 9 are marked with points corresponding to τ^*, which

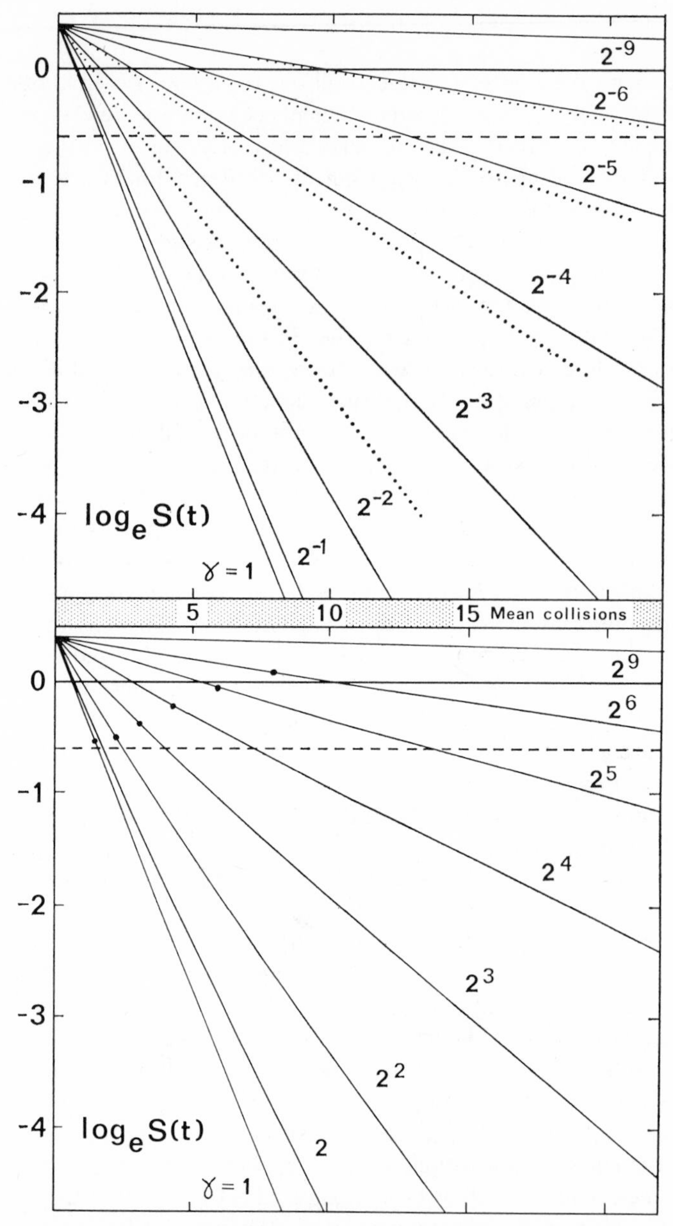

Fig. 9. The equilibrium energy-autocorrelation function $S(t)$ for the hard-sphere gas. The time is scaled in units of the mean equilibrium collision time Z_{00}^{-1}. The horizontal dotted line measures a factor e^{-1} in the decay of fluctuations, and the others show the Gaussian results for small mass ratios predicted by the energy Fokker-Planck equation. From Hoare and Kaplinsky.[25] The dots in the lower half indicate the characteristic time $\tau^* = 1/Z(0)$ in the mean-collisional time-scale.

will be a rough measure of the stage where the continuum modes cease to contribute. Even if the discrete spectral resolution were most unrealistic in the region $t < \tau^*$ this could hardly modify the conclusions just drawn.

X. CONCLUSION

Anyone following the development this far will be very conscious that the fragmentary nature of the subject matter still frustrates any effort to give a really unified treatment in standard statistical-mechanical language. This is of course in part due to the many gaps which remain to be filled in the theory and numerical investigations alike, but it is equally a reflection of the very different practical needs and possibilities governing the several fields which touch on each other in this work. These practical differences probably determine mathematical attitudes more than may appear from the bare equations. While there can be little doubt that technological pressures have played their part in neutron transport theory, many advances can be seen to be inspired by the simple experimental fact that there exist flux-detecting probes for neutrons. This brings to the transport equations a directness hardly possible in "ordinary" kinetic theory. Nature is also kind to the neutron theorist in other ways: the linearity of equations is virtually guaranteed, and scattering is by interactions which are as short-range as could be desired. The confidence which these fortunate conditions give to the theoretician is very apparent in the literature.

Several topics will be seen to be lacking which might have filled out the picture presented. A full spectral theory and numerical treatment of the Rayleigh piston is still incomplete, but is in hand at this time. It is curious indeed that we should still know less about the one-dimensional case than the three-dimensional one, but this hardly justifies further delay in presenting this article. It might also have been natural to continue with the theory of spatially dependent processes in molecular systems. However there seems to be a case for treating this as a separate topic since it joins rather than overlaps with the material presented here. Spatial transport theory is still growing fast and is perhaps now at the crucial stage reached several years ago in the scalar theory. We hope therefore to develop this subject in a later article which will again focus on results lying published but unpublicized in the reactor physics journals.

In putting this work into perspective with better-known areas of statistical mechanics, the overriding impression is that, for every subtlety of the scalar, spatially homogeneous, infinite-medium, linear-transport problem exhibited here, there must necessarily arise a whole multitude of complications in the corresponding vector, spatially dependent, and finite-medium boundary-value problems. And when each of these complications is reflected in turn into the profundities of the full nonlinear Boltzmann equation, which can surely be no less than an order of magnitude still more difficult, then one is conscious, to say the very least, of a new humility towards the latter. For the author the considerable advances described here have the curious effect of shedding light in the direction of the full Boltzmann equation, while at the same time causing its monstrous form to recede still further into the distance. Nevertheless it would be wrong to neglect the outer fringes which are open to illumination in this way and have every claim to interest in their own right.

Acknowledgment

The author is particularly grateful to M. M. R. Williams for personal advice and encouragement and to others, particularly J. H. Ferzinger, J. U. Koppel, and M. Rahman, for helpful communications. Acknowledgment is also made of financial support from the Science Research Council, London during the completion of this work.

References

1. J. W. Strutt (Baron Rayleigh), *Phil. Mag.*, **32,** 424 (1891); *Scientific Papers*, Vol. 3, Cambridge University Press, London, 1902, p. 473.
2. L. Waldmann, *Handbuch der Physik* S. Flügge, Ed., Vol. 12, Springer, Berlin, 1958.
3. H. Grad, *Handbuch der Physik*, S. Flügge, Ed., Vol. 12, Springer, Berlin, 1958, p. 205.
4. M. M. R. Williams, *The Slowing Down and Thermalization of Neutrons*, North Holland, Amsterdam, 1966.
5. P. F. Zweifel, and K. M. Case, *Linear Transport Theory*, Addison-Wesley, Reading, Mass., 1967.
6. J. H. Ferzinger and P. F. Zweifel, *Theory of Neutron Slowing-down in Nuclear Reactors*, M.I.T. Press, Boston, 1967.
7. M. D. Kostin, *J. Chem. Phys.*, **43,** 2679 (1965).
8. B. S. Rabinovitch and R. Diesen, *J. Chem. Phys.*, **30,** 735 (1959).
9. J. L. Lebowitz and P. G. Bergmann, *Ann. Phys.* (*N.Y.*), **1,** 1 (1957).
10. T. A. Bak and J. L. Lebowitz, *Phys. Rev.*, **131,** 1138 (1963).
11. S. Chapman and T. G. Cowling, *The Mathematical Theory of Non-uniform Gases*, 2nd Ed., Cambridge University Press, London, 1952.
12. E. P. Wigner and J. E. Wilkins, *A.E.C. Rept. D*-2275, 1944.
13. K. Andersen and K. E. Shuler, *J. Chem. Phys.*, **40,** 633 (1964).
14. S. E. Nielsen and T. A. Bak, *J. Chem. Phys.*, **41,** 665 (1964).
15. I. Kuščer and M. M. R. Williams, *Phys. Fluids*, **10,** 1922 (1967).
16. H. Grad, *Rarefied Gas Dynamics*, J. A. Laurmann, Ed., Academic Press, New York, 1963, p. 26.
17. R. J. Glauber, *Phys. Rev.*, **87,** 189 (1952).
18. L. Van Hove, *Phys. Rev.*, **95,** 249 (1954).
19. P. A. Egelstaff, *An Introduction to the Liquid State*, Academic Press, London, 1967, Chaps. 8 and 9.
20. R. C. Desai and M. Nelkin, *Nuclear Sci. and Eng.*, **24,** 142 (1966).
21. M. M. R. Williams, *J. Phys.*, **2,** D389 (1969).

22. F. G. Tricomi, *Integral Equations*, Interscience, New York, 1957.
23. D. Bohm and E. P. Gross, *Phys. Rev.*, **75,** 1851 (1949).
24. C. S. Shapiro and N. Corngold, *Phys. Rev.*, **137A,** 1686 (1965).
25. M. R. Hoare and C. H. Kaplinsky, *J. Chem. Phys.*, **52,** 3336 (1970).
26. I. Oppenheim, K. E. Shuler, and G. Weiss, *Adv. Mol. Relaxation Processes*, **1,** 13 (1967).
27. K. M. Case, *Ann. Phys. (N.Y.)* **7,** 349 (1959); **9,** 1 (1960).
28. J. R. Mika, *Nuclear Sci. and Eng.*, **11,** 415 (1961).
29. R. Zelzany and A. Kurszell, *Ann. Phys. (N.Y.)*, **16,** 81 (1962).
30. N. G. Van Kampen, *Physica*, **21,** 949 (1955).
31. P. A. M. Dirac, *The Principles of Quantum Mechanics*, 3rd Ed., Clarendon Press, Oxford, 1947, p. 195.
32. C. Cercignani, *Mathematical Methods in Kinetic Theory*, Plenum Press, New York, 1969.
33. G. E. Uhlenbeck and G. W. Ford, *Lectures in Statistical Mechanics*, American Mathematical Society, Providence, R.I., 1963.
34. J. U. Koppel, *Nuclear Sci. and Eng.*, **16,** 101 (1963).
35. J. H. Ferzinger, *Phys. Fluids*, **8,** 426 (1965).
36. J. K. Bruckner and J. H. Ferzinger, *Phys. Fluids*, **9,** 2309 (1966).
37. Y. Shizuta, *Progr. Theoret. Phys. (Kyoto)*, **32,** 489 (1964).
38. (a) L. Schwartz, *Mathematics for the Physical Sciences*, Addision-Wesley, Reading, Mass., 1966. (b) M. J. Lighthill, *Introduction to Fourier Analysis and Generalized Functions*, Cambridge University Press, London, 1958.
39. (a) A. H. Zemanian, *Distribution Theory and Transform Analysis*, McGraw-Hill, London, 1965. (b) W. F. Donoghue, *Distributions and Fourier Transforms*, Academic Press, London, 1969.
40. Masamichi Tokizawa, *J. Math. Phys.*, **10,** 1834 (1960).
41. M. D. Kruskal and I. B. Bernstein, *Phys. Fluids*, **7,** 457 (1964).
42. A. Joseph, *Intern. J. Quant. Chem.*, **1,** 615 (1967).
43. I. Kuščer and N. Corngold, *Phys. Rev.*, **139,** A981 (1965); **140,** AB5 (1966).
44. N. Corngold, P. Michael, and W. Wollman, *Nuclear Sci. and Eng.* **15,** 13 (1963).
45. M. M. R. Williams, *Nuclear Sci. and Eng.*, **26,** 262 (1966).
46. M. M. R. Williams, *Nuclear Sci. and Eng.*, **33,** 262 (1968).
47. R. P. Feynmann, *Phys. Rev.*, **56,** 340 (1939).
48. R. Courant and D. Hilbert, *Methods of Mathematical Physics*, Vol. 1, Interscience, New York, 1953.
49. T. Wood, *Proc. Phys. Soc.*, **85,** 805 (1965).
50. B. J. Alder and T. Wainwright, in *Transport Processes in Statistical Mechanics*, I. Prigogine, Ed., Interscience, New York, 1958.
51. C. S. Shapiro, *Report BNL* 8433, Brookhaven National Laboratory, Upton, N.Y., 1964.
52. M. Cadilhac, *Report CEA-R* 2368, Centre d'Etudes Nucleaires de Saclay, 1964.
53. G. Schaeffer and K. Allsop, *Proc. B.N.L. Conference*, **2,** 614 (1962).
54. J. Heading, *An Introduction to Phase-integral Methods*, Methuen, London, 1962.
55. N. I. Muskhelishvili, *Singular Integral Equations*, P. Nordhoff, Grönigen, 1953. See also W. Pogorzelski, *Integral Equations*, Pergamon, London, 1966.
56. M. Rahman and M. K. Sundaresan, *Phys. Letters*, **A25,** 705 (1967).
57. M. Rahman and M. K. Sundaresan, *Can. J. Phys.*, **46,** 2287 (1968).
58. M. Rahman, *Can. J. Phys.*, **48,** 151 (1970).

59. M. S. Green, *J. Chem. Phys.*, **19,** 1036 (1951).
60. G. E. Uhlenbeck and L. S. Ornstein, *Phys. Rev.*, **36,** 823 (1930).
61. Ming Chen Wang and G. E. Uhlenbeck, *Rev. Mod. Phys.*, **17,** 323 (1945).
62. Ref. 73, Chap. 15, Section 12.
63. H. A. Kramers, *Physica*, **7,** 284 (1940).
64. J. E. Moyal, *J. Roy. Statist. Soc.*, **B11,** 150 (1949).
65. J. Keilson and J. E. Storer, *Quart. J. Appl. Math.*, **10,** 243 (1952).
66. N. G. Van Kampen, *Can. J. Phys.*, **39,** 551 (1961).
67. R. F. Pawula, *Phys. Rev.*, **162,** 186 (1967).
68. H. C. Anderson, I. Oppenheim, K. E. Shuler, and G. Weiss, *J. Math. Phys.*, **5,** 522 (1964).
69. F. H. Ree and R. E. Kidder, *Phys. Fluids*, **6,** 857 (1963).
70. J. E. Wilkins, *Ann. Math.*, **49,** 189 (1948).
71. H. Oser, K. E. Shuler, and G. Weiss, *J. Chem. Phys.*, **41,** 2661 (1964).
72. M. Lax, *Rev. Mod. Phys.*, **32,** 25 (1960).
73. F. Reif, *Fundamentals of Statistical and Thermal Physics*, McGraw-Hill, New York, 1965, Section 15.15.

LOW-ENERGY ELECTRON DIFFRACTION*

G. A. SOMORJAI AND H. H. FARRELL†

Inorganic Materials Research Division, Lawrence Radiation Laboratory and Department of Chemistry, University of California, Berkeley, California

CONTENTS

* This work was done under the auspices of the U.S. Atomic Energy Commission.

† Current address: Department of Applied Science, Brookhaven National Laboratory, Upton, N.Y. 11973.

I. INTRODUCTION

In recent years there has been remarkable progress in our understanding of the structure of surfaces. Most of the structural investigations have been carried out using low-energy electron diffraction (LEED). Just as X-ray diffraction may be used to study the bulk structure, low-energy electron diffraction probes the structure of surfaces.

The lack of structural information in surface reactions has long impeded the progress of surface science. Using LEED one can determine the structure of the clean surface and monitor the structure of adsorbed gases during the different stages of chemical surface reaction. Thus, correlation between the structure and chemistry of surfaces can be established. Surface phase transformations of many kinds (order-order, order-disorder, etc.) can be studied by LEED. Finally, the dynamics of surface atoms, their mean square displacements, or their diffusion along the surface can be investigated.

The application of low-energy electron diffraction has led to the discovery of several new surface phenomena. It has been found that the arrangement of surface atoms in clean solid surfaces could be different from the arrangement of atoms in the bulk unit cell. Solid surfaces may undergo structural rearrangements or changes of chemical composition, while no corresponding changes may occur in the bulk of the crystal. It has been found that atoms chemisorbed on solid surfaces form ordered surface structures. The nature of the surface structure depends on the crystal orientation, the chemistry and the concentration of adsorbed gas atoms, and the temperature. Low-energy electron diffraction studies revealed that the mean square displacement of surface atoms is larger than the mean square displacement of bulk atoms.

The major obstacle in the path of surface structural studies using low-energy electron diffraction is the lack of a simple theory which could explain the scattered low-energy electron beam intensities. The application of such a theory in model calculations where the important variables are the atomic positions should lead, just like the use of the kinematic theory in X-ray diffraction, to complete description of the surface structure. It is hoped that such a theory will become available in the very near future. Until then the assignment of atomic positions in surface structures, solely on the basis of the diffraction pattern, is not unambiguous. This explains the different interpretations which may be given to the same diffraction pattern and the concentration of LEED studies on only simple monatomic or diatomic surfaces. Frequently, however, the available supplementary chemical information using other experimental techniques permits one to identify the surface structure correctly and to eliminate most of the alternative models.

This review attempts to present the state of the field of low-energy electron diffraction. We shall describe the theory and the experiment and then review the structural studies which have been carried out using disordered surfaces, clean ordered surfaces, adsorbed gases, and condensable vapors on single-crystal surfaces.

In order to carry out a low-energy electron diffraction experiment one needs (a) ultra-high vacuum ($<10^{-8}$ torr), (b) one face of a pure single crystal, and (c) a well-focused electron beam in the energy range 1–500 eV. At the present state of our technology such an experiment can be carried out with relative ease.

II. SYMMETRY AND NOMENCLATURE OF TWO-DIMENSIONAL STRUCTURES

One of the most notable features of low-energy electron diffraction from ordered single-crystal surfaces is the symmetry of the diffraction pattern. This symmetry is a direct consequence of the periodic arrangement of the atoms or molecules in the surface of the crystal.

It is convenient to regard the structure of the surface to be arbitrarily constructed of a lattice and a basis. A lattice is an array of points in space such that the arrangement of atoms around any lattice point is identical to that around every other lattice point. The basis represents the arrangement of atoms around the lattice points. A basis may be as simple as a single metal atom placed on a lattice point for many metal crystals, or it may be as complex as a set of DNA molecules. A unit cell which contains lattice points only at the corners is called a primitive cell.

All lattice points are related by the translation operations

$$\mathbf{T} = n_a \mathbf{a} + n_b \mathbf{b} + n_c \mathbf{c} \tag{1}$$

where n_a, n_b, and n_c are integers and **a**, **b**, and **c** are translation vectors whose dimensions are those of the sides of unit cell and whose directions are parallel to the sides of the unit cell. The two-dimensional (2D) lattice of a surface may be characterized by two-dimensional translation operations

$$\mathbf{T} = n_a \mathbf{a} + n_b \mathbf{b} \tag{2}$$

Note that the translational operations are symmetry operations that leave the surface invariant. The regular arrangement of a reasonably perfect single-crystal surface frequently allows for the application of other symmetry operations such as rotations and mirror reflections that will also leave the surface invariant.

It may be shown that perfect two-dimensional symmetry allows for only a finite number of different types of rotations (even though each allowed rotation may be applied an infinite number of times). Only those rotations through an angle of $2\pi/n$ where $n = 1, 2, 3, 4$, and 6 are allowed.[1] It is easily seen that if rotations for $n = 4$ are allowed, then the lattice must be square. Similarly, rotations through $2\pi/n$ for $n = 3$ or 6 must be associated with hexagonal lattices. In this manner, the allowed rotations place restrictions on the types of primitive translations that may occur. Mirror operations will also restrict the types of primitive translations that are allowed. As a consequence of these mutual restrictions, there are only five two-dimensional Bravais or space lattices that are possible. These are shown in Figure 1.

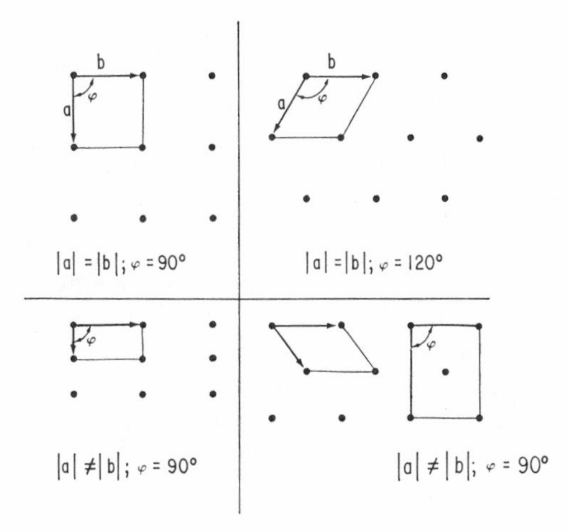

Fig. 1. The five two-dimensional Bravais lattices.

The symmetry of the surface is only partially described by its Bravais lattice. The basis or arrangement of atoms around each lattice point will itself remain invariant under certain symmetry operations even if it is only the trivial operation of rotation through 360°. The collection of symmetry operations that leave a basis invariant is called a crystallographic point group. There are 10 two-dimensional crystallographic point groups. Five of them are characterized by the permissible rotations $2\pi/n$ for $n = 1, 2, 3, 4$, and 6. The other five are characterized by the permissible rotations and by mirror reflections. If one mirror plane is allowed, then the rotations will generate a set of equivalent mirror planes (except for $n = 1$, of course).

The total symmetry of a crystal surface is described by the combination of the Bravais lattice and the crystallographic point group of the basis. There are 17 unique and allowed combination of the five Bravais lattices and ten crystallographic point groups. These are called two-dimensional space groups. The reader is referred to an excellent discussion of these space groups by Wood.[2]

In the energy range usually employed in low-energy electron diffraction, the de Broglie wavelength associated with the electron will be of the order of angstroms [$\lambda(\text{Å}) = (150/\text{eV})^{1/2}$]. This length is similar to the inter-row spacing of the atoms on most single-crystal surfaces. Since the atoms in the crystal are arranged in an orderly fashion, there will be only certain regions in space where the reflections from parallel rows of atoms will interfere constructively. When the scattering takes place from a two-dimensional array, the regions in space of allowed constructive interference will be rods rather than points, as in the three-dimensional case. These rods, or diffraction beams, can be characterized by the equivalent parallel crystallographic planes from which the constructive interference that formed a given diffraction beam originated.

A convenient parameter for indexing a diffraction beam is the reciprocal lattice vector, $\mathbf{G}_{hk}$. This reciprocal lattice vector contains information about both the magnitude of the interplanar spacing under consideration and the direction of the normal to that set of crystallographic planes. For a given two-dimensional Bravais lattice with primitive translational vectors, **a** and **b**, the corresponding primitive reciprocal lattice vectors $\mathbf{G}_a$ and $\mathbf{G}_b$ are defined by the following relationships:

$$\mathbf{a} \cdot \mathbf{G}_a = \mathbf{b} \cdot \mathbf{G}_b = 2\pi \tag{3a}$$

$$\mathbf{a} \cdot \mathbf{G}_b = \mathbf{b} \cdot \mathbf{G}_a = 0 \tag{3b}$$

and

$$\mathbf{G}_a = 2\pi \frac{\mathbf{b} \times \mathbf{c}}{\mathbf{a} \cdot [\mathbf{b} \times \mathbf{c}]} \qquad \mathbf{G}_b = 2\pi \frac{\mathbf{a} \times \mathbf{c}}{\mathbf{b} \cdot [\mathbf{a} \times \mathbf{c}]} \tag{3c}$$

where **c** is a unit vector perpendicular to the surface. Just as a real space translational vector, **T**, may be constructed from an integral number of primitive real space translational vectors $n_a \mathbf{a} + n_b \mathbf{b}$, a reciprocal lattice vector $\mathbf{G}_{hk} = h\mathbf{G}_a + k\mathbf{G}_b$ may be constructed from the primitive reciprocal lattice vectors $\mathbf{G}_a$ and $\mathbf{G}_b$.

The reciprocal lattice vector $\mathbf{G}_{hk}$ has the following interesting properties. Its direction is normal to the 2D crystallographic planes with (*hk*) Miller indices and its length is equal to 2π times the reciprocal of the spacing of these (*hk*) planes.

The indexing of the various diffraction beams is a relatively simple matter. The specularly reflected beam does not involve any change in the component of the electron momentum that is parallel to the surface of the crystal. Therefore, it is associated with only the null parallel reciprocal lattice vector, $\mathbf{G}_{00} = 0 \cdot \mathbf{G}_a + 0 \cdot \mathbf{G}_b$ and is customarily indexed as the (00) beam. In a similar fashion, the first-order diffraction beams for the square, rectangular, or oblique lattices may be indexed as (01), (10), (01), or (10) as they are associated with the reciprocal lattice vectors $\mathbf{G}_{01} = \mathbf{G}_b$, $\mathbf{G}_{10} = \mathbf{G}_a$, $\mathbf{G}_{0\bar{1}} = -\mathbf{G}_b$, or $\mathbf{G}_{\bar{1}0} = -\mathbf{G}_a$, respectively. Higher-order beams may be indexed in an analogous manner. The unit cell vectors of the primitive two-dimensional unit cell and indexing of the three densest crystal faces in the face-centered cubic structures are given in Figures 2a–j.

Some caution must be exercised in the indexing of low-energy electron diffraction patterns. The choice of indices is dependent upon the choice of unit cell. For example, the first-order diffraction beams from the (100) face of face-centered cubic crystals may be indexed as either (10) or (11), depending upon whether the primitive two-dimensional lattice (Fig. 3a) or the full X-ray unit cell containing a centered atom is used (Fig. 3b). Both notational systems are used in the literature.

A common observation in low-energy electron diffraction patterns is the occurrence of "extra" or "fractional-order" spots. In addition to the normal diffraction beams associated with the lattice spacing of the bulk crystal, there frequently appear other diffraction beams that often may be

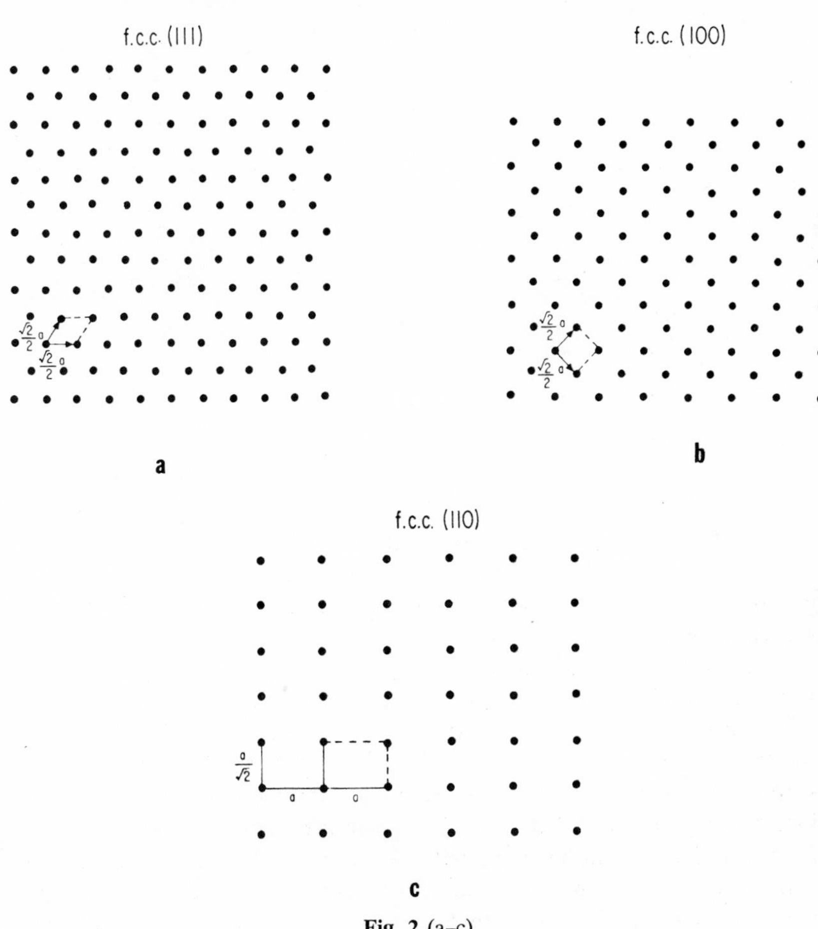

Fig. 2 (a–c)

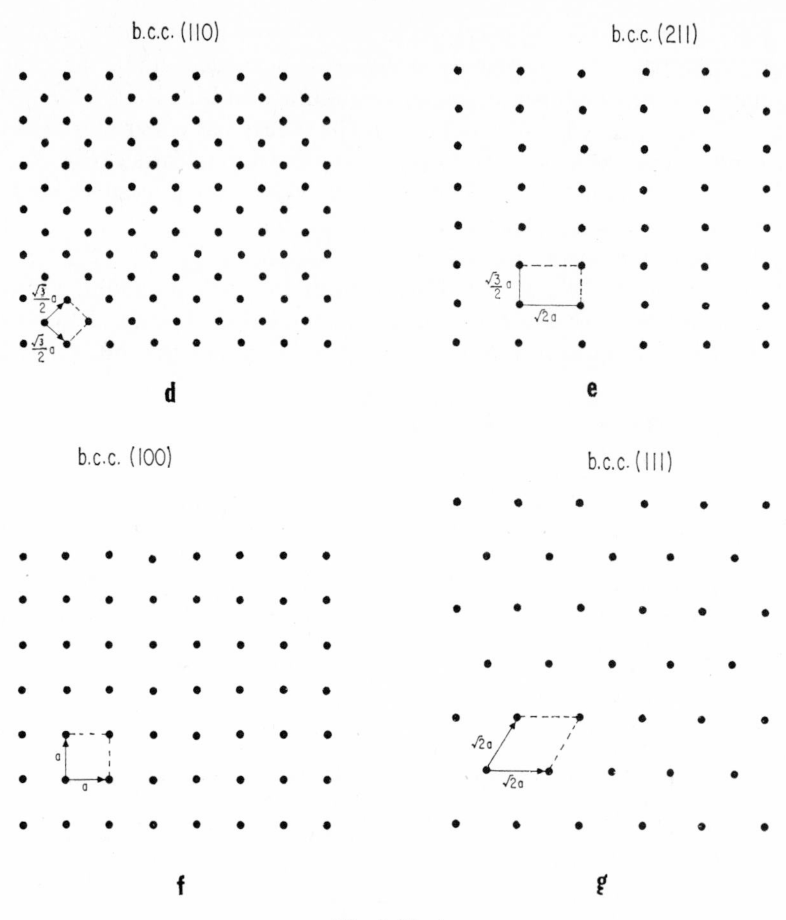

Fig. 2 (d–g)

indexed with fractional indices. These extra diffraction beams are usually associated with structures on the surface which are characterized by larger unit cells than the projection of the bulk unit cell onto the crystal surface. For example, if molecules from the ambient were to condense out onto every other lattice site of the surface, then this new structure on the surface of the crystal would have a periodicity twice that of the "clean" substrate surface. As the reciprocal lattice vectors are inversely proportional to the interplanar spacings, then the new primitive reciprocal lattice vectors for the surface would have one-half the length of those in the original set. Therefore, in addition to the old set of reciprocal lattice vectors parallel to the surface, there would then exist a new set with fractional-order

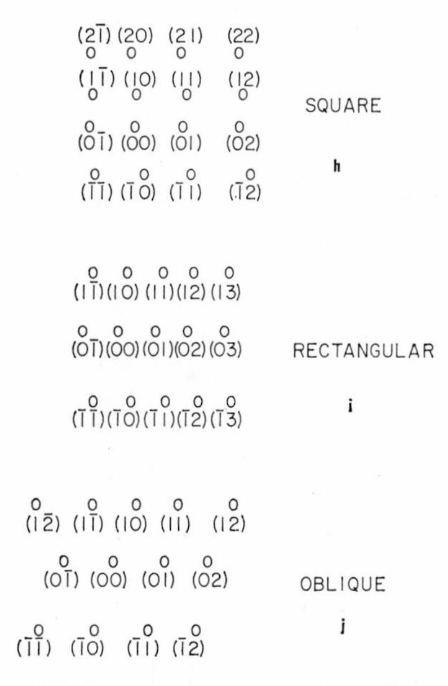

Fig. 2. Unit cell vectors (a–g) of the primitive two-dimensional unit cell and indexes (h–j) for the three densest f.c.c. and b.c.c. crystal faces.

indices. Associated with this new structure and these new parallel reciprocal lattice vectors, there would also be a new set of diffraction beams.

Surface structures are formed not only by the adsorption of gases onto the surface, but also by the segregation of bulk impurities onto the surface of the crystal and by the reconstruction of clean crystal surfaces. As with the simple surfaces, their Bravais lattices may be either centered or primitive. Though perhaps less common, the dimensions of the structured surface may bear a nonintegral relationship to the dimensions of the simple surface. Furthermore, the structured surface may have a unit cell that is rotated relative to that of the clean surface.

Often, a surface structure will exist that has the usual dimensions along one translation, but a large dimension along the other translation direction. These structures are frequently denoted as being $(1 \times n)$ where the 1 indicates the usual bulk cell dimension along the x-direction while the n indicates n times the bulk unit cell dimension along the y-direction. This will give rise to a diffraction pattern with the usual number of diffraction spots along one reciprocal lattice direction and n times the usual number along the other direction (Fig. 4a–c). When the unit cell vectors

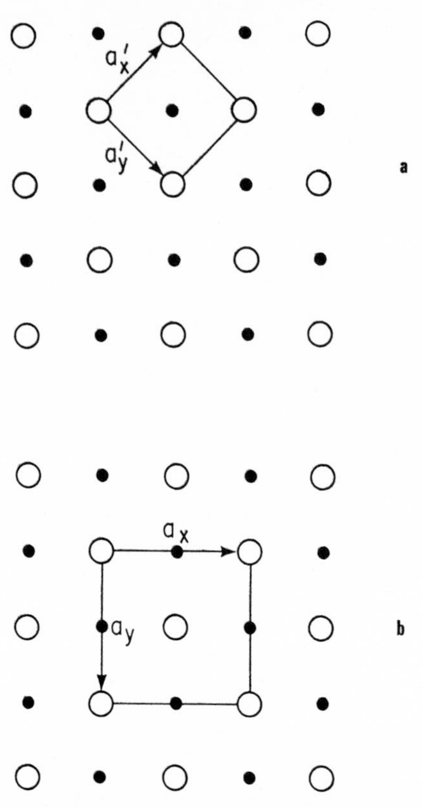

Fig. 3. (a) The primitive two-dimensional unit cell and (b) the X-ray unit cell projection. ○, atoms in surface; ●, atoms in second layer.

of the substrate in both directions are identical on the original surface (as on the (100) face of f.c.c. or b.c.c. solids), then it is possible to have two types of domains, one set of the $(1 \times n)$ and one set of the $(n \times 1)$ kind. When this occurs, it may be observed that the diffraction pattern will have n times the normal number of spots along both directions. This type of diffraction pattern is not to be confused with that arising from a true $(n \times n)$ structure where the surface structure has n times the usual dimension along both directions on all portions of the crystal. For example, a (1×2) structure on a square surface may contain two types of domains rotated relative to one another by 90° and giving rise to $(0, \frac{1}{2})$ and $(\frac{1}{2}, 0)$ spots. A true (2×2) structure, however, will also give rise to $(\frac{1}{2}, \frac{1}{2})$ spots in addition to those which appear for the domain structure (Fig. 4a–c).

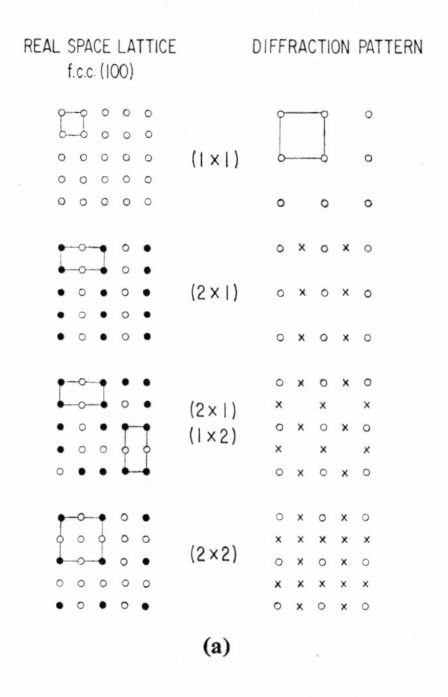
REAL SPACE LATTICE
f.c.c. (100)
DIFFRACTION PATTERN
(1 x 1)
(2 x 1)
(2 x 1)
(1 x 2)
(2 x 2)

(a)

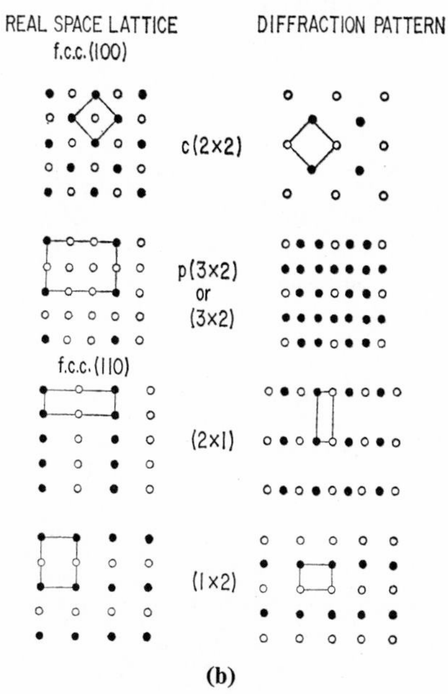
REAL SPACE LATTICE
f.c.c.(100)
DIFFRACTION PATTERN
c(2x2)
p(3x2)
or
(3x2)
f.c.c.(110)
(2x1)
(1x2)

(b)

Fig. 4

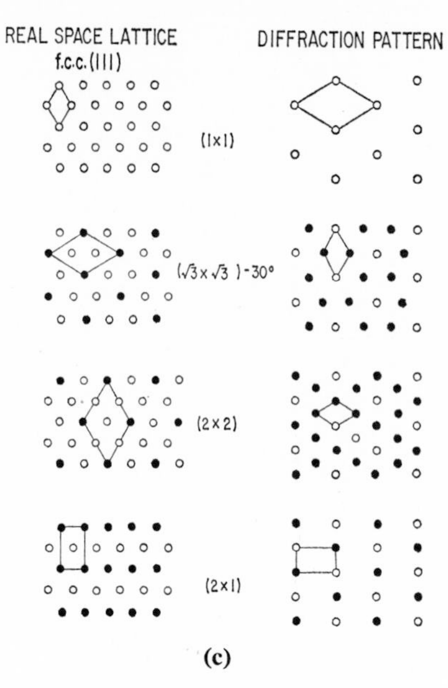

Fig. 4. Schematic diagram of surface structures on the (100), (110), and (111) crystal faces of a face-centered cubic crystal.

Surface structures of the type $(n \times m)$ where $n \neq m$ are frequently formed. For example, carbon monoxide on the palladium (100) face gives a $c(4 \times 2)$ structure.[3] The indices used to characterize the structure of the surface need not be integer. If every third lattice site on a hexagonal face is distinguished from the other sites, then a $(\sqrt{3} \times \sqrt{3})$-30° surface structure may arise. The angle given after the $(n \times m)$ notation indicates the orientation of the new unit cell relative to the original unit cell. If every other lattice site on a square face is unique, then a $(\sqrt{2} \times \sqrt{2})$-45° surface structure could be formed. To avoid the noninteger notation, this structure is usually labeled a $c(2 \times 2)$ where the c indicates that this is a centered (2×2) structure. Occasionally, the notation $p(n \times m)$ is used where the p indicates that a primitive unit cell has been taken. This p is frequently deleted either when it is understood that the unit cell is primitive, or when the detailed geometry of the unit cell is unknown. If a surface structure is known to be associated with some contaminant or adsorbed gas, it is customary to denote the adsorbate material in the description of the surface structure as $(n \times m)$-S where S is the chemical symbol or formula for the adsorbate. Perhaps one of the simplest examples of this would be the oxygen surface structure on molybdenum where the oxygen atoms or molecules on the surface of the metal have the same unit mesh as the clean metal surface. This structure would be denoted as the Mo(100)-(1 × 1)-O structure where the chemical symbol and crystallographic face of the substrate are given first, then the unit mesh of the surface structure relative to that of the substrate, and finally the chemical symbol of the impurity.

A useful and simple method for determining the real space lattice of a surface structure from its reciprocal space lattice vectors as displayed in its diffraction pattern has been developed by Park and Madden.[3] Noting the reciprocal and the real space vector obey the relation

$$\mathbf{a} \cdot \mathbf{G}_a = 2\pi$$

it is possible to construct two matrices, **A** and **G**, such that

$$\underset{\approx}{\mathbf{A}} \cdot \underset{\approx}{\mathbf{G}} = \underset{\approx}{\mathbf{1}} \tag{4a}$$

or

$$\underset{\approx}{\mathbf{A}} = \underset{\approx}{\mathbf{G}}^{-1} \tag{4b}$$

If the components of **G** are taken as the indexes of the partial-order spots and are expressed in terms of the basis vectors of the "clean" diffraction pattern, then its inverse, $\underset{\approx}{\mathbf{A}}$, will give the coordinates of the basis vectors of the real space lattice for the surface structure in terms of the primitive lattice vectors for the clean or unreconstructed surface. For example, a (2 × 2) surface structure on a rectangular or square surface will give rise to a diffraction pattern characterized by diffraction spots with the indexes $n/2$ and $m/2$. In fact, the total diffraction pattern, or reciprocal lattice space net, may be regarded as being generated from the basis vectors $(0\ \tfrac{1}{2})$ and $(\tfrac{1}{2}\ 0)$ where the first-order diffraction spots from the clean surface would have the indices (10) and (01). We may therefore construct a reciprocal lattice matrix $\underset{\approx}{\mathbf{G}} = \begin{vmatrix} 0 & \tfrac{1}{2} \\ \tfrac{1}{2} & 0 \end{vmatrix}$ which has as its inverse the real space matrix $= \underset{\approx}{\mathbf{A}} \begin{vmatrix} 0 & 2 \\ 2 & 0 \end{vmatrix}$ which may be regarded as being constructed of the real space vectors (0 2) and (2 0) which are simply the primitive translational vectors of the (2 × 2) surface structure expressed in terms of translational vectors for the clean surface. Similarly, a $c(2 \times 2)$ surface give rise to a diffraction pattern or reciprocal lattice net which can be generated from the vectors $(\tfrac{1}{2}\ \tfrac{1}{2})$ and $(\tfrac{1}{2} - \tfrac{1}{2})$. The resulting reciprocal lattice matrix $\underset{\approx}{\mathbf{G}} = \begin{vmatrix} \tfrac{1}{2} & \tfrac{1}{2} \\ \tfrac{1}{2} & -\tfrac{1}{2} \end{vmatrix}$ has as its inverse the real space matrix

$\underset{\approx}{\mathbf{A}} = \begin{vmatrix} -1 & 1 \\ 1 & 1 \end{vmatrix}$ which displays the real space translational vectors for the $c(2 \times 2)$ surface structure as $(\bar{1}\ 1)$ and $(1\ 1)$. This method is very powerful for the analysis of complicated surface structures. Note, however, that any analysis of the geometry of the diffraction pattern will give information only about the two-dimensional space lattice. In order to determine the arrangement of the basis around the lattice points and in a direction perpendicular to the surface, an analysis of the intensities of the diffraction features must be performed. The problems inherent in such an analysis will be discussed in a later section.

III. THE NATURE OF LOW-ENERGY ELECTRON DIFFRACTION

Many of the unique characteristics of low-energy electron diffraction in the energy range 0–500 eV are due to the large scattering cross sections of atoms to low-energy electrons. Particularly at very low electron energies, 0–100 eV, these cross sections may be of the order of square angstroms. As a consequence, there will be substantial amplitudes scattered into the nonforward directions, and the probability that the electron will be found in the transmitted beam will be significantly less than unity. This results in a high probability that an electron will be incapable of penetrating very deeply into a solid under these conditions before it is scattered, either elastically or inelastically, out of the forward-scattered beam. Therefore, most of the intensity that is back-scattered out of the crystal comes from either the surface or the neighborhood of the surface. This, of course, makes low-energy electron diffraction an ideal tool for studying the structure of surfaces.

Unfortunately, the very aspect that makes low-energy electron diffraction valuable for surface structure analysis also complicates this analysis. That is, because the scattering cross sections are large, not only will the electron be scattered predominantly from the vicinity of the surface, but it will also have a significant probability of being scattered more than once. This phenomenon is known as multiple scattering, and its importance vitiates the applicability of the kinematic theory of diffraction which has been used so successfully in the X-ray case, where only single-scattering or kinematic events are important.

One of the interesting consequences of the fact that scattering is confined to the vicinity of the surface is that the full three-dimensional periodicity of the crystal is not experienced by the electron. We therefore are dealing with a potential which has essentially perfect periodicity in the two dimensions parallel to the surface but has imperfect periodicity perpendicular to the surface. This perfect two-dimensional periodicity insures that diffraction will occur and that the electron will be scattered only into certain discrete rods or beams, destructive interferences having taken place along all other directions in space.

More concisely, as has been noted by Boudreaux and Heine,[4] the only exact quantum number in the system is that component of the wave vector $\mathbf{K}_{\parallel}$ or $\mathbf{K}_{xy}$, that is parallel to the surface, and this is indeterminant to the extent of adding any reciprocal lattice vector that is parallel to the surface in the usual sense of the Bloch theorem. Due to the imperfect periodicity perpendicular to the surface, however, that component of the wave vector, $\mathbf{K}_{\perp}$ or $\mathbf{K}_z$, that is perpendicular to the surface is not constrained to take on only certain discrete values as it would be in the X-ray diffraction case.

However, when only elastic scattering is considered, this perpendicular component is defined by the parallel component and the condition that the total magnitude of the wave vector must be conserved. If the incident electrons are characterized by a total wave vector, $\mathbf{K}^{\circ}$, then the components parallel to and perpendicular to the surface may be denoted as $\mathbf{K}_{\parallel}^{\circ}$ and $\mathbf{K}_{\perp}^{\circ}$, respectively. In a similar manner, a diffraction beam may be characterized by $\mathbf{K}'$ with components $\mathbf{K}_{\parallel}'$ and $\mathbf{K}_{\perp}'$. Now, the constraint on the parallel component may be written as

$$\mathbf{K}_{\parallel}' = \mathbf{K}_{\parallel}^{0} + \mathbf{G}_{\parallel} \tag{5}$$

where $\mathbf{G}_{\parallel}$ is some reciprocal lattice vector parallel to the surface. If the surface has rectangular or square symmetry, $\mathbf{G}_{\parallel} = 2\pi(\hat{x}h/a_x + \hat{y}k/a_y)$ where h and k are integers, $\hat{x}$ and $\hat{y}$ are unit vectors in the x and y directions, and $\mathbf{a}_x$ and $\mathbf{a}_y$ are the primitive translational vectors of the surface lattice net in the x and y directions, respectively. The z direction has been taken as being perpendicular to the surface.

In free space, the energy of the electron is directly proportional to the square of the total wave vector as $E = \hbar^2 |\mathbf{K}|^2/2m$, where $\hbar$ is Planck's constant divided by 2π and m is the mass of the electron. Therefore, the constraint that the scattering must be elastic may be written as

$$|\mathbf{K}'|^2 = |\mathbf{K}^{\circ}|^2 \tag{6a}$$

or

$$|\mathbf{K}_{\parallel}'|^2 + |\mathbf{K}_{\perp}'|^2 = |\mathbf{K}^{\circ}|^2 \tag{6b}$$

Rearranging Equation (2b), $\mathbf{K}_{\perp}'$ may be determined as

$$\mathbf{K}_{\perp}'^{hk} = \pm\hat{z}\sqrt{|K^{\circ}|^2 - |\mathbf{G}_{\parallel}{}^{hk}|^2 - |\mathbf{K}_{\parallel}^{\circ}|^2} \tag{7}$$

Note that $\mathbf{K}'_\perp$ may be either positive or negative, corresponding to a diffraction beam directed either into or out of the crystal. Real values of $\mathbf{K}'_\perp$ correspond to traveling waves or allowed states in the crystal, while complex or imaginary values correspond to damped or evanescent waves at the surface and forbidden states in the bulk of the crystal.

There are actually an infinite number of solutions to Equation (7).[5] First, there are those within the Ewald sphere where $|\mathbf{K}^\circ|^2 > |\mathbf{K}'_\parallel|^2$. The Ewald sphere is that surface in reciprocal space with a radius of $|\mathbf{K}^\circ|$. When within this sphere, $\mathbf{K}'_\perp$ is real, at least when not in a band gap where it may assume complex values.[1] In the following, the states characterized by real values of $\mathbf{K}'_\perp$ will be referred to as "allowed" states. Secondly, there are those solutions that lie outside of the Ewald sphere where $|\mathbf{K}^\circ|^2 < |\mathbf{K}'_\parallel|^2$. For these cases, $\mathbf{K}'_\perp$ is purely imaginary and the associated eigenfunctions are strongly damped. Note that $\mathbf{K}'_\perp$ may be either positive or negative, corresponding to diffraction beams directed both into and out of the crystal (Fig. 5). As most low-energy electron diffraction studies are made as a function of electron energy, it is of value to inspect Equations (5), (6), and (7) for their energy dependence. Equation (6) states the necessity that the diffracted beam have a wave vector of the same magnitude as the incident beam for elastic scattering whereas Equation (5) states that the parallel component may contain some reciprocal lattice vector. It may therefore be seen that at a low-enough beam voltage these two equations may not be fulfilled simultaneously with real values of $\mathbf{K}'_\perp$ except for the null parallel reciprocal lattice vector. In this region, only the transmitted and the specularly reflected beams are allowed. All other beams will be forbidden, or evanescent. Upon going to higher energies, the magnitude of the wave vector becomes large enough to accommodate the smallest reciprocal lattice vector, and the first-order diffraction beams will be allowed in addition to the transmitted and specularly reflected beams. At still higher voltages, higher-order diffraction beams will come into existence. When a diffraction beam first appears, the component of its wave vector perpendicular to the surface will have zero magnitude, and the emergent beam will lie in the surface. At a slightly higher energy, $|\mathbf{K}'_\perp|$ will have a finite value and diffraction beams directed both into and

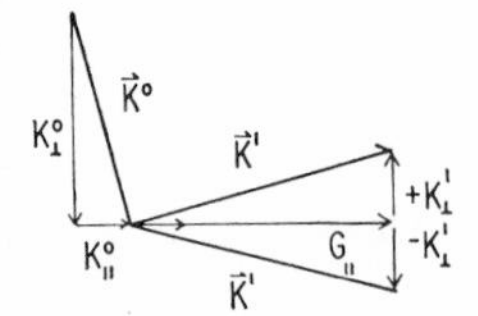

Fig. 5. Wave vectors for the incident beam and two diffraction beams showing their components parallel and perpendicular to the surface.

out of the surface will appear. As the energy is increased, the angle that these new beams will make with the surface increases and these beams will asymptotically approach the axis of the incident or specularly reflected beam. Viewing only the back-scattered beams, upon increasing the electron energy one would first see new diffraction beams appear parallel to the surface of the crystal and then rise up out of this surface and sweep through space towards the specularly reflected beam. These considerations arise solely from the symmetry, that is, the two-dimensional periodicity parallel to the surface. They are completely independent of the nature of the surface other than its symmetry and the dimensions of its unit cell parallel to the surface.

Information about dimensionalities perpendicular to the surface and about the type of scattering centers involved is, however, contained in the intensities of these diffraction beams. To appreciate the possible variations in these beam intensities, let us consider two limiting cases.

A. The Two-Dimensional Diffraction Limit

The first case is that in which there are no periodic modulations in the potential in the direction perpendicular to the surface that are experienced by the electron. This is essentially the two-dimensional grating problem. Here, if one were to monitor the intensities of the back-diffracted beams as a function of electron energy, one would find, at best, a monotonic variation (Fig. 6a). Conceptually, this situation could occur if the scattering cross sections were sufficiently large that the electrons never penetrated the first atomic layer of the surface.

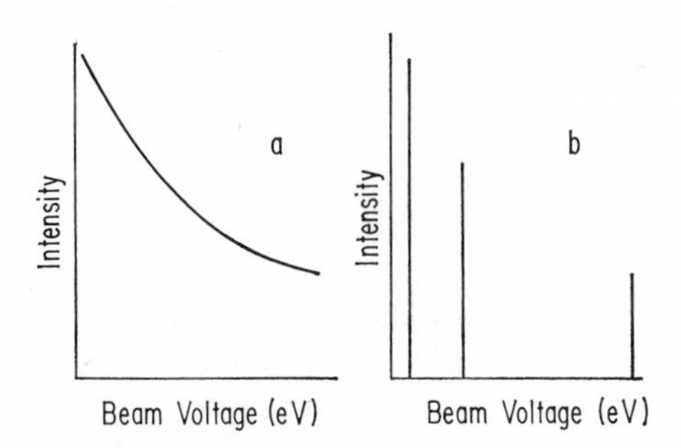

Fig. 6. Intensity of the (00) beam as a function of electron energy in (a) the pure two-dimensional diffraction limit and in (b) the pure three-dimensional diffraction limit.

B. The Three-Dimensional Diffraction Limit

The second limiting case arises in the opposite limit where the cross sections for back scattering are quite small so that the electron can penetrate deeply into the crystal before being scattered. In this case, the effect of the surface can be ignored and the electron will be diffracted predominantly in an environment where it is subjected to the full three-dimensional periodicity of the crystal. Now the perpendicular component of the reciprocal lattice vector is no longer free to assume a continuum of values but is limited to certain discrete values by this periodicity in the z direction. This constraint on $\mathbf{K}'_{\perp}$ may be expressed in a manner similar to Equation (5) as

$$\mathbf{K}'_{\perp} = \mathbf{K}^{\circ}_{\perp} + \mathbf{G}_{\perp} \tag{8}$$

where $\mathbf{G}_{\perp}$ is some reciprocal lattice vector perpendicular to the surface. Note that the combination of Equations (6) and (8) is just the Bragg equation for X-ray diffraction expressed in reciprocal space. This can be seen in the following manner. We may write $|\mathbf{K}' - \mathbf{K}^{\circ}| = 2K \sin(\theta/2)$, $|K| = 2\pi/\lambda$, and $|\mathbf{G}| = 2\pi\, n/d$ where θ is the angle between the initial and the final directions of travel and d is the interplanar spacing perpendicular to the scattering vector $\mathbf{G}$. Substituting these real space expressions into the reciprocal space expression obtained by combining Equations (6) and (8), the classical Bragg expression

$$n\lambda = 2d \sin(\theta/2) \tag{9}$$

is obtained.

If one were to look at the intensities of the diffracted beams in this limit, it would be observed that they were zero except at those points where Equations (6) and (8) were met simultaneously (Fig. 6b).

C. Low-Energy Electron Diffraction

We now have two extreme cases, one where the intensity varies smoothly with electron energy, and the other where the intensity varies abruptly being zero except at certain discrete energies and points in space. Reality for LEED is, of course, somewhere in between. In Figure 7 we show the intensities of the different (hk) diffraction beams, I_{hk} as a function of electron energy, eV, which is obtained for the Al(100) surface at normal incidence. There are modulations in the beam intensities, some, but not all, corresponding to maxima predicted by Equation (8). Furthermore, particularly at low beam voltages, there is usually finite intensity in these diffraction beams at energies that do not correspond to any diffraction condition. These observations may be explained by the fact that, even

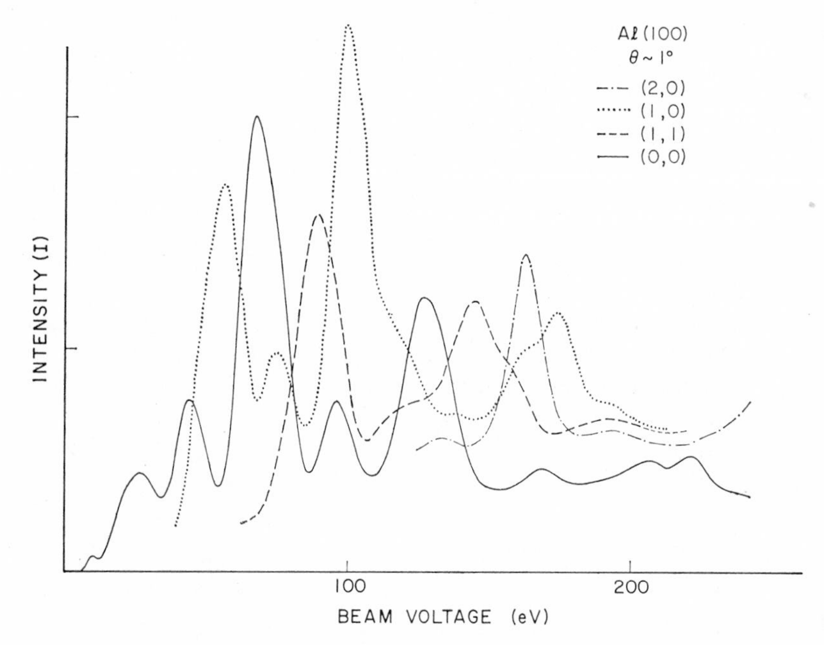

Fig. 7. Intensity of the low-index diffraction beams as a function of electron energy, eV, from the (100) face of aluminum.

though the scattering cross sections are rather large, they are not so large that the electron does not have a finite probability of penetrating the first and even several of the topmost atomic layers parallel to the surface. Consequently, the electron may experience some degree of the full three-dimensional periodicity of the crystal.

However, the observation of intensity maxima at energies other than those predicted from Equation (8) indicate that the situation is not so simple as outlined above. As mentioned before, the very fact that the scattering cross sections are reasonably large can lead to multiple-scattering events. These may be envisioned in the following manner. As the amplitudes of the nontransmitted diffraction beam are substantial, and as the cross sections are large, the diffracted beams themselves may act as primary beams or electron sources. Consequently, we must consider diffraction conditions of the form of Equation (8), but between diffracted beams rather than only between the primary, or incident, beam and a diffraction beam.

We therefore have the new condition

$$\mathbf{K}''_{\perp} = \mathbf{K}'_{\perp} + \mathbf{G}_{\perp} \tag{10}$$

where both $\mathbf{K}'_{\perp}$ and $\mathbf{K}''_{\perp}$ are wave vector components corresponding to diffraction beams. Note that Equation (8) may be considered as a special case of Equation (10). The analogous condition for the parallel components

$$\mathbf{K}''_{\parallel} = \mathbf{K}'_{\parallel} + \mathbf{G}_{\parallel} \tag{11}$$

is always met. This is guaranteed by Equation (5).

For sufficiently large cross sections, still more phenomena can be observed. For example, when the condition expressed in Equation (10) is met between two diffraction beams, a subsidiary maximum may be observed in a third beam, even though no appropriate diffraction condition is met. This is because all the beams are more or less coupled for sufficiently large cross sections, and an increase in the intensity of one of them may result in an increase in the intensity of another.

The actual intensity maxima that are observed may be arbitrarily categorized into three different types on the basis of the associated diffraction conditions.

(1) Kinematic or Single Diffraction: The first group is comprised of those maxima whose positions are predicted by Equation (8). This is the kinematic or single-scattering case, and peaks should appear at these positions even in the limit of negligible multiple scattering.

(2) Double Diffraction: In this case, we have those peaks whose positions are predicted by Equation (10) rather than Equation (8). This is a simple multiple-scattering situation and may be called the double-diffraction case as it necessitates only two successive scattering events.

(3) Tertiary and High-Order Scattering: This case contains all intensity maxima not directly predicted by Equations (8) and (10). Observation of these phenomena should be limited to those situations where multiple scattering is quite strong. One would expect when inelastic scattering was important that maxima of this type would be experimentally observed only with difficulty.

Although the division of intensity maxima into these three different categories presents a useful classification scheme, it is rather artificial as higher-order scattering events may contribute to the intensities of maxima classified as either kinematic or double diffraction even though only one or two events need be considered to predict their positions.

How well the position (in electron energy) of the diffraction maxima is predicted just by taking single- and double-diffraction events into account is indicated from the work of Farrell and Somorjai.[6] They have measured the intensities of several diffraction beams [(00), (10), (11), (20), and (22)] from the (100) surface of several face-centered cubic metals as a function of electron energy at normal incidence. Under these conditions the diffraction beams with the same indices and the same sign of K's are degenerate. This is one of the simplest conditions which facilitates the analysis of the data since the number of diffraction beams to be considered is much less than under conditions of non-normal electron beam incidence. First of all they have found that when the intensity data from the different (100) f.c.c. surfaces which were available in the literature are plotted on a "reduced" electron energy scale, eV $d^2 \cos^2 \theta$, the peak positions for the different materials seem to fall at the same corrected electron energies. This is shown for two different beams in Figures 8a and 8b. By plotting the data on a "normalized" energy scale (I_{hk} vs. eV $d^2 \cos^2 \theta$) one can compensate for the variations of the lattice parameter among the metals. Thus, these results indicate that the same diffraction processes are operative in all the metal surfaces with the same crystal structure and surface orientation. However, the intensities of these peaks vary considerably from material to material, presumably reflecting variations in the characteristics of the atomic potentials.

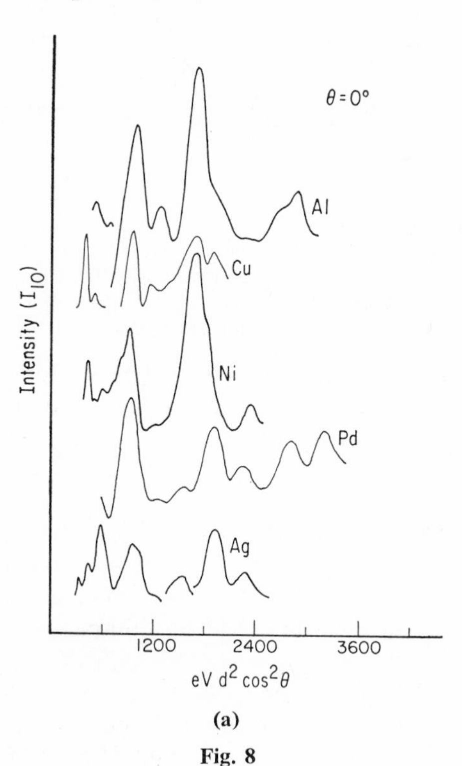

(a)

Fig. 8

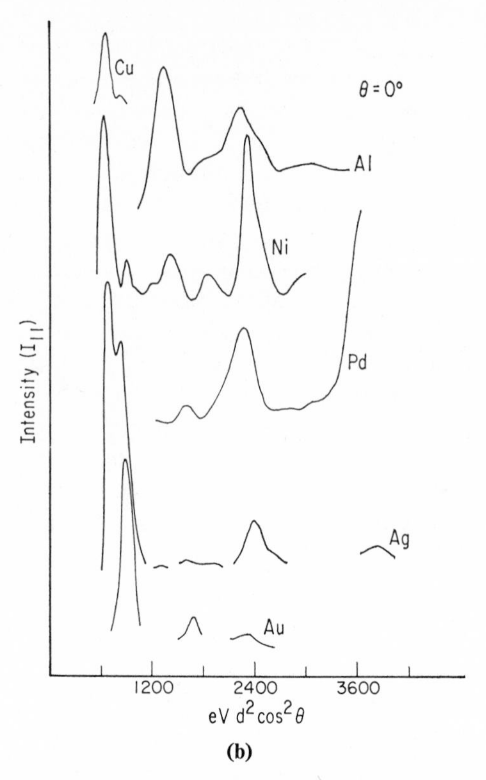

(b)

Fig. 8. The intensities of (a) the (10) and (b) the (11) diffraction beams as a function of normalized electron energy for the (100) faces of aluminum, copper, nickel, palladium, silver, and gold at normal incidence.

Then Farrell and Somorjai[6] calculated the peak positions that could be predicted by assuming that only single and double diffraction take place in the (100) surfaces, and compared the calculated peak positions with those found by experiments for six different metals (Al, Pd, Ag, Au, Cu, and Ni).

Most of the experimental and calculated values of peak positions showed coincidence within the accuracy of the measurements. This result seems to substantiate that single- and double-diffraction events are the dominant scattering processes in low-energy electron diffraction from f.c.c. metal surfaces.

The double-diffraction condition, $2\mathbf{K}_z' = \mathbf{G}_z$, appears to be particularly dominant in the electron energy region just above the appearance energy of the diffraction beam under consideration. There also appears to be a general tendency for diffraction conditions with relatively small magnitudes of **G** to dominate. As most atomic potentials would favor forward scattering this is physically reasonable.

IV. IMPORTANT PARAMETERS OF THE LOW-ENERGY ELECTRON DIFFRACTION EXPERIMENT AND OF THE SURFACE STRUCTURE CALCULATIONS

A. The Inner Potential

A crystalline solid is composed of ordered arrays of atoms which are themselves made up of negative electrons and positive nuclei. These create a symmetric, but complex, potential which is naturally quite different from that experienced by an electron in free space. Consequently, when an incident electron strikes a crystal, the change in potential which it experiences brings about a corresponding change in the de Broglie wavelength of the electron. This increase in the kinetic energy of the electron entering the solid is commonly described as being due to the "inner potential." The electron is accelerated as it enters the solid since at interatomic distances the nuclei are only imperfectly shielded by the core and valence electrons. The average inner potential experienced by a primary electron will be dependent upon the energy of that electron as the degree of shielding of the positive nuclei will to some extent be dependent upon the electron-electron correlation.

If the potential of the crystal is expressed as a Fourier expansion $V(\mathbf{r}) = \sum_G V_G e^{-i\mathbf{G}\cdot\mathbf{r}}$ then the inner potential is well represented by the first or static term, V_0, in the expansion. This term is essentially the matrix element of the potential taken between identical initial and final eigenstates that represent the incident electron (i.e., $V_0 = \langle \mathbf{k}^\circ | V(\mathbf{r}) | \mathbf{k}^\circ \rangle$). As the energy of the incident electron changes, its eigenstate will change causing a corresponding change in the value of the matrix element that approximates the inner potential. This matrix element is the diagonal term of the potential in the secular determinant that describes the interaction of the electron with the crystal. When the determinant is properly diagonalized, other off-diagonal terms will appear along the diagonal and will represent contributions to the effective inner potential experienced by an incident electron. However, as higher-order Fourier coefficients are usually at least an order of magnitude smaller than the zeroth-order term, these contributions may be regarded as refinements on a reasonable zeroth-order approximation.

Pendry has used the pseudo-potential approach to calculate the average "inner potential" (i.e., the diagonal matrix element of the potential) for niobium and nickel.[7] In these calculations, the effects of screening, correlation, surface dipole, inelastic, and incoherent processes have been neglected on the grounds that at higher energies their contributions will be less than about 2 eV. In Pendry's calculations, the largest contribution to the inner potential came from the Hartree term which includes the effect of the nuclear potentials and an *averaged* core state contribution. The Hartree contribution is independent of the primary energy of the incident electrons and is the high-energy limit of the inner potential. For nickel, Pendry calculated this term to be about 14 eV and for niobium, about 19 eV. Similar calculations gave a high-energy limit of about 12 eV for the inner potential of graphite.

In the range of primary energies intermediate between the Fermi energy and that region where the inner potential approaches the high-energy limit (between 5 and 100 eV), the calculated inner potentials for nickel and niobium show considerable energy dependence (Fig. 9). They both exhibit a minimum in absolute value that is a consequence of the partial cancellation of an exchange term and a pseudo-potential term in Pendry's calculations. Although it is of electrostatic origin, the exchange term has no classical analog. It expresses the difference in the coulomb interaction energy of systems where the electron spins are parallel or antiparallel, and is a consequence of the Pauli exclusion principle.[8] The pseudo-potential term expresses the deviations from a plane-wave nature due to nodes at the atom centers in the eigenfunction of the electron. Both of these terms diminish in importance at higher energies. It is tempting to extrapolate from these calculations the expectation that all material will

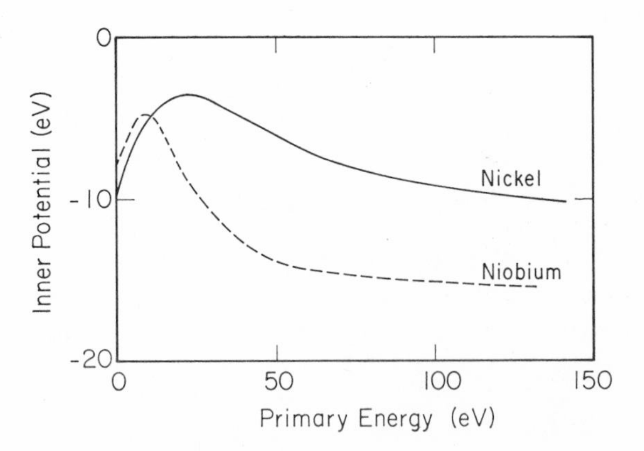

Fig. 9. Inner potential as a function of the energy of the incident electron beam for nickel and for niobium.

have an energy-dependent inner potential of minimum absolute value at some energy between the Fermi energy and the high-energy limit. It is to be hoped that calculations will be performed in the near future for other materials. This information would greatly simplify the interpretation of low-energy electron diffraction data.

A large number of experimental estimates have been made of the effective inner potential experienced by an incident electron. One of the recent publications dealing with the problems involved in an accurate experimental determination of inner potential is that by Stern and Gervais.[9] They observed for the (110) face of tungsten by how much a diffraction condition was shifted from its theoretical position calculated with zero inner potential. The difference between the calculated and the observed values was assigned to an inner potential correction. They discuss the necessity of choosing a diffraction condition for which there are not significant multiple reflections. When this condition is not met, then several diffraction conditions may be strongly coupled, making the experimental extrication of an average inner potential a difficult and uncertain process. Further, they have shown that the experimentally determined average inner potential has an effective angular dependence of $1/\sin^2 \theta$ leading to large corrections at glancing angles of incidence. Careful measurements far from normal incidence for diffraction conditions that did not excite strong multiple scattering gave a value of 20 eV $\pm$ 1 eV for W(110). Because all the measurements were made about 100 eV, this may be taken as the high-energy limit for the inner potential of tungsten. Note that this value is commensurate with that calculated for niobium in the high-energy limit.

Seah has experimentally determined a value of 14 eV for the inner potential of Ag(111) films on mica in the high-energy region.[10] This may be compared with a value of 12 eV determined by Segall[11] and a value of 10 eV at the Fermi surface obtained by adding the work function of silver to its Fermi energy. Other representative values for experimentally determined inner potential corrections are 9 eV for graphite,[12] 16 eV for nickel,[13] 22 eV for nickel,[14] and 11–12 eV for lithium fluoride.[15] Other values are 16 eV for tantalum,[16] 8–19 eV for vandium,[17] and 19 eV for iron.[18]

B. The Atomic Scattering Factor

One of the most important parameters which enters into all calculations of surface structure from the intensities of the diffracted low-energy electron beams is the amplitude scattered by a single atom in the crystal surface. The scattered amplitude, $V_{\mathbf{q}}$, is called the form factor or the atomic scattering factor. The amplitude in any given diffraction beam is dependent upon the probabilities that electrons will be scattered out of the primary beam (or other diffraction beams) into that beam from various points in the crystal. These scattering probabilities are dependent upon the atomic potential. For single-scattering events, the scattering amplitude may be regarded in the first approximation as being proportional to the Fourier coefficient of the potential that is characterized by the scattering vector between the initial and the final state of the electron—that is,

$$f_{\mathbf{q}} \sim \langle \mathbf{K}' | \, V(r) \, | \mathbf{K}^{\circ} \rangle = V_{\mathbf{q}} \tag{12}$$

where $\mathbf{q} = \mathbf{K}' - \mathbf{K}^{\circ}$ is the scattering vector. One of the simplest model potentials used in LEED calculations is the isotropic *s*-wave scattering potential. With this potential, scattering in all directions has the same probability. While this is a particularly convenient potential to apply in computations, it is rather unrealistic in that most experiments indicate that atomic potentials tend to be forward scattering. That is, on the average, the probability that an electron will be scattered into a new direction that is considerably different from the original direction will be significantly less than the probability that the electron will either continue along its original direction or be deflected through only relatively small scattering angles.

A somewhat more realistic atomic potential that illustrates this forward-scattering tendency is the shielded coulombic potential. Here, the positive nuclear charge is regarded as being uniformly shielded by the surrounding electrons and one may write

$$V(\mathbf{r}) = \frac{Ze^2}{r} \exp\left[-\lambda r\right] \tag{13}$$

where Ze^2/r is the nuclear coulombic potential and λ is the "screening length" of the core and the valence electrons. Upon Fourier transformation this potential gives form factors of the form

$$V_q = \frac{4\pi e^2 Z}{\lambda^2 + q^2} \tag{14}$$

Note that when the scattering vector, $\mathbf{q}$, is very small and the electron has been scattered through only a small angle, that V_q will be larger than when q is large and the electron has been back-scattered away from the original direction. As the electron amplitude in any given direction is proportional to the corresponding form factor, it may be seen that forward scattering is more probable than back scattering for such a potential.

It is worth noting that the effective screening length of the electron is dependent upon the energy of the incident electron and actually decreases when the velocity of the scattered electron increases. This is because the core and valence electrons are less effective in shielding the incident electron from the coulombic nuclear charge at higher energies. It may be shown that,

$$\lambda = \frac{4\pi e^2 n}{(2/3)\varepsilon_f}\left[1/2 + \frac{4k_f^2 - q^2}{8k_f q}\ln\left|\frac{2k_f + q}{2k_f - q}\right|\right] \tag{15}$$

where n is the average electron density, e is the charge of the electron, ε_f is the Fermi energy, k_f is the wave vector.[19] For aluminum, λ is on the order of 2 Å below the Fermi surface. Note that the decrease in the screening length with the increase in electron energy increases the relative probability of forward scattering. This is reasonable as one would expect the more energetic electrons to be less easily deflected than those with relatively low velocity.

Although the shielded coulombic potential leads to form factors that are an improvement over those obtained from an isotropic potential, it is still inadequate for an accurate description of low-energy electron diffraction. As the electrons in an atom are not arranged uniformly around the nucleus, real form factors will not have as simple a form as that in Equation (14). Further, those considerations such as exchange and correlation that contribute to the voltage dependence of the inner potential will also effect the off-diagonal matrix elements of the potential, that is, the form factors. A detailed consideration of these effects leads to form factors that are not necessarily monotonic functions of the scattering vector or electron energy. The form factors may actually become zero and change in sign under the proper circumstances.

Few calculations have been performed that attempt to calculate the actual form factors from a detailed consideration of the electronic structure of crystals. One of the more promising approaches is by the pseudo-potential method. Unfortunately, most available pseudo-potential form factors have been evaluated near the Fermi surface and the extension of these calculations to the energy range of interest in LEED has only recently been undertaken.

Figure 10 shows the form factor, f_q, for aluminum calculated by Heine and Animalu at the Fermi surface by the pseudo-potential method.[20] This may be compared with that calculated from an isotropic s-wave potential and with that calculated from a simple screened coulombic potential. Note that the pseudo-potential form factor varies markedly

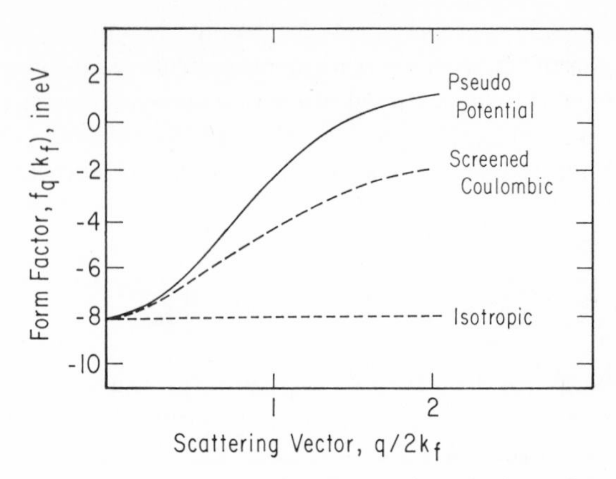

Fig. 10. Atomic scattering factor, f_q, calculated for aluminum.

with the magnitude of the scattering vector. However, it still has its largest absolute value in the forward-scattering direction.

There is a body of experimental evidence that indicates that there is a considerable amount of structure in the form factor, or atomic scattering factor, for the scattering of low-energy electrons. If an electron beam is scattered from a completely random arrangement of atoms, then, to a first approximation, the intensity is proportional only to the number of scattering centers and to the squared modulus of the atomic scattering factor. The spatial disorder of the atoms would preclude any periodic modulation of the intensity, as discussed in Section VII. Therefore, if the intensity scattered by a perfectly disordered surface is measured, experimental information may be obtained about the form factors. It is, however, very difficult to prepare a solid surface which shows uncorrelated disorder as atoms tend to prefer to be at some average distance from one another. Atomic arrangements that show no long-range order may be obtained using liquid surfaces or by depositing amorphous layers of material on foreign substrates.

Experimental Determination of the Atomic Scattering Factor. Goodman and Somorjai have studied the background intensity from low-energy electrons scattered from liquid lead, tin, and bismuth surfaces.[21] They have found that there are definite nonmonotonic intensity variations with both energy and scattering angle that cannot be correlated with the bulk radial distribution function due to density fluctuations in the liquids. These intensity variations are most likely due primarily to variations in the atomic scattering factor.

Lander and Morrison have studied the intensity back-scattered from disordered layers deposited on ordered substrates.[22] The possibility of orienting influences and scattering by the ordered-substrate materials could conceivably effect their results. Unquestionably, one of the greatest present needs in low-energy electron diffraction structure calculations is information, both experimental and theoretical, on the nature of the atomic scattering factors in the energy region of interest.

C. Inelastic Scattering of Low-Energy Electrons

An electron may be subjected to several different types of scattering interactions in a crystal. These scattering mechanisms may be classified as being either elastic or inelastic depending upon whether or not the energy of the electron has been changed during the interaction. The elastic mechanisms lead to the usual diffraction phenomena investigated with LEED. The inelastic mechanisms are also extremely important in the interpretation of low-energy electron diffraction information as inelastic processes are generally far more probable than elastic processes in the energy range where most LEED studies are performed (10–500 eV).[23]

The inelastic scattering mechanisms may be further subdivided into two categories: those involving energy loss through coupling into the thermal motions of the atoms in the crystal; and those involving energy loss through electronic excitation of the core and/or the valence or conduction electrons in the crystal. The thermal effects, that is, those involving energy exchange between the electron and the phonons in the crystal, lead to thermal diffuse scattering, the Debye-Waller effect, and other related phenomena.[19] These will be discussed below. The electron-electron interactions may be classified according to the types of electrons in the crystal that are excited by the incident electron.

At sufficiently low energies, the incident electrons are not energetic enough to excite the electrons in the core states of the crystal. In this region, the only electronic processes that are important are those involving the valence or the conduction electrons. Here, energy may be lost either by the excitation of a valence electron to a higher state out of the crystal, or by the excitation of plasma oscillations. For incident electrons with energies less than the plasmon energy (≈ 10 eV), inelastic processes are relatively unimportant and the scattering is mostly elastic.

Armstrong finds between 10% and 50% of the electrons back-scattered from tungsten are elastic in the range between 2 and about 16 eV incident beam energy.[24] Lander notes that the total elastic reflectivity averages about 20% for most materials in the region below 10 eV and then drops to about 1% at about 100 eV.[25] This percentage will be considerably less for the lighter elements and more for the heavier elements. At energies above 100 eV, there is, on the average, only a very slow decrease in the fraction of elastically scattered electrons. The percentage of elastically scattered electrons as a function of electron energy is plotted for the (100) face of platinum in a typical LEED experiment in Figure 11.

The very-low-energy region (<5 eV) of high elastic reflectivity is not easily accessible with the conventional commerically available apparatus. Most investigations of elastic scattering processes have been carried out above 20 eV. At these energies, surface and bulk plasmons may be excited. The energy associated with the surface plasmon mode is about $1/\sqrt{2}$ that for the bulk.[26] The probability that an electron will lose energy through excitation of a plasmon mode rises steeply at energies just above the excitation threshold.[27] This probability has a maximum at several times the excitation energy and then decreases slowly at higher energies.

A convenient quantity for characterizing inelastic losses through electron-electron collisions in a crystal is the "mean free path," λ. Quinn has calculated that λ is on the order of 1000 Å at 1 eV and falls rapidly to about 100 Å at 3 eV for aluminum. The mean free path for plasmon emission has been calculated to be about 10 Å above about twice the excitation energy of the plasmon mode.[27] Recently, Duke and Tucker have performed calculations employing values of 4 and 8 Å for electron-electron mean free path lengths.[28]

Figure 12 shows the effect of damping the scattering amplitude with an arbitrary inelastic loss factor of the form $e^{-d/\lambda}$ for several values of λ. It may be seen that the penetration of electrons more than several monolayers into a crystal is severely curtailed by inelastic losses for values of λ less than about 6 Å. Though the employment of an inelastic loss factor of the form $e^{-d/\lambda}$ is not mathematically rigorous, it does give some physical insight into the constraint placed on the penetration of an incident electron by inelastic collisions with conduction electrons.

Proceeding to higher energies, the primary electrons eventually become sufficiently energetic to excite the bound electrons in the core states of the atoms in the lattice into higher states. As with the plasmon modes, the probability of exciting these core states rises steeply at energies just above the excitation energy and then reaches a maximum at several times the excitation energy.[29] The probability then decreases very gradually as the primary energy is increased further. The exact value of the probability of exciting a core state will, of course, depend upon the material and the transition under consideration. Similarly, the threshold value for excitation of core states will also depend upon the same considerations, and will range from several electron volts for many elements with filled core states near

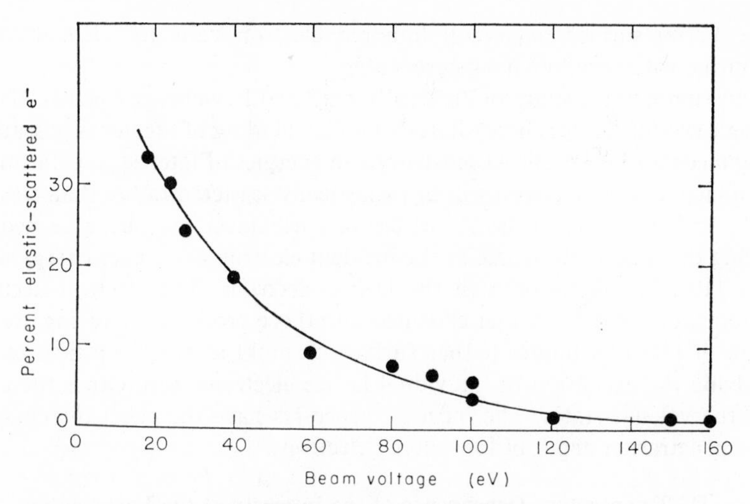

Fig. 11. Fraction of elastically scattered electrons as a function of electron energy for the (100) face of platinum.

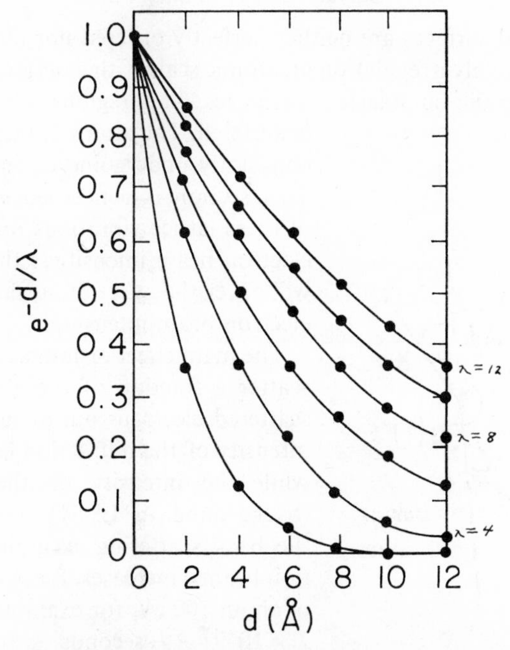

Fig. 12. Damping of the scattering amplitude by an inelastic loss factor, $e^{-d/\lambda}$ as a function of distance, d, from the surface.

the Fermi surface to several hundred electron volts for elements like flourine with very low lying core states.

In summary, a study of elastically scattered low-energy electrons from single crystal surfaces necessitates an understanding of the inelastic scattering mechanism as well. At most electron energies of interest, it is far more probable that an electron will be inelastically scattered rather than elastically scattered. The inelastic scattering mechanism may be classified as being thermal or electronic as the incident electron exchanges energy with the lattice phonons or with the lattice electrons. The electron-electron interactions may be further classified into those processes involving excitation of plasmon modes (either surface or bulk) and those processes involving the excitation of individual lattice electrons from either the core or the valence states. The inelastic processes have the effect of reducing the penetration depth of low-energy electrons.

D. Temperature Dependence of the Intensity of the Low-Energy Electron Diffraction Beams

1. *Surface Debye-Waller Factor*

Real crystal surfaces are neither perfectly ordered nor ideally flat. Real surfaces are highly irregular on an atomic scale with emerging dislocations, steps, pits, grain boundaries, vacancies, and regions where atoms are disordered. The atoms in these surfaces are constantly undergoing thermal vibrations. This section is concerned with the effect of these lattice vibrations on the scattered electron beam intensities; the next section will cover the effect of surface disorder on electron beam intensities.

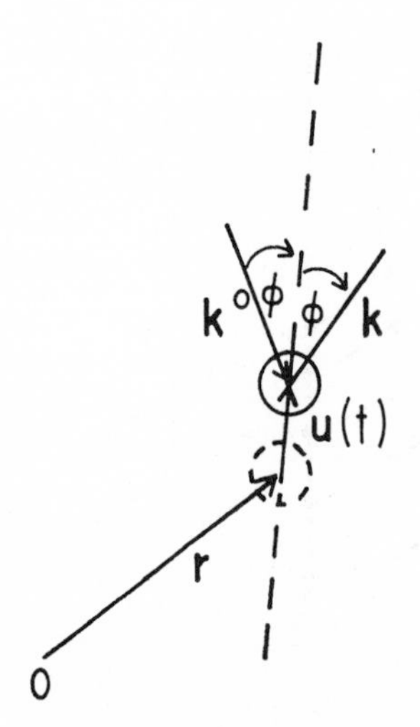

Fig. 13. Definition of $\mathbf{u}(t)$, $\mathbf{r}$, $\mathbf{k}_0$, and $\mathbf{k}$ for the scattering process.

The main effect of lattice vibrations is to scatter a fraction of the elastically backscattered electrons out of phase. Thus the intensity of the diffraction beams decrease while the intensity of the background (background in LEED is defined as all the back scattering excluding the diffraction beams) increases. Electrons of energies of about 100 eV, for example, spend about $2 \times 10^{-17} \times l$ seconds scattering (where l = distance in Å traversed; t = transit time $= l\sqrt{m_e/2\,\text{eV}} \times (\sqrt{10^{-16}}/1.6 \times 10^{-12})$

where m_e = mass of electron, eV = electron energy). Since characteristic vibrational frequencies are as "slow" as 10^{-12} sec, the electron "sees" a disordered "snapshot" of the lattice. However, in the laboratory frame we monitor intensities for times of one-tenth of a second or longer and thus obtain an average of a great number of "snapshots," of the disordered lattice. We can calculate the effect of lattice vibrations on laboratory measurements of intensity. Define an arbitrary atom position at 0°K by a vector **r** as shown in Figure 13. At any finite temperature the atom will be displaced by an amount $\mathbf{u}(t)$, a time-dependent function. In the kinematic approximation the scattered intensity from an array of such scattering centers is

$$I = |f|^2 \left\{ \sum_{ll'} \exp\left[i(\mathbf{K}' - \mathbf{K}^\circ) \cdot (\mathbf{r}_l - \mathbf{r}_{l'}) + i(\mathbf{K}' - \mathbf{K}^\circ) \cdot (\mathbf{u}_l - \mathbf{u}_{l'})\right] \right\} \tag{16}$$

where we are summing over all pairs of scattering centers l, l'.[30] Only the first term in the exponential would appear for a static lattice.

Without any loss of generality we can expand the displacements in a complete set of the normal lattice mode coordinates:[31]

$$\mathbf{u}_l(t) = \sum_{qj} \mathbf{u}_{qj} a_{qj} \cos(\omega_{qj} t - \mathbf{q} \cdot \mathbf{r}_l - \psi_{qj}) \tag{17}$$

where the summation is over all the lattice modes q and polarizations, j. The u_{qj} are unit vectors in the direction of the phonon of wave vector, **q** of frequency ω_{qj}, amplitude a_{qj}, and arbitrary phase angle ψ_{qj}.

Following the derivation given by James[30] for X-ray diffraction, the effect of this "phonon" scattering on the scattering intensity can be determined. The assumption used in this calculation are as follows: (1) the ergodic hypothesis, that is, the time average over all the thermal motions (what is actually observed experimentally) is equivalent to an ensemble average of the thermal motions; (2) that thermal motions are symmetric, that is, the net (or average) motion along any coordinate is zero; (3) that the thermal motions are small. The result of James's calculations is

$$I = |F_{hkl}|^2 e^{-2W} - |f|^2 e^{-2W} \sum_{ll'} 2 \cos\{\mathbf{q} \cdot (\mathbf{r}_l - \mathbf{r}_{l'})\} \exp i\, \Delta\mathbf{K} \cdot (\mathbf{r}_l - \mathbf{r}_{l'}) \tag{18}$$

where $|F_{hkl}|^2$ is the kinematic diffraction intensity for a perfectly ordered lattice, $2W$ is the so-called Debye-Waller factor and is equal to $\sum_q |\Delta\mathbf{K} \cdot \mathbf{u}_q|^2$ where $\Delta\mathbf{K}$ is the scattering vector.

In the high-temperature limit of the Debye model[19,32] ($T > \theta_D$) the mean square displacement is given by

$$\langle u^2 \rangle = \frac{3N\hbar^2}{Mk} \cdot \frac{T}{\theta_D{}^2} \tag{19}$$

where N = Avogadro's number, $\hbar$ = Planck's constant divided by 2π, k = Boltzmann's constant, M = atomic mass in grams, $T = {}^\circ$K, θ_D = Debye temperature. For the specularly reflected beam,

$$|\Delta \mathbf{K}|^2 = K^2 \cos^2 \phi = (4\pi^2/\lambda^2) \cos^2 \phi$$

thus

$$-2W = \frac{4\pi^2 \cos^2 \phi \langle u^2 \rangle}{\lambda^2} = \frac{12Nh^2}{Mk} \cdot \frac{\cos^2 \phi}{\lambda^2} \cdot \frac{T}{\theta_D{}^2} \tag{20}$$

or (by substitution of $\lambda = \sqrt{150.4/\text{eV}}$ and collecting constants),

$$\exp_{10}(-2W) = \exp_{10} \frac{-CVT \cos^2 \phi}{M\theta^2_{D,\,\text{EFF}}}, \qquad C = \text{const.} = 66.6 \frac{g\ {}^\circ K}{mole\ eV} \tag{21}$$

where ϕ = angle of incidence and $\theta_{D,\text{EFF}}$ = effective Debye temperature.

Equation 21 combined with Equation 18 suggests that at a given beam voltage and angle of incidence the intensity of a diffraction feature decreases as an exponential function of temperature. From this result an effective Debye temperature for the atoms involved in the scattering can be derived. From results using LEED[32,33,34] the values of θ_D for the surface layers are smaller than for bulk layers. However, as indicated in Section VI. A.2, LEED samples an increasing amount of the bulk as the energy increases. Thus at different voltages, the beam penetrates a different number of layers and the measured Debye temperature which we designate as $\theta_{D,\text{EFF}}$ is some average of the surface and bulk layers. In the limit of low voltages $\theta_{D,\text{EFF}} \to \theta_{D,\text{surf}}$ and at high voltages $\theta_{D,\text{EFF}} \to \theta_{D,\text{BULK}}$. Studies of $\theta_{D,\text{EFF}}$ as a function of beam voltage provide a means of studying the surface dynamics of crystals. However, one must be very cautious in applying Equation 21. First, the use of the Debye model may not be appropriate to describe the surface motions where anharmonic effects could be large. Second, the second term in Equation 18, usually referred to as the thermal diffuse scattering, must be evaluated. For certain values of $2W$ the effect of the thermal diffuse scattering on the Debye-Waller results may be significant.

Maradudin[35] has evaluated the cubic and quartic contributions to anharmonic motion and their effect on the Debye-Waller factor. He obtains the result:

$$2W = \frac{16\pi^2 \cos^2 \phi \alpha_0{}^2}{\lambda^2} \cdot \frac{T}{\theta_\infty} \cdot 1.861 \times 10^{-4}\left[1 + 0.0483\frac{T}{\theta_\infty}\right] \tag{22}$$

where α_0 = lattice parameter and θ_∞ is a parameter determined independently in Maradudin's model. Using lead as an example, $\alpha_0 = 4.95$ Å, $\theta_\infty = 143.4°$K. One may fit his value for $2W$ into the form of Equation 21 if a temperature-dependent $\theta(T)$ is defined as

$$\frac{1}{\theta_D{}^2(T)} = \frac{Mk\alpha_0{}^2}{Nh^2\theta_\infty} 0.6205 \times 10^{-4}\left[1 + 0.0483\frac{T}{\theta_\infty}\right] \tag{23}$$

Using the values for lead, this becomes

$$\frac{1}{\theta_D{}^2(T)} = \frac{1}{154(143.4)}\left[1 + 0.0483\frac{T}{143.4}\right] \tag{24}$$

which for $T \sim \theta_\infty$, gives $\theta_D \sim \theta_\infty$. Maradudin's model indicates that anharmonic effects should be expected to increase linearly with temperature (being about 9% anharmonic at 0°C and about 20% at the melting point) and that to first-order anharmonicity affects only the magnitude of the θ_D but not the form of the Debye-Waller factor.

Another effect which is important in LEED studies of the Debye-Waller factor is the second term in Equation 18, called the thermal diffuse scattering. Thermal diffuse scattering arises from the independence of the phonon modes from each other. Webb et al.,[36] have shown that the thermal diffuse scattering intensity, the second term in Equation 18 is (let I_2 = second term)

$$I_2 = \frac{|f_0|^2}{4} e^{-2W}[2WI_0(\mathbf{\Delta K} \pm \mathbf{q})] \tag{25}$$

where $|I_0(\mathbf{\Delta K})|^2 \cdot |f_0|^2 = |F_{hkl}|^2$, that is, $I_0(\mathbf{\Delta k})$ is referred to as the interference function and is nonzero only where the argument $\mathbf{\Delta K} = \mathbf{G}$, a reciprocal lattice vector. From Equation 25, I_2 has significant magnitude only where $\mathbf{\Delta K} \pm \mathbf{q} = \mathbf{G}$. Webb[36] shows that the thermal diffuse intensity falls off in a manner inversely proportional to the distance in reciprocal space from the nearest reciprocal lattice rod. Studying the ratio R of the thermal diffuse intensity to the kinematic intensity Webb finds $R = 2W/4$ $(1 + \Delta)$ where Δ is a small correction factor, less than unity, and of order $|\mathbf{q}|^2/|\mathbf{G}^2|$ which decreases to zero for large $|\mathbf{G}|$.

2. *Effect of Multiple Scattering*

We have, thus far, neglected multiple-scattering effects, though they are definitely prominent, especially at low energies. Work in this laboratory[6] indicates that the double-diffraction mechanism is the most likely. Figure 14 indicates a possible double-diffraction process. An incident beam $\mathbf{K}_0$ making an angle ϕ with the normal to the surface may scatter in two ways: part of the beam is specularly reflected into the beam $\mathbf{K}$, another part scatters into vector $\mathbf{K}_1$. The angle between $\mathbf{K}_0$ and $\mathbf{K}_1$ is 2ϕ. The $\mathbf{K}_1$ beam may then be rescattered into $\mathbf{K}_2$ beam (actually identical to $\mathbf{K}$) where the angle between $\mathbf{K}_1$ and $\mathbf{K}_2$ is $2\phi_2$. From simple geometrical considerations: $\Delta\mathbf{K} = \Delta\mathbf{K}_1 + \Delta\mathbf{K}_2$. Physically, the constructive interference between $\mathbf{K}_2$ and $\mathbf{K}$ could contribute to a diffraction maximum. The Debye-Waller factor for the double-scattered case is:

$$2W = \frac{CVT}{2M}\left[\frac{\cos^2\phi}{\theta_D{}^2} + \frac{\cos^2\phi_1}{\theta_{D_1}{}^2} + \frac{\cos^2\phi_2}{\theta_{D_2}{}^2}\right] \tag{26}$$

where θ_{D_n} refers to the effective Debye temperature for thermal motions in the direction $\Delta\mathbf{k}_n$. Comparing Equations 21 and 26, if

$$\frac{\cos^2\phi_1}{\theta_{D_1}{}^2} + \frac{\cos^2\phi_2}{\theta_{D_2}{}^2} = \frac{\cos^2\phi}{\theta_D{}^2} \tag{27}$$

then the results interpreted in terms of kinematic diffraction would be in agreement with this dynamical result. Assuming $\theta_D = \theta_{D_1} = \theta_{D_2}$,

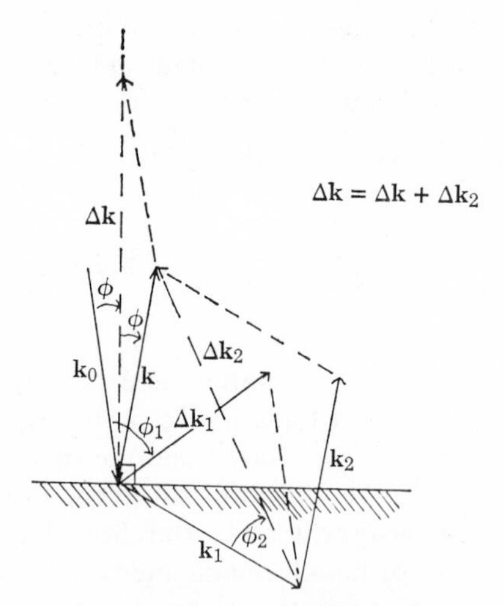

Fig. 14. Vector diagram for double-diffraction mechanism.

this condition is met whenever $\phi = 0°$. For ϕ near 0° and $\theta_D \sim \theta_{D_1} \sim \theta_{D_2}$, the most usual case experimentally, the results interpreted in terms of kinematic diffraction do not differ significantly from the dynamical result. However, especially at lower energies, where the differences in surface and bulk θ_D's are most significant and at large angles of incidence, the interpretation of Debye-Waller experiments not allowing for multiple-scattering effects could lead to discrepancies. The problem is tractable, however, since dynamical theory does predict the exact multiple-diffraction mechanisms applicable, and by a form of iterative procedure the θ_{D_n}'s could be determined.

V. COMPUTATIONAL PROCEDURES TO EVALUATE THE SCATTERED LOW-ENERGY ELECTRON INTENSITIES

From the preceding considerations, we see that the geometry of the scattered beams is uniquely defined by the dimensions and two-dimensional symmetry of the crystal surface, and by the energy and angle of incidence of the primary beam. Further, we now know that intensity maxima may appear in these diffraction beams when certain diffraction conditions are met. However, the relative magnitude of these intensity maxima and their precise relationship to the chemical nature and exact positions of the scattering centers can only be determined through a more quantitative investigation of the scattering phenomena.

There are a number of different approaches currently popular in the literature, but they all involve either explicitly or implicitly finding a solution, in some degree of approximation, to the Schrödinger equation. It should be emphasized that while many of these approaches appear formalistically different, they are all concerned with the same physical phenomena. They differ primarily in their viewpoint and in the nature of their approximations. The current literature on theoretical calculations of the intensity of LEED beams may be roughly subdivided into two parts on the basis of their starting points.

The first group begins with the differential form of the Schrödinger equation

$$(\nabla^2 + K^2)\psi(\mathbf{r}, \mathbf{K}) = u(\mathbf{r})\psi(\mathbf{r}, \mathbf{K}) \tag{28}$$

where K is the magnitude of the wave vector, and $u(\mathbf{r})$ is $2m/\hbar^2$ times the potential. In general, both the potential and the eigenfunction are expanded in a Bloch or Fourier series and the resulting set of linear inhomogeneous equations are then solved for the coefficients of the eigenfunctions. Frequently, these solutions are obtained for the eigenfunctions within the crystal and those in free space are determined by matching $\psi(\mathbf{r}, \mathbf{K})$ and its first derivative at the surface. Variations on this approach have been employed by Hirabayashi and Takeishi,[37] Boudreaux and Heine,[4] Hoffman and Smith,[38] Jepsen and Marcus,[5] and Ohtsuki[39] among others. Historically, this method has its roots in the works of Bethe[40] and Von Laue.[41]

The second basic approach begins with the integral form of the Schrödinger equation

$$\psi(\mathbf{r}, \mathbf{K}) = \psi^\circ(\mathbf{r}, \mathbf{K}) - 1/4\pi \int_{\mathbf{r}'} G(\mathbf{r}, \mathbf{r}')u(\mathbf{r}')\psi(\mathbf{r}', \mathbf{K})\, d^3r' \tag{29}$$

where $\psi^\circ(\mathbf{r}, \mathbf{K})$ is the incident beam, $G(\mathbf{r}, \mathbf{r}')$ is a Green function, and $u(r')$ is the potential defined above. An excellent description of the transformation of the differential form of the Schrödinger equation to its integral form is given by Merzbacher.[42] The effect of the integral operator, $\int_{\mathbf{r}'} d^3r'\, G(\mathbf{r}, \mathbf{r}\ )u(\mathbf{r}')$ on the eigenfunction $\psi(\mathbf{r}', \mathbf{K})$ may be regarded as a projection or an evolution of this eigenfunction from a point $\mathbf{r}'$ to another point $\mathbf{r}$. As the solution appears also on the right-hand side of Equation 29 under the integral sign, an iterative procedure is often followed. Alternatively, the quantities involved in the integral may be expanded in some appropriate basis set, such as partial waves, the integral solved, and the resulting set of coupled linearly dependent equations resolved as in the case of the differential Schrödinger equation approach. The integral equation, or Green function, approach has been utilized by McRae,[43] Kambe,[44,45] and Beeby[46] among others. Historically it is similar to the dynamical theory of X-ray diffraction developed by Darwin.[47]

Regardless of the starting point, there are several basic assumptions employed by most authors. The first is that the incident or primary beam of electrons may be represented as a plane wave. As the actual wave is presumably coherent for hundreds to thousands of angstroms,[3] this is probably not a bad approximation.

The second assumption is that the crystal has perfect periodicity parallel to the surface. The degree of perfection required perpendicular to the surface varies from paper to paper. The neglect of the existence of ledges and other surface imperfections is not important in a qualitative discussion, though there is some evidence that surface damage can change the results in actual situations.[3]

It is frequently assumed that the electrons that are elastically scattered into the region exterior to the crystal are contained in a number of discrete beams whose wave vectors are defined by Equations 5 and 7. While this is definitely true far away from a scattering center, it is not necessarily true in its immediate vicinity. However, calculations performed by McRae[43] for the case of isotropic scatterers indicate that deviations from a plane-wave nature may be negligible.

Further, the lattice is generally assumed to be static. This assumption is not valid except perhaps for those materials having a large atomic weight and a high Debye temperature.

Inelastic scattering is usually either ignored or considered on as simple a basis as possible. When considered, it is usually represented as atomic excitations, and collective phenomena such as plasma resonances are usually neglected. The lack of a detailed consideration of inelastic scattering is somewhat dangerous, particularly as it is frequently the dominant scattering mechanism.[25]

The last assumption is that the scattering is nonrelativistic. This is a reasonably good assumption for low energies and light atoms, but further investigation into its validity under other situations is necessary. From the first assumption, we may write the incident beam as

$$\psi^{\circ}(\mathbf{r}, \mathbf{K}) = e^{i\mathbf{K}^{\circ} \cdot \mathbf{r}} \tag{30}$$

From the second assumption, that of perfect two-dimensional periodicity parallel to the surface of the crystal, we may express the potential of the crystal as a Fourier expansion

$$u(\mathbf{r}) = \sum_{G_{\parallel}} V_{G_{\parallel}}(z) e^{-i\mathbf{G}_{\parallel} \cdot \mathbf{r}_{\parallel}} \tag{31}$$

where $\mathbf{G}_{\parallel}$ is a reciprocal lattice vector parallel to the surface and z is the coordinate perpendicular to the surface. There are several alternative expressions for $V_{G_{\parallel}}(z)$ which will be used below. The first is in the limit of perfect three-dimensional periodicity,

$$V_{G_{\parallel}}(z) = \sum_{G_{\perp}} V_G e^{-i\mathbf{G}_{\perp} \cdot \mathbf{z}} \tag{32}$$

where $\mathbf{G}_{\perp}$ is some reciprocal lattice vector perpendicular to the surface and $\mathbf{G} = \mathbf{G}_{\parallel} + \mathbf{G}_{\perp}$. This expansion will be used in the X-ray or kinematic limit, and for the matching calculations.

The second expansion of $V_{G_{\parallel}}(z)$ is an integral Fourier expansion

$$V_{G_{\parallel}}(z) = \int_{-\infty}^{+\infty} d\gamma V_G e^{-i\gamma \cdot \mathbf{z}}$$

where $\mathbf{G} = \mathbf{G}_{\parallel} + \gamma$, and γ is some continuously varying parameter in the z direction. This expression is useful near the surface where the periodicity in the z direction is weak or nonexistent. Further, when the potential is expressed as a sum of scattering centers

$$u(\mathbf{r}) = \sum_{s} W_s(\mathbf{r} - \mathbf{R}_s) \tag{34}$$

where the summation is over either layers or atomic sites, then

$$u(\mathbf{r}) = \sum_{G,s} W_{s,G}\, e^{-i\mathbf{G}\cdot(\mathbf{r}-\mathbf{R}_s)} \tag{35}$$

and

$$V_G = \sum_{s} W_{s,G}\, e^{+i\mathbf{G}\cdot\mathbf{R}_s} \tag{36}$$

where the summation over G can be taken to formally include an integration over γ. This expansion is important when there are two or more atoms per unit cell. Below, the structure factor terms, $e^{i\mathbf{G}\cdot\mathbf{R}_s}$, will usually be carried implicitly in V_G, which may also contain the Debye-Weller factor.

Another result of the second assumption is that, like the potential, the wave function may be expressed as a Fourier expansion

$$\psi(\mathbf{r}, \mathbf{K}) = \sum_{G_{\|}} f_{G_{\|}}(z) e^{i(\mathbf{K}_{\|}^{\circ} + \mathbf{G}_{\|})\cdot\mathbf{r}_{\|}} \tag{37}$$

This is a result of Bloch's theorem for two-dimensional periodicity. As for $V_{G_{\|}}(z)$, $f_{G_{\|}}(z)$ may be expressed as a discrete or continuous Fourier expansion as the situation warrants. Again $\mathbf{G} = \mathbf{G}_{\|} + \mathbf{G}_{\perp}$ or $\mathbf{G} = \mathbf{G}_{\|} + \boldsymbol{\gamma}$ will be used, and $\sum_G$ will be taken to contain an implicit summation over $\mathbf{G}_{\perp}$ or integration over γ as is appropriate to the circumstances. With this in mind, we may write

$$\psi(\mathbf{r}, \mathbf{K}) = \sum_{G} f_G\, e^{i(\mathbf{K}^{\circ} + \mathbf{G})\cdot\mathbf{r}} \tag{38}$$

It is instructive to consider the calculational procedures in the limit of negligible multiple scattering. More extensive calculations must reduce to these solutions for very small cross sections and these results serve as a basic frame of reference within which multiple-scattering phenomena may be discussed. In addition, general techniques can be outlined with a minimum of detail.

In the kinematic limit, when multiple scattering is insignificant, it may be assumed that the electron has a much greater probability of being found in the primary, or transmitted beam than in any other. In addition, it may be assumed that the electron penetrates deeply enough into the crystal to experience its full three-dimensional periodicity, and that surface effects are negligible. We therefore may use the three-dimensional expansion of the potential and wave function.

Substituting Equations 31, 32, and 38 into the differential form of the

Schrödinger equation, there results

$$(\nabla^2 + K^2) \sum_G f_G\, e^{i(\mathbf{K}^\circ + \mathbf{G}) \cdot \mathbf{r}} = \sum_{G''} V_{G''}\, e^{-\mathbf{G}'' \cdot \mathbf{r}} \sum_{G'} f_{G'}\, e^{i(\mathbf{K}^\circ + \mathbf{G}') \cdot \mathbf{r}} \tag{39}$$

which becomes

$$\sum_G \{(K^2 - |\mathbf{K}^\circ + \mathbf{G}|^2) f_G - \sum_{G'} V_{G'-G}\, f_{G'}\} e^{i(\mathbf{K}^\circ + \mathbf{G}) \cdot \mathbf{r}} = 0 \tag{40}$$

As the functions $e^{i(\mathbf{K}^\circ + \mathbf{G}) \cdot \mathbf{r}}$ form a linearly independent basis set, we need only consider the set of equations

$$(K^2 - |\mathbf{K}^\circ + \mathbf{G}|^2) f_G - \sum_{G'} V_{G'-G}\, f_{G'} = 0 \tag{41}$$

In this limit, it is not necessary to solve simultaneously this total set of equations in order to determine the amplitudes f_G, as we have made the assumption that $f_0 \gg f_G$. Therefore Equation 41 becomes

$$(K^2 - |\mathbf{K}^\circ + \mathbf{G}|^2 - V_0) f_G - V_G f_0 = 0 \tag{42}$$

or

$$f_G = \frac{V_G}{(K^2 - |\mathbf{K}^\circ + \mathbf{G}|^2 - V_0)} f_0 \tag{43}$$

This is essentially the X-ray result, that the amplitude of the various diffraction beams is proportional to a Fourier coefficient in the expansion of the potential.

We obtain a similar result by using the integral equation approach. There, assuming that $f_0 \gg f_G$ is the same as making the first Born approximation. That is, we may substitute $\psi^\circ(\mathbf{r}, \mathbf{K})$ for $\psi(\mathbf{r}, \mathbf{K})$ under the integral sign on the right-hand side of Equation 29. Making this approximation and substituting Equations 31 and 32 into Equation 29 we obtain

$$\psi(\mathbf{r}, \mathbf{K}) \cong -\int_{\mathbf{r}'} \int_{\mathbf{K}'} d^3K' \frac{e^{i\mathbf{K}' \cdot (\mathbf{r} - \mathbf{r}')}}{|\mathbf{K}'|^2 - |\mathbf{K}^\circ|^2} \sum_G V_G\, e^{-i\mathbf{G} \cdot \mathbf{r}} e^{i\mathbf{K}^\circ \cdot \mathbf{r}'} d^3r' \tag{44}$$

where the spectral form of the Green function is used (see the following section). Utilizing the following two equations

$$\int_{\mathbf{r}} e^{i\mathbf{Q} \cdot \mathbf{r}}\, d^3\mathbf{r} = \delta(\mathbf{Q}); \; \mathbf{Q} = \mathbf{K}^\circ - \mathbf{G} - \mathbf{K}' \tag{45}$$

and

$$\int_{\mathbf{K}'} d^3K' \frac{e^{i\mathbf{K}' \cdot \mathbf{r}}}{|\mathbf{K}'|^2 - |\mathbf{K}^\circ|^2} \delta(\mathbf{K}', \mathbf{K}^\circ - \mathbf{G}) = \frac{e^{i(\mathbf{K}^\circ - \mathbf{G}) \cdot \mathbf{r}}}{|\mathbf{K}^\circ - \mathbf{G}|^2 - |\mathbf{K}^\circ|^2} \tag{46}$$

we obtain

$$\psi(\mathbf{r}, \mathbf{K}) = \sum_G \frac{V_G}{|\mathbf{K}^\circ|^2 - |\mathbf{K}^\circ - \mathbf{G}|^2} e^{-i(\mathbf{K}^\circ - \mathbf{G}) \cdot \mathbf{r}} \tag{47}$$

where, by a comparison with Equations 38 and 43 it may be seen that the relative amplitudes are identical with those from the differential form of the Schrödinger equation.

A. Differential Equation Approach

One of the earliest nonkinematic LEED calculations was performed by Hirabayashi and Takeishi.[37] They used the differential equation approach in an extension of Von Laue's[41] dynamical theory. An explicit accounting of the termination of the crystal periodicity at the surface was made by utilizing the forms for the potential and wave function given in Equations 31 and 37. This set of coupled first-order differential equations in $f_{G_{\parallel}}(z)$ could conceivably be solved for the amplitudes of the various diffraction beams. However, Kirabayashi and Takeishi did not attempt a completely self-consistent solution, but rather made the approximation $|f_0(z)| \gg |f_G(z)|$, that is, the intensity of the incident beam is much stronger than that of any of the diffracted beams. Numerical calculations were performed for the specularly reflected beam in the case of graphite and were compared with experimental results. The agreement is not bad in the region above 100 eV but becomes progressively worse at lower voltages. This is not unexpected as the approximation $|f_0(z)| \gg |f_G(z)|$ becomes less valid at lower energies. This paper is of significance as it was the first to attempt a dynamical treatment of low-energy electron diffraction. Not only did it illustrate that reasonable agreement with experimental data could be obtained at higher energies by considering only a limited number of beams, but it further underlined the fact that the amplitudes of the diffracted beams are not negligible relative to that of the transmitted beam in the very-low-energy region. The condition that $f_G(z) \gg f_0(z)$ is precisely that which is associated with multiple scattering, and it is this condition which necessitates a more self-consistent treatment of the problems.

A related but more complete method has gained considerable popularity recently, particularly among the solid state physicists. This is the wave-matching approach where the wave equation is first solved within the perfectly infinite crystal and then the eigenfunctions outside the crystal are determined by matching these wave functions and their normal derivatives at the surface. In this approach, the primary problem is identical with that of determining the energy band structure within the crystal, but only for that energy and that component of the wave vector, K, parallel to the surface which characterize the incident beam. This method has the advantage that it may draw upon much of the knowledge accumulated about energy band calculations. It is particularly applicable to uncontaminated and unreconstructed surfaces, and leads to a clear insight into the relationship between reflected intensities and the band structure of the solid. The wave function inside the solid may be expressed as a linear combination of the Bloch functions for the perfect bulk crystal, as in Equation 38. The first phase of the problem within the framework of this approach is to solve the wave equation within the crystal. Inserting Equations 31, 32, and 38 into the differential form of the Schrödinger equation, there results

$$(\nabla^2 + K^2) \sum_G f_G \, e^{i(\mathbf{K}+\mathbf{G})\cdot\mathbf{r}} = \sum_{G''} V_{G''} e^{-i\mathbf{G}''\cdot\mathbf{r}} \sum_{G'} f_{G'} \, e^{i(\mathbf{K}+\mathbf{G}')\cdot\mathbf{r}} \qquad (48)$$

which, upon performing the indicated differentiation and then rearranging, becomes

$$\sum_G \left[(K^2 - |\mathbf{K} - \mathbf{G}|^2 - V_0) f_G - \sum_{G' \neq G} V_{G'-G} f_{G'} \right] e^{i(\mathbf{K}+\mathbf{G})\cdot\mathbf{r}} = 0 \qquad (49)$$

or, since the traveling wave terms, $e^{i(\mathbf{k}+\mathbf{G})\cdot\mathbf{r}}$, are linearly independent

$$(K^2 - |\mathbf{K} - \mathbf{G}|^2 - V_0) f_G - \sum_{G' \neq G} V_{G'-G} \, f_{G'} = 0 \qquad (50)$$

This set of linearly dependent equations in the amplitudes, f_G, has solutions if and only if the secular determinant is equal to zero, that is,

$$\begin{vmatrix} (K^2 - |\mathbf{K} - \mathbf{G}_{11}|^2 - V_0) & -V_{G_{12}} \cdots \\ -V_{G_{21}} & (K^2 - |\mathbf{K} - \mathbf{G}_{22}|^2 - V_0) \\ \vdots & \vdots \end{vmatrix} = 0 \qquad (51)$$

The relative values of the amplitudes f_G, may be determined as co-factors of the secular matrix,[48] and their absolute values then determined from the normalization condition

$$\sum_G |f_G|^2 = 1 \qquad (52)$$

Again, the above solution is essentially identical to that for an energy band problem with the exception that those solutions which attenuate or are damped are also considered.

The second phase of the problem is to match the wave function and its first derivative with respect to the surface normal within the crystal to that wave function and its derivative that are exterior to the crystal. In this manner, the amplitude of the diffracted beams in free space may be determined. The matching equations are

$$\psi(\mathbf{r}, \mathbf{K}) = \psi_B(\mathbf{r}, \mathbf{K}); \quad z < z_s \tag{53a}$$

$$\psi(\mathbf{r}, \mathbf{K}) = \psi_E(\mathbf{r}, \mathbf{K}); \quad z > z_s \tag{53b}$$

$$\psi_B(\mathbf{r}, \mathbf{K})\Big|_{\mathbf{z}=\mathbf{z}_s} = \psi_E(\mathbf{r}, \mathbf{K})\Big|_{\mathbf{z}=\mathbf{z}_s} \tag{54}$$

and

$$d\psi_B(\mathbf{r}, \mathbf{K})/dz\Big|_{\mathbf{z}=\mathbf{z}_s} = d\psi_E(\mathbf{r}, \mathbf{K})/dz\Big|_{\mathbf{z}=\mathbf{z}_s} \tag{55}$$

$\psi_B(\mathbf{r}, \mathbf{K})$ is the wave function in the bulk of the crystal, $\psi_E(\mathbf{r}, \mathbf{K})$ is that exterior to the crystal, and $\mathbf{z}_s$ is the coordinate of the crystal surface.

The simplest case, the two-beam case at normal incidence where only the transmitted and the specularly reflected beams are allowed, has been discussed in detail by Boudreaux and Heine.[4] The development is as follows. Within the crystal, the Bloch function is given by

$$\psi_B(\mathbf{r}, \mathbf{K}) = f_0\, e^{i\mathbf{K}\cdot\mathbf{r}} + f_{2k}\, e^{i(\mathbf{K}-\mathbf{G}_\perp)\cdot\mathbf{z}} \tag{56}$$

where K is given by Equation 51. When $\mathbf{K} = \mathbf{G}_\perp/2$, we are at the end of a Brillouin zone and consequently in an energy gap. For the smallest $G_\perp$ this corresponds to the first Bragg reflection. Away from the gap, the wave function within the crystal is predominantly that of a traveling wave direction into the crystal, and $f_0 > f_{2k}$. The coefficient of the back-reflected wave, $f_{G_\perp}$, is given, to a first order, by Equation 43 for the kinematic case.

However, within the band gap the waves are strongly coupled and a simple perturbation approach is no longer valid. It may be shown that within the gap, f_0 and f_{2k} have the same magnitude, and differ at most only by a phase factor, 2ϕ; i.e.

$$|f_0|\, e^{+i\phi} = |f_{2k}|\, e^{-i\phi} \tag{57}$$

where ϕ varies from 0 to $+\pi/2$ from one edge of the gap to the other.[4] The sign of ϕ depends upon the sign of $V_{G_\perp}$. Further, at energies inside the gap, there are no corresponding real values of K. This is a direct consequence of Equation 51 and has the physical significance that there are no traveling waves allowed within the crystal at these energies. There are, however, complex values of K that are allowed that correspond to evanescent or damped waves that are localized at the surface of the crystal. It follows then that

$$\mathbf{K} = \mathbf{K}_R + i\mathbf{K}_M \tag{58}$$

where $\mathbf{K}_M$ is the imaginary part of $\mathbf{K}$ and $\mathbf{K}_R$ is the real part. $|\mathbf{K}_R|$ is equal to $|\mathbf{G}_\perp|$ within the gap and $|\mathbf{K}_M|$ is zero at the edges of the gap. Within the gap, the Bloch function inside the crystal is

$$\psi_B(\mathbf{r}, \mathbf{K}) = |f|[e^{i\phi}e^{-K_M z}e^{i\mathbf{G}_\perp \cdot \mathbf{z}/2} + e^{-i\phi}e^{-K_M z}e^{-i\mathbf{G}_\perp \cdot \mathbf{z}/2}] \tag{59a}$$

$$\psi_B(r, K) = |f|\, e^{-K_M z} \cos (G_\perp \cdot z/2 + \phi) \tag{59b}$$

The wave function outside the crystal is

$$\psi_E(\mathbf{r}, \mathbf{K}) = e^{i\mathbf{K}^\circ \mathbf{z}} + f'_{2k} e^{-i\mathbf{K}^\circ \cdot \mathbf{r}} \tag{60}$$

By matching ψ_B and ψ_E and their first derivatives at the surface, the value of f'_{2k}, the amplitude of the specularly reflected beam, may be determined. It is found that $|f'_{2k}|$, the magnitude of back-reflected amplitude, is equal to unity. This is not unexpected as all the electrons striking the crystal must be back-reflected at energies within the band gap since there are no allowed traveling waves within the crystal in this region. When inelastic scattering is taken into account, $|f'_{2k}|$ of course will be less than unity. As the band gap is of width V_G, it follows that, to a first approximation this also will be the width of the Bragg peak. Similar arguments hold at higher beam voltages and for other diffraction beams. Consider the case where a higher-order diffraction beam characterized by $\mathbf{K}' = \mathbf{K}'_\parallel + \mathbf{K}'_\perp$ meets a diffraction condition of the form

$$2\mathbf{K}'_\perp = \mathbf{G}_\perp \tag{61}$$

The higher-order diffraction beam will behave in a similar manner to the specularly reflected beam discussed above. At this point, there is a band gap, and no traveling waves with K' are allowed in the crystal.[4] Consequently, the electron must be either reflected out of the crystal or, alternatively, scattered into some beam for which there is an allowed state. Actual calculations have been performed using variations on this wave-matching technique. Hoffman and Smith[38] have applied this approach to the problem of calculating the intensities of the (00), (01), and (11) diffraction beams from the (100) face of aluminum at normal incidence. They used a 27-term Fourier expansion of the potential with a 10 eV inner potential correction and a constant 2.5 V imaginary part of the potential to simulate inelastic scattering. In addition to Bragg peaks predicted by kinematic theory, they found secondary peaks associated with multiple-scattering phenomena. While the agreement with experimental data is imperfect, it does illustrate the validity of this approach for real problems.

Model calculations using the wave-matching approach have been performed by Carpart[49] and by Marcus and Jepsen[5] for simple cubic crystals. Marcus and Jepsen effected the solution of the one-dimensional linear differential equations through the use of a formal propagation matrix. They used a potential of point ions of charge Z in a sea of uniform negative charge. The calculations were performed for non-normal incidence. Their published results show both the band structure and the reflected intensities. The strong correlation between the band structure and the intensities is quite obvious. The several types of multiple-scattering phenomena discussed above are well represented. Carpart has used a pure wave-matching approach. His calculations are particularly important as they were performed for a cubic ensemble of s-wave scatterers. This same model potential was used by McRae[43] in the first self-consistent dynamical LEED calculations using the integral equation approach. The strong agreement between the results of these two approaches substantiates their fundamental similarities. It is of interest to note that while the s-wave scatterer potential is an easy model in the integral equation approach, it is a particularly difficult model within the differential equation approach. This is because all the Fourier coefficients have the same magnitude and consequently, a large number of terms must be carried. Consequently, the claim is made that the achieved agreement constitutes rather important evidence that the method can be used for real situations. Carpart's work also includes a band structure calculation, and again, there is a definite relationship between the band structure and the beam intensities.

Recently, Pendry has used the pseudo-potential method and the wave-matching approach to calculate intensities for niobium and nickel that show a fair agreement with the experimental data.[7] This is one of the few calculations where a realistic potential has been employed and promises to be one of the more fruitful approaches.

The reader is referred to a review article by Boudreaux, Perry, and Stern for a more detailed discussion of the differential equation approach.[50]

B. Integral Equation Approach

While the differential form of the Schrödinger equation has been employed in a number of different approaches that are related to the determination of the band structure of solids, the integral form is conceptually more concerned with the scattering mechanisms from a number of different scattering centers. Further, the previous approach is most easily handled when the crystal has perfect three-dimensional symmetry right up to the surface, while the following method initially assumed nothing about the periodicity of the system in the direction normal to the surface.

Assuming the potential to be formally expressible as a sum of individual scattering centers as in Equation 34, the integral form of the Schrödinger equation becomes a sum of integral equations

$$\psi(\mathbf{r}, \mathbf{K}) = \psi^\circ(\mathbf{r}, \mathbf{K}) - 4\pi \sum_s \int_{\mathbf{r}'} G(\mathbf{r}, \mathbf{r}') W_s(\mathbf{r}' - \mathbf{R}_s) \psi(\mathbf{r}', \mathbf{K}) \, d^3r' \quad (62)$$

where, if all the centers are identical, only one integral need be evaluated. The formal solution is now independent of the total symmetry, or lack thereof, of the problem. However, as most LEED problems do have a two-dimensional symmetry parallel to the surface, it is useful to introduce this as it results in some simplification of the problem. This symmetry is explicitly assumed when Equations 35 and 38 for the potential and the wave function are substituted into Equation 62, which then becomes

$$\psi(\mathbf{r}, \mathbf{K}) = \psi^\circ(\mathbf{r}, \mathbf{K}) - 4\pi \sum_{g, g'} \int_{\mathbf{r}'} [G(\mathbf{r}, \mathbf{r}') V_{g-g'}(z') \times e^{-i(\mathbf{g}-\mathbf{g}')\cdot \mathbf{r}'_{\|}} f_{g'}(z') e^{i(\mathbf{K}^\circ + \mathbf{g}')\cdot \mathbf{r}'_{\|}} d^3r'] \quad (63)$$

Here, g has been used to indicate $G_{\|}$ in order to avoid confusion with the Green function $G(r, r')$. The terms of the structure factor, $e^{i(\mathbf{g}-\mathbf{g}')\cdot \mathbf{R}_s}$, have been absorbed into $V_{s,g-g'}(z)$. The Green function has several different acceptable forms; among others, it may be used as an expansion of spherical harmonics or in its spectral form

$$G(\mathbf{r}, \mathbf{r}') = \int_{\mathbf{K}'} d^3K' \frac{e^{i\mathbf{K}\cdot(\mathbf{r}-\mathbf{r}')}}{|\mathbf{K}'|^2 - |\mathbf{K}^\circ|^2}. \quad (64)$$

Substituting this spectral form of the Green function into Equation 6 and integrating over $\mathbf{r}_{\|}$, one obtains

$$\psi(\mathbf{r}, \mathbf{K}) = \psi^\circ(\mathbf{r}, \mathbf{K}) - 4\pi \sum_{g, g'} \int_{z'} dz' V_{g-g'}(z') f_{g'}(z') \times \int_{\mathbf{K}'} d^3K' \frac{e^{i\mathbf{K}_{z'}\cdot(\mathbf{z}-\mathbf{z}')}}{|\mathbf{K}'|^2 - |\mathbf{K}^\circ|^2} e^{i\mathbf{K}_{\|}'\cdot \mathbf{r}_{\|}} \delta(\mathbf{K}'_{\|}, \mathbf{K}^\circ_{\|} + \mathbf{g}) \quad (65)$$

Using the properties of the delta function to first integrate over $K_{\|}$ and then over $\mathbf{K}'_z$, one obtains

$$\psi(\mathbf{r}, \mathbf{K}) = \psi^\circ(\mathbf{r}, \mathbf{K}) - 4\pi^2 \sum_{g, g'} \left[\frac{e^{-i(\mathbf{K}^\circ_{\|} + \mathbf{g} + \mathbf{k}_g)\cdot r}}{k_g} \times \int_{z'} V_{g-g'}(z') f'_g(z') e^{ik_g z'} \, dz' \right] \quad (66)$$

Here, $|\mathbf{k}_g| \equiv \sqrt{|\mathbf{K}^\circ|^2 - |\mathbf{K}_\parallel + \mathbf{g}|^2}$ is the component of $\mathbf{K}'$ perpendicular to the surface.

This formal solution illustrates several points about the integral equation approach. The use of symmetry and the expansion of the potential into a sum of individual potentials have been mentioned above. Further, the solution may usually be expressed as a sum of plane-wave states characterized by the appropriate parallel reciprocal lattice vector. The amplitudes in these states are, of course, dependent upon the nature of the potential and the geometry of the crystal. Moreover, they are inversely proportional to the perpendicular component, $|\mathbf{K}_\perp'| = |\mathbf{k}_g|$ of the diffracted wave vector. This is a direct consequence of the imposition of perfect two-dimensional symmetry on the Green function.

Kambe[44] has shown how to derive a specific form of the Green function that is particularly tailored to this problem as

$$G(\mathbf{r}, \mathbf{r}') = \sum_g e \frac{i(\mathbf{K}_\parallel{}^\circ + \mathbf{g} + \mathbf{k}_g) \cdot (\mathbf{r} - \mathbf{r}')}{2ik_g} \tag{67}$$

In addition, he has given an excellent discussion of the relationship between the Green function and the integral equation approach.

Kambe has also developed a solution to the integral form of the Schrödinger equation.[45] The key to the whole approach is the particular choice of the form of the potential. As in Equation 34, it is assumed that the potential can be expressed as a sum of potentials centered at particular atomic positions. Further, it is assumed that these potentials are of the "muffin-tin" type, specifically, that the total potential is contained in a series of spherically symmetric nonoverlapping globes and that there is a constant potential between these spheres of zero value. As the wave function and its first derivative must both be continuous, it follows that at the surface of these spheres, the wave functions that are inside any given sphere must match those that are external to it. Moreover, because there is no potential between the spheres, any outgoing wave that leaves a sphere must travel unperturbed, at least until it enters another sphere. Therefore, if $\psi(\mathbf{r}, \mathbf{K})$ is known at the surface of the sphere, its value in free space may be calculated.

The first self-consistent dynamical theory of LEED to be published was that of E. G. McRae.[43a] This paper was particularly significant not only for the mathematical formalism, but also for the model calculations that it contained. These calculations qualitatively illustrated many of the important aspects of multiple scattering such as its dependence upon cross section, angle of incidence, and so forth.

In many ways, McRae's derivation of a solution for the wave equation is similar to that of Kambe's. They both employ a Green function approach and a "muffin-tin" potential, and both expand into spherical harmonics to perform the integration. However, McRae's approach differs in that the potential between the spheres is not constrained to have a zero value. In addition, Green's theorem is not evoked and only volume integrals are used. Further, $G(\mathbf{r}, \mathbf{r}')$ is utilized in its real space expression rather than as an expansion of Blochlike functions.

The salient feature of McRae's theory is the concept of the effective field $\psi^s(r, K)$. The total field is considered to be composed of the primary field, $\psi^\circ(r, K)$, and the fields emitted by all the atoms, $\sum_s \psi^s(r, K)$. Within this viewpoint, the effective field incident on any given atom is the sum of the primary field and all the fields emitted by all the other atoms. This is the basis of this self-consistent approach. The field emitted by any atom is a function of all the fields emitted by all the other atoms and, for sufficiently large cross sections, multiple scattering of all orders is a logical consequence of this interdependence.

This formalism was used to calculate the intensity of back-diffracted electron beams from the (100) face of a hypothetical simple cubic crystal. A number of different intensity maxima were observed in the calculated plots. McRae has studied the behavior of these intensity maxima, or peaks, as a function of cross section.[43] He has found that as the cross section is reduced, those peaks that are nonkinematic in nature diminish in intensity more rapidly than do those that are allowed in the kinematic limit. This is reasonable as the nonkinematic peaks have their origin in multiple scattering in contrast to the single-scattering kinematic peaks. The smaller that one makes the cross sections, the more improbable multiple scattering will be relative to single scattering, all else being equal. In addition to changes in the ratios of peak heights, McRae has found that the peak positions may move when the cross sections are reduced. In the limit of small cross sections, the positions approach those predicted from the free electron model. This is to be expected, as reducing the cross sections is essentially the same as reducing the interactions of the electron with the crystal. Therefore, the band gaps become more narrow and the coupling between different beams is diminished.

Both McRae[43] and Marcus and Jepsen[5] have considered the effect of non-normal incidence on the intensity vs. energy curves. In general, those beams that are strongly coupled to other beams in a given energy range developed very pronounced fine structure when the degeneracy was broken by deviating from normal incidence. This is in sharp contrast to the kinematic case where maxima would be expected to move, but would not be expected to split and develop fine structure when the angle of incidence is varied. The development and variation of fine structure with changes in the angle of incidence have been observed experimentally.[51]

McRae has also studied the effect of introducing inelastic scattering by assigning a complex value to the scattering phase shift.[43] The effect was to change the shape and reduce the height of the peaks without changing their position or their base width. Ohtsuki has also considered the effect of inelastic scattering.[39] He has formally developed a theoretical approach to the LEED problem in the limit of strong absorption, that is, when the diffraction potential is small compared with the inelastic potential. His qualitative conclusions are similar to those of McRae. His formalism is sufficiently general to include bulk phenomena that are not well represented by individual atomic excitations.

Jones and Strozier[23] have also considered the effect of inelastic scattering on low-energy electron diffraction. They have performed calculations that indicate that the inclusion of inelastic effects leads to low reflectivities, asymmetric peak shapes, and in contrast to McRae, broad (10–20 eV) peak widths.

Duke and Tucker have derived a single-electron propagator formalism that includes inelastic processes.[28] Their calculations indicate that inelastic electron-electron interactions in the solid limit the penetration of the incident elastic beam to a depth of 10 Å or less for electrons between 15 and 150 eV incident energy.

When a surface structure is present, that is, when the surface layer is different from all the underlying bulk layers, it is more convenient to use a detailed scattering approach such as the integral equation method rather than the differential equation, or band structure approach. The formalisms of McRae[43] and Kambe[45] may be used to effect a solution to this problem. In addition, several authors have approached this problem through the use of a scattering or transfer matrix.

Beeby[46] has developed a method where the amplitude of the diffracted beam is expressed as an infinite summation. This form is particularly interesting because of the physical interpretation of his result. The first term in the summation is the single-scattering term. It represents the electron being scattered only once before leaving the crystal and would be the dominant term in the kinematic limit. The second term is a double-scattering term. The electron is first scattered at a point r_1 and is then scattered again at a second point r_2 before leaving the crystal. The following terms correspond to higher-order multiple scattering events. This approach is of course similar to an iterative Born expansion. The stepwise picture leads to a fairly direct interpretation of the physical significance of the various terms.

McRae[52] has considered the problem in a similar manner. He has approached the problem as a generalization of Darwin's theory of diffraction.[47] Unlike Darwin, however, he has considered all beams to be coupled and has allowed for the possibility that the surface may differ from the bulk of the crystal. McRae[53] has considered in particular the case where only single and double diffraction are important. Like Beeby, he has expressed the amplitude of the diffracted beam as a summation

$$\underset{\approx}{\mathbf{b}}_0 \cong \underset{\approx}{\mathbf{b}}_1 + \underset{\approx}{\mathbf{b}}_2 \tag{68}$$

where $\underset{\approx}{\mathbf{b}}_0$ is a column vector whose components are the amplitudes of the plane-wave components of the total wave field emitted by the crystal. The term $\underset{\approx}{\mathbf{b}}_1$ contains those contributions from single-scattering events and may be regarded as a modified kinematic expression for the diffraction amplitude. The term $\underset{\approx}{\mathbf{b}}_2$ corresponds to double-diffraction events where the electron has been scattered twice before leaving the crystal.

The physical meaning of the various terms is illustrated in Figure 15. The heavy line indicates the unique surface layer. The bulk layers that are chosen are to be considered as representative terms.

This approach has been suggested by Bauer,[54] among others, and should be useful where multiple scattering is weak, but not so weak as to place the problem in the kinematic limit. This situation could conceivably arise

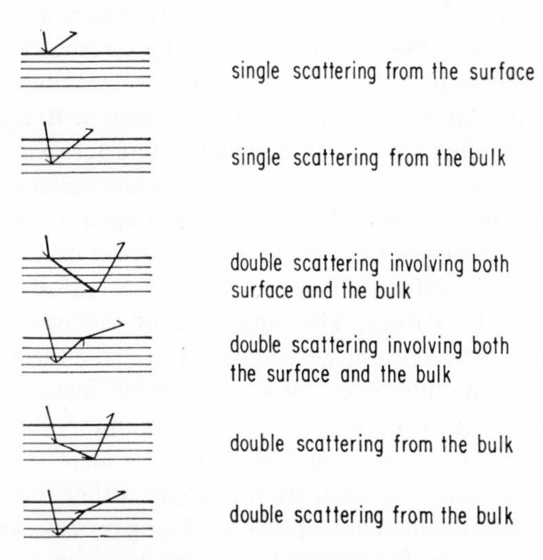

Fig. 15. Schematic representation of several simple scattering processes involving the surface layer of atoms and/or bulk atomic layers.

when inelastic scattering is strong, or when the number of diffraction beams is sufficiently large that the amplitude in any given beam is small.

When the surface layer has a periodicity that bears an integral multiple relationship to the periodicity of the bulk, fractional-order beams will be diffracted back from the crystal. The only nonvanishing contributions from b_1 to the intensity of these fractional-order beams will come from the surface layer. This contribution will contribute little to the modulation of the intensity of these fractional-order beams. Therefore, the contributions primarily from b_2 will determine the structure of the intensity curves. Furthermore, according to McRae,[53] the peak position should resemble a superposition of intensity curves for the integral-order beams. Physically, one may regard this process in the following manner. The diffraction beams that are formed within the crystal have large amplitudes in the back direction in the neighborhood of band gaps. As these large-amplitude integral-order diffraction beams leave the crystals, they impinge upon the surface layer. Part of their intensity is lost by scattering at the surface layer into the fractional-order beams. Thus, the surface layer serves to mix the intensities of the various beams. From these considerations, it is to be expected that surface structures with the same periodicity, but different chemical nature, should give rise to peaks in the same positions. The intensities of these peaks should of course be dependent upon the detailed nature of the scattering centers.

McRae and Winkler[55] have considered the case where a gas is adsorbed in register on a crystal. They find that when the surface layer differs significantly from the bulk, that the secondary or fractional-order Bragg peaks are damped relative to the kinematically allowed Bragg peaks. This result may be interpreted in terms of destructive interference in the double-diffraction terms because of the disparity between the surface and the bulk. The stepwise diffraction picture formally developed by McRae[53] and Beeby[46] has been used earlier in a more intuitive form by Gafner.[56] He has carried out a multiple diffraction calculation for several of the diffraction beams from the Ni(111) face. The amplitude of the waves which were formed at each diffraction event were adjusted to make their sum equal to the incident amplitude multiplied by an absorption factor to account for inelastic scattering. This is in contrast to the usual method of normalizing through intensities rather than amplitudes. The relative scattering factor was assumed to be unity for all scattering angles other than zero, where it was given the value of nine. The stepwise scattering process was considered in the following manner. The normally incident beam was diffracted into the several allowed diffraction beams at the first layer. The beams scattered into the crystal were allowed to undergo oscillatory diffraction between the first and second layer until all the beams had amplitudes less than some prescribed value. The beams that were scattered out of the crystal in this process were gathered up with those scattered back out of the incident beam. The beams that were scattered forward in this process were combined vectorially, and oscillatory diffraction between layers 2 and 3 was allowed to proceed as in the preceding case. This process was continued until all beam amplitudes in the crystal had fallen below the prescribed limit. Despite the approximations and assumptions within this model (or perhaps because of them), the agreement between the calculated and the experimentally observed intensity curves is quite encouraging.

C. Effect of Multiple Scattering on the Temperature Dependence of the Diffraction Beam Intensities

Let us finally consider, in general, the effect of multiple scattering on the temperature dependence of the diffraction beam intensities. Using an approach similar to that of Beeby[46] and McRae,[53] one may express the total eigenfunction for a diffraction beam in a Born-type expansion as

$$\psi_T(\mathbf{r}, \mathbf{K}) = \psi_0(\mathbf{r}, \mathbf{K}) + \psi_1(\mathbf{r}, \mathbf{K}) + \psi_2(\mathbf{r}, \mathbf{K}) + \cdots \tag{69}$$

where

$$\psi_n(\mathbf{r}, \mathbf{K}) = 1/4\pi \int G(\mathbf{r}, \mathbf{r}')V(\mathbf{r}')\psi_{n-1}(\mathbf{r}', \mathbf{K})\, d^3\mathbf{r}'. \tag{70}$$

Here $\psi_0(\mathbf{r}, \mathbf{K})$ is the eigenfunction of the incident beam. The term $\psi_1(\mathbf{r}, \mathbf{K})$ corresponds to that portion of the total eigenfunction that has been kinematically or singly scattered. Double-scattering events would be contained in $\psi_2(r, K)$ and higher-order events would be represented by the other terms in the expansion. It may be shown that

$$\psi_1(\mathbf{r}, \mathbf{K}') = 4\pi(K_z')^{-1} e^{i(\mathbf{K}^\circ + \mathbf{G}) \cdot \mathbf{r}} \sum_s e^{i\mathbf{G} \cdot \mathbf{r}_s} V_{s, G} \tag{71a}$$

or

$$\psi_1(\mathbf{r}, \mathbf{K}) = 4\pi(K_z')^{-1} e^{i(\mathbf{K}^\circ + \mathbf{G}) \cdot \mathbf{r}} \sum_s e^{i \cdot \mathbf{G} \cdot \mathbf{r}_s^\circ} e^{-W_s} V_{s, G} \tag{71b}$$

If one makes the assumption that all the layers are identical, then

$$\psi_1(\mathbf{r}, \mathbf{K}') = f_G\, e^{i(\mathbf{K}^\circ + \mathbf{G}) \cdot \mathbf{r}} \tag{72}$$

where

$$f_G = 4\pi e^{-W} \frac{V_{s, G}}{2K_z'} \sum_s e^{i\mathbf{G} \cdot \mathbf{r}_s^\circ} \tag{73}$$

In a similar manner, it may be shown that the double-diffraction term can be written as

$$\psi_2(\mathbf{r}, \mathbf{K}) = \sum_{G'} f_{G-G'} f_G e^{i(\mathbf{K}^\circ + \mathbf{G})\cdot \mathbf{r}} \tag{74}$$

In general,

$$\psi_n(\mathbf{K}', \mathbf{r}) = \left[\prod_{i=1}^{n} \left(\sum_{G_i} f_{G_i} e^{i\mathbf{G}_i \cdot \mathbf{r}}\right)\right] e^{i\mathbf{K}^\circ \cdot \mathbf{r}} \tag{75}$$

The intensities of the various beams characterized by G are given by

$$\begin{aligned} I &= \psi_T^*(\mathbf{r}, \mathbf{K}^\circ + \mathbf{G})\psi_T(\mathbf{r}, \mathbf{K}^\circ + \mathbf{G}) \\ &= (\sum_n \psi_n^*(\mathbf{r}, \mathbf{K}'))(\sum_m \psi_m(\mathbf{r}, \mathbf{K}')) \\ &= \sum_{n,m} \psi_n^*(\mathbf{r}, \mathbf{K}')\psi_m(\mathbf{r}, \mathbf{K}') \end{aligned} \tag{76}$$

where the eigenfunctions have been expanded in a Born-type series. On the basis of this expansion, the intensity itself may be expressed as a series as

$$I_T = \sum_n I_n \tag{77}$$

Here, the first term

$$I_1 = |f_G|^2 \tag{78}$$

is the kinematic term and would contribute to the intensity even in the absence of multiple scattering. This term, of course, carries the kinematic temperature dependence as

$$I_1 = I_1^\circ e^{-2W} \tag{79}$$

where W is a linear function of temperature in the Debye approximation. I_1 arises as the product of the amplitudes for single-scattering events.

The second term, I_2, is generated as the product of the amplitudes for single-scattering events with the amplitude for double-scattering events and may be written as

$$I_2 = \sum_{G'} |f_G^* f_{G-G'} f_{G'}|. \tag{80}$$

This term will be referred to as the double-diffraction contribution to the total intensity. It has the temperature-dependent form

$$I_2 = I_2^\circ e^{-W_1} e^{-W_{12}} e^{-W_2} \tag{81}$$

where W_1 is proportional to $|\mathbf{G}|^2$, W_{12} to $|\mathbf{G}' - \mathbf{G}|^2$ and W_2 to $|\mathbf{G}'|^2$ in the approximation that the crystal is isotropic and that all the layers are identical.

There will be two contributions to the third term in the expansion of the intensity. The first will arise as a product of the single-scattering amplitude and the triple-scattering amplitude. The second term comes from a product of the double-scattering amplitudes. Consequently, this intensity term will have the form

$$I_3 = \sum_{G_1, G_2} [|f_G^* f_{G-G_1-G_2}^* f_{G_1-G_2} f_{G_2}| + |f_{G-G_1}^* f_{G_1}^* f_{G-G_2} f_{G_2}|] \qquad (82)$$

Its temperature dependence may be determined in a manner similar to that for I_2. Higher-order scattering contributions to the intensity will have increasingly complex forms and will bring correspondingly more complicated temperature-dependent terms into the total intensity. As f_G must be less than or equal to unity, these higher-order terms should be generally less important. However, there do exist cases where a term may be more important than the preceding lower-order terms. For example, there are observed "secondary" Bragg peaks in the specularly reflected intensity that do not correspond to kinematic diffraction conditions. When multiple scattering is reasonably strong, diffraction conditions of the form

$$K_z^{00} + K_z^{hk} = G_z \qquad (83)$$

can lead to intensity maxima in the (00) beam. Note that, even though this condition is kinematic for the (h, k) beam, it must involve at least double diffraction to produce an intensity maximum in the (00) beam. Consequently, the double-diffraction contribution to the total intensity, I_2, may be expected to be larger than the kinematic contribution, I_1. Higher-order contributions may also be significant. Therefore, it may be expected that the experimentally determined quantity, $T(d \ln I/dT)$, will more closely resemble $W_1 + W_{12} + W_2$ rather than the kinematic $2W$, assumed in the simple model. Because one might expect terms like $W_1 + W_{12} + W_2$ to be larger than $2W$, it would seem at first glance that multiple scattering alone could lead to the apparent determination of lower "effective" Debye temperatures or higher "effective" root-mean-square (r.m.s.) displacements for the surface. This is however not necessarily true in all cases. For simplicity, let us retain the assumption that the crystal is isotropic and that all the layers are identical. We may then write

$$W = \frac{3}{2} \frac{\hbar^2 T}{mk} \left[\frac{G_x^2}{\theta_x^2} + \frac{G_y^2}{\theta_y^2} + \frac{G_z^2}{\theta_z^2} \right] \qquad (84)$$

as

$$W = B\,|\mathbf{G}|^2 \qquad (85)$$

Within this approximation

$$2W_1 = 2B\,|\mathbf{G}_1|^2 \tag{86}$$

and

$$W_1 + W_{12} + W_2 = B(|\mathbf{G}_1|^2 + |\mathbf{G}_1 - \mathbf{G}_2|^2 + |\mathbf{G}_2|^2) \tag{87}$$

It may easily be seen that three cases arise. When

$$|\mathbf{G}_2|^2 + |\,\mathbf{G}_1 - \mathbf{G}_2|^2 < |\mathbf{G}_1|^2 \tag{88}$$

then $W_1 + W_{12} + W_2$ will be less than $2W_1$. In this case the experimentally determined effective Debye temperature derived from the simple model would be less than the actual effective Debye temperature. Alternatively, the apparent r.m.s. displacements would be greater than those actually contributing to the temperature dependence of the intensity.

In the second case, $|\mathbf{G}_2|^2 + |\,\mathbf{G}_1 - \mathbf{G}_2|^2 > |\mathbf{G}_1|^2$ and $W_1 + W_{12} + W_2$ is greater than $2W_1$. Here, the experimentally determined value for the r.m.s. displacements would be less than the real value.

In the third case $|\mathbf{G}_2|^2 + |\,\mathbf{G}_1 - \mathbf{G}_2|^2 = |\mathbf{G}_1|^2$ and $W_1 + W_{12} + W_2$ is equal to $2W_1$. In this case, the use of the simple kinematic model to determine the effective Debye temperature and the atomic displacements would lead to the same results as the use of a more complicated multiple-scattering model.

At normal incidence, double-diffraction contributions to the specularly reflected beam fall in the last case. Thus, if one neglects any possible asymmetry of the surface, one would expect that the contributions from this type of mechanism would give results that were experimentally indistinguishable from those arising from kinematic scattering. Away from normal incidence, double-diffraction contributions will no longer fall into the third case, but will give rise to contributions of both the first and the second type. Whether the experimentally determined r.m.s. displacements will be greater than or smaller than the actual displacements will depend upon the detailed nature of the scattering potential. For simple forward-scattering potentials, such as the screened coulombic potentials, one would expect those terms giving smaller apparent r.m.s. displacements to dominate.

Higher-order scattering events can also lead to apparent displacements that are either greater than or less than the real displacements. In the limit of an isotropic crystal with identical layers, the relationship between $\sum_i W_i$ and $2W$ may be determined in a manner similar to that for the double-diffraction situation. Again, one would expect that those terms leading to smaller apparent r.m.s. displacements would dominate when the scattering potential was of a smooth, forward-scattering type. Similar arguments may be made concerning the effect of multiple scattering on the temperature dependence of the intensity of the higher-order diffraction beams.

The assumption that all the layers of the crystal are identical is unrealistic, particularly in the presence of a surface structure.

Let us then consider the case where the first layer is different from all the other layers. For simplicity, the factors V_G/K_z' will be taken to be unity. The kinematic contribution to the eigenfunction for a given diffraction beam may then be written as

$$\psi_1(\mathbf{r}, \mathbf{K}') = e^{i(\mathbf{K}^\circ + \mathbf{G})\cdot r} \sum_{S=1}^{\infty} e^{i\mathbf{G}\cdot\mathbf{r}_s^\circ} e^{-W_s} \tag{89}$$

or

$$\psi_1(\mathbf{r}, \mathbf{K}') = e^{i\mathbf{K}'\cdot\mathbf{r}}(e^{-W_0} e^{i\mathbf{G}\cdot r_1^\circ} + e^{-W} \sum_{S=2}^{\infty} [e^{i\mathbf{G}\cdot\mathbf{R}}]^{S-1}) \tag{89b}$$

where W_0 is the Debye-Waller factor for the first layer, W is the Debye Waller factor for all the other layers, $\mathbf{r}_1^\circ$ is the coordinate of the surface, and R is the translational vector between layers. Making the definitions that $\alpha = e^{-W_0}$, $\beta = e^{-W}$, $\phi_0 = e^{i\mathbf{G}\cdot\mathbf{r}_1^\circ}$ and $\phi = e^{i\mathbf{G}\cdot\mathbf{R}}$, it may be shown that

$$\psi_1(\mathbf{r}, \mathbf{K}') = e^{i\mathbf{K}'\cdot\mathbf{r}}[\alpha\phi_0 - \beta\phi/(1 - \phi)] \tag{90}$$

The corresponding single-scattering contribution to the intensity may be written as

$$I_1 = \left| \alpha\phi_0 - \frac{\beta\phi}{1 - \phi} \right|^2 \tag{91}$$

When all the interplanar spacings are equivalent, this reduces to

$$I_1 = \alpha^2 - \alpha\beta + \frac{\beta^2}{2(1 + \cos(\mathbf{G}\cdot\mathbf{R}))} \tag{92}$$

This, of course, is essentially Darwin's result with the inclusion of the Debye-Waller factor previously considered by Lyon and Somorjai.[31,57]

Proceeding to higher-order scattering events, the double-scattering contribution to the total eigenfunction may be written as

$$\psi_2(\mathbf{r}, \mathbf{K}) = e^{i(\mathbf{K}^\circ + \mathbf{G})\cdot\mathbf{r}}(\sum_s e^{i(\mathbf{G} - \mathbf{G}_1)\cdot\mathbf{r}_s} e^{-W_{12}})(\sum_t e^{i\mathbf{G}_1\cdot\mathbf{r}_t^\circ} e^{-W_2}) \tag{93}$$

or

$$\psi_2(\mathbf{r}, \mathbf{K}) = e^{i(\mathbf{K}^\circ + \mathbf{G})\cdot\mathbf{r}} \left(\alpha_1 - \frac{\beta_1\phi_1}{1 - \phi_1}\right)\left(\alpha_2 - \frac{\beta_2\phi_2}{1 - \phi_2}\right) \tag{94}$$

where it has been assumed that all the interplanar spacings are equivalent and α_i and β_i have been defined in a manner similar to that for the kinematic case. The double-diffraction contribution to the total intensity partakes of the form

$$I_2 = \left|\left(\alpha - \beta\frac{\phi^*}{1-\phi^*}\right)\left(\alpha_1 - \beta_1\frac{\phi_1}{1-\phi_1}\right)\left(\alpha_2 - \beta_2\frac{\phi_2}{1+\phi_2}\right)\right| \tag{95}$$

where α and β correspond to the singly scattered amplitude and the α_i and β_i correspond to the doubly scattered amplitude. This term has a particularly interesting form when applied to fractional-order beams arising from the presence of a surface structure. These beams are forbidden in the bulk of the crystal. One may therefore make the simplifying assumption that scattering into these beams can occur only at the surface. When this is the case, I_2 reduces to

$$I_2 = \left|\alpha\alpha_2\beta_1\frac{\phi_1}{1-\phi_1}\right| \tag{96}$$

The terms α and α_2 correspond to scattering events at the surface layer where the electron is diffracted into back-scattered fractional-order beams. The term $(\beta_1\phi_1/1-\phi)$ corresponds to scattering events that can occur in the bulk of the crystal between the incident and some intermediate integral-order beam. It may be seen that if the surface species is loosely bonded relative to the bulk species, then the intensities of the fractional-order beams should exhibit a stronger temperature dependence when double diffraction occurs than would be observed for the integral-order beams. This of course is also true for the kinematic contribution where

$$I_1 = |\alpha|^2 \tag{97}$$

It may be shown that higher-order contributions to the total intensity will be of the form

$$I_n = \left|\prod_{i=1}^{n+1}\left(\alpha_i - \beta_i\frac{\phi_i}{1-\phi_i}\right)\right| \tag{98}$$

VI. THE LOW-ENERGY ELECTRON DIFFRACTION EXPERIMENT

In order to carry out surface studies by low-energy electron diffraction one needs (a) a well-defined electron beam, (b) a single crystal surface, and (c) ultra-high vacuum in order to keep the surface under conditions desired in the experiment. We shall now discuss these three experimental parameters in some detail.

A. Electron Optics

1. *Thermal Spread and Coherence Length*

The electrons are obtained by thermionic emission from a hot cathode. For a barium oxide cathode the operating surface temperature is about 800°C. If the cathode material is a refractory metal such as tungsten which may be coated with lanthanum hexaboride, the surface temperature of the cathode is of the order of 2000°C. For cathodes made out of barium silicates an operating surface temperature of the order of 1200°C is used. The temperature of the cathode determines the initial thermal energy spread of the electron beam. Assuming a Maxwellian distribution of electron velocities, the average thermal energy of the electrons is $\frac{3}{2}kT$. If we consider the energy spread ΔE equal to $\frac{3}{2}kT$, approximately 95% of the electron beam is contained in the thermal energy spread of $2\Delta E = 3kT$. Thus, the energy spread at 800°C is of the order of 0.2 eV while at 2000°C it is of the order of 0.6 eV. These electrons are then focused electrostatically and allowed to impinge on the crystal surface which is held at ground potential. The electron beam is on the order of 1 mm in diameter. For a completely ordered surface, the coherence length of the electron to a large extent is determined by the size of the source of the electron beam. Incoherence within the electron wave packet sets the upper limit to the number of atoms which can contribute to coherent scattering. This incoherence arises from the finite size of the electron source and the incoherence due to the spreading of the wave pockets over the distance between two scattering centers. This latter distance is usually referred to as the Fresnel zone. If r = width of Fresnel zone $= (R\lambda/2)^{1/2}$, where R = distance from scattering center to detector, and λ is the wavelength of the incident wave, then using the appropriate values for LEED, $R \cong 7 \times 10^8$ Å, gives $r = 2 \times 10^4$ Å However, because of the need for high intensities in LEED, the instrumental incoherence introduced by the use of a large electron source is much more significant. The coherence width of the electron beam at the scattering object,

$$\Delta\chi = \frac{\lambda}{2(1 + \Delta E/2E)\beta_s} \tag{99}$$

where β_s = half-angle indeterminacy in the angle of incidence for an incident electron due to the size of the electron source, ΔE = thermal spread of electron beam, and E is the energy of the electron beam. In LEED, $\beta_s \sim 0.001$ radians, $\lambda = 1$ Å (for $E = 150$ eV), $\Delta E \sim 0.2$ eV. These values give a coherence width of about 500 Å, that is, much smaller than the width of the Fresnel zone. Thus, in LEED, no area larger than $\sim(\Delta\chi)^2$

can contribute coherently to the diffraction pattern since no area larger than this receives coherent radiation.

The question of what is the minimum area necessary to give a coherent diffraction pattern has not been definitely answered experimentally. However, if one assumes that ordered arrays of 25–100 atoms are sufficient to give coherent diffraction, best agreement with present results is obtained.

Thus, considering all these factors, we can characterize coherence in LEED by the following description: minimum order necessary to give a coherent diffraction beam consists of ordered patches on about 10% of the crystal surface. As the surface is further ordered, the intensity of the diffraction spots should increase, and the sharpness of the spots improve until the surface consists almost entirely of regions of ordered arrays of about 10,000 atoms. Beyond this degree of ordering, the experimental factors prevent any improvement in the pattern, the macroscopic beam width (about 1 mm) limiting the sharpness of the spots, and the source incoherence limiting the intensity.

2. *Penetration of the Low-Energy Electron Beam and Energy Analysis of the Back-Scattered Electrons*

In order to obtain experimental information about the structure of the surface we would like to have most of the low-energy electrons which are elastically scattered, and thus contain diffraction information, to scatter from the topmost layer of atoms at the surface without any further penetration into the bulk of the crystal. As expected, the actual penetration depth depends on the energy of the low-energy electron beam. The penetration depth of the electron beam has been probed experimentally by depositing an epitaxial layer of one metal over another metal, detecting the amount deposited as a function of time, and correlating the result with the gradual disappearance of the diffraction features from the underlying substrate. It was concluded that a reasonable estimate for the penetration of low-energy electrons is 2–5 atomic layers in the energy range 5–100 eV. Studies of the deposition of amorphous silicon on silicon single-crystal surfaces by Jona have also indicated that the intensity of the diffraction spots of the underlying substrate could be reduced by over 95% upon the deposition of two monolayers of amorphous material.[58] An empirical correlation between the penetration depth and the energy of electrons is given by Heidenreich.[59]

The equation, L (penetration depth) $= 2 + (\text{eV}/150)^2$, gives a reasonable estimate of the electron beam penetration.[60]

LEED experiments indicate that the number of elastically scattered electrons which are back-reflected from crystal surfaces varies with the incident electron beam energy. About 20% of the incident electrons are back-scattered elastically at 10 eV, about 1% at 100 eV, and less at higher energies.[25] Figure 11 shows the representative result obtained for scattering from a number of face-centered metal (100) surfaces. Although the inelastically scattered electrons contain much valuable information about the composition of surfaces (which has been brought to light by recent advances in Auger spectroscopy), in a diffraction experiment the elastic and inelastic components of an electron beam have to be separated as the primary diffraction information is contained only in the elastic component. Since the elastic fraction is a small part of the total scattered electron beam the energy separation of these two components is a prerequisite of a successful low-energy electron diffraction experiment. This separation can be carried out in a number of different ways. Perhaps the most popular and the most prominent at the present is the so-called post-acceleration technique. The scheme of the electron optics is shown in Figure 16; the back-scattered electrons travel a fieldfree path to the first grid which is held at ground potential, as is the crystal. Energy analysis takes place at the second grid which is held at cathode potential. This grid, in principle, repels all the electrons which have lost energy in the collision with the surface and allows only the elastically scattered electrons to penetrate. The elastic component which has penetrated the grid system is then accelerated by the application of a large positive potential, 5,000–7,000 eV, onto a fluorescent screen where, due to radiative recombinations after excitation by the electron beam, light is emitted where the electron hits. The light intensity is proportional to the number of electrons hitting the screen. This post-acceleration technique is an excellent means of instantaneously displaying the

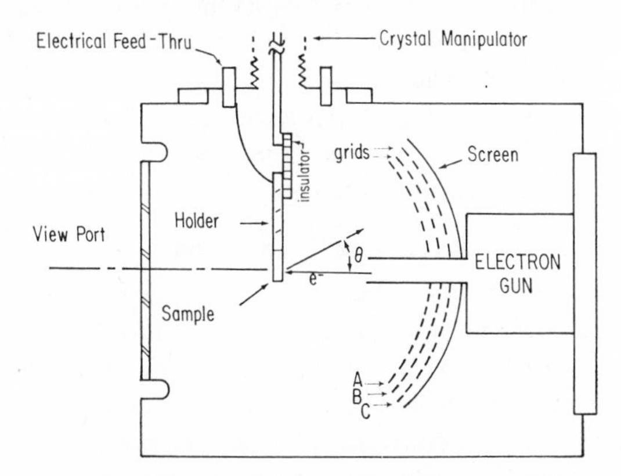

Fig. 16. Scheme of the low-energy electron diffraction apparatus of the post-acceleration type.

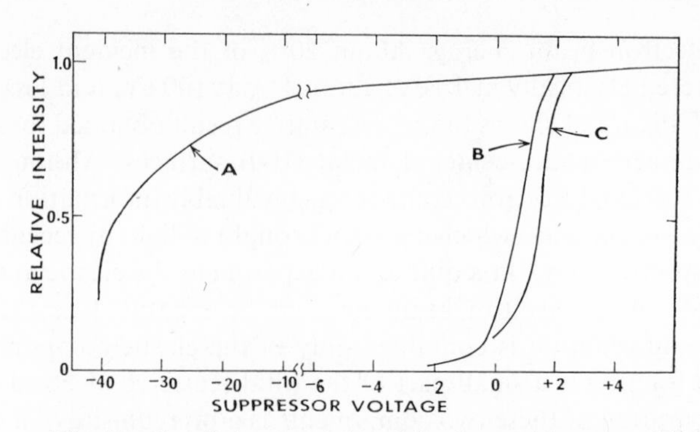

Fig. 17. Cutoff characteristics of the three-grid system with varying repeller grid potential with respect to the cathode potential.

diffraction pattern. In order to improve the energy selection of the repelling grid, often two grids instead of one are used for rejecting the inelastically scattered electrons. The cutoff characteristics of a three-grid system are shown in Figure 17. If the single-crystal surface is placed at the center of curvature of the three grids and the fluorescent screen, one should obtain an undistorted low-energy electron diffraction patterns. The solid angle subtended by the fluorescent screen is ~95°. There are other detection techniques which are often used in low-energy electron diffraction studies. A Faraday cup which can be rotated 180° is frequently used to monitor the low-energy electron beam intensity. While the fluorescent screen intensities allow one to measure relative intensities, the Faraday cup detection allows absolute intensity measurements which are necessary in some experiments. Another experimental geometry for low-energy electron diffraction studies uses a magnetically deflected low-energy electron beam. An advantage of this geometry is that the (00) or specular beam is not shadowed by the electron gun under conditions of normal electron beam incidence as it is with the "post-acceleration" technique or the Faraday cup detection technique.

B. Crystal Preparation

The preparation of well-ordered, clean, single-crystal surfaces is a very important phase of the low-energy electron diffraction experiment.[60] The crystals are oriented by the Laue X-ray diffraction technique using a precision goniometer, to within ±1° of the desired crystal face. The crystals are then cut to a convenient shape and are polished and etched by suitable chemical or electrochemical techniques. For soft single crystals such as lead or bismuth, spark cutting rather than mechanical cutting should be used to prepare the sample. Furthermore, for these crystals mechanical polishing should be minimized because of the extensive damage such a mechanical treatment might introduce into the surface order. For harder metals such as silver, gold, nickel, and platinum, spark cutting can also be used for preparation, followed by mechanical polishing with diamond or carbide powders of successively finer mesh. Mechanical polishing can easily correct the small deviations from the desired crystal orientation which are due to erroneous orientation of the original single-crystal sample. The chemical etching or electrochemical polishing serves to remove the damage introduced by the mechanical polishing of the crystal surface, which can often be as deep as one micron. The chemical etchants which are used vary from crystal to crystal; the etchants used for several metal and semiconductor surfaces are published in the literature. The samples are then mounted on holder assembly and placed into the diffraction chamber. The chamber is closed, pumped down, and baked at a temperature of roughly 250°C in order to obtain ultra-high vacuum conditions. Further cleaning of the crystal surfaces is then carried out by *in situ* ion bombardment and heating cycles until reproducibly clean, ordered, single-crystal surfaces are obtained. An ion bombardment gun is an essential part of a low-energy electron diffraction apparatus. Suitable ambient conditions for ion bombardment are $1\text{–}4 \times 10^{-5}$ torr gas pressure (argon and xenon are used frequently), ion accelerating potential in the range 140–350 V, and ion currents in the range of 1–2 $\mu A/cm^2$. Such an ion bombardment treatment removes traces of contamination introduced by the etching procedure and by exposure to the ambient during the mounting of the crystal. After ion bombardment the single crystal is heated to anneal out the surface disorder introduced by the ion bombardment treatment and the crystal face may then exhibit an excellent diffraction pattern. If not, the bombardment-annealing cycle is repeated until a diffraction pattern of the desired quality is obtained.

Since the coherence length of the electron is of the order of 500 Å a sharp diffraction pattern could be formed from microscopically rough surfaces. The diffraction pattern would be due to ordered domains approximately 500 Å or larger in diameter, and the intensity of the diffraction spots would depend upon the number of these domains which all contribute independently to the total scattered elastic amplitude. In several experiments where the vaporization of surfaces have been studied by low-energy electron diffraction, extremely rough surfaces have given excellent diffraction patterns. One question which may continually be asked in low-energy electron diffraction studies is, what fraction of the surface has to be ordered to obtain a diffraction pattern. A partial answer to this

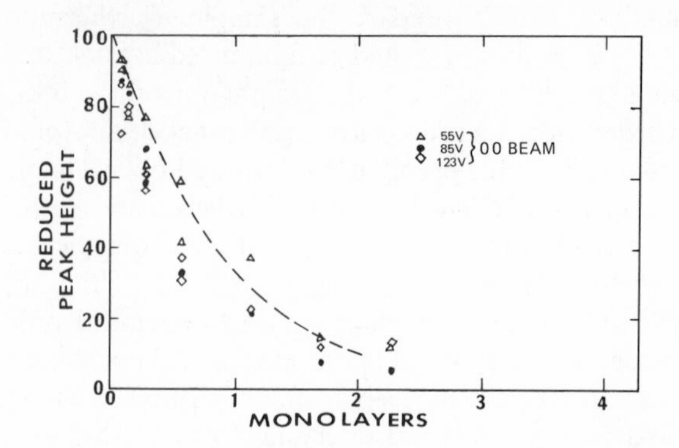

Fig. 18. Intensity of the (00) beam as a function of deposited amorphous silicon.

question was provided by the experiment in which amorphous silicon was deposited on a silicon single-crystal surface. It was found that as the deposition of the amorphous material continued, the intensity was reduced to about 50% of the initial intensity with the deposition of one-half of a monolayer. The reduced heights of selected specular intensity maximum as a function of the deposited amorphous silicon is shown in Figure 18. Thus, we find that even though the surface may be covered by 5–10% of amorphous material or disordered atoms the diffraction patterns would not be very sensitive to the presence of this surface concentration.

Another difficulty in low-energy electron diffraction experiments is to ascertain the cleanliness of the surface. Although the low-energy electron diffraction patterns may change or the intensity of the diffraction spots may reflect the presence of ordered impurities by the appearance of new diffraction features, the presence of amorphous impurities even in concentrations as high as 10% of the total number of surface atoms (that is, of the order of 10^{14} atoms per square centimeter) may not be easily detectable by low-energy electron diffraction. For example, the decomposition of hydrocarbons can lead to the deposition of amorphous carbon on the surface which may go undetected. In certain experiments, the deposition of amorphous impurities must be ascertained. In these situations, it is usually necessary to resort to auxiliary experimental techniques which, in combination with low-energy electron diffraction, can be used to give detailed information about the concentration of impurities and the nature of impurities on the surface. Recent advances in Auger spectroscopy, which analyzes the energy distribution of inelastically scattered electrons, frequently allow one to detect impurity concentrations on the surface of less than 5% of the total number of surface atoms. The detection and identification of impurities remains one of the continual problems in low-energy electron diffraction studies.

C. Ultra-High Vacuum

The low-energy electron diffraction experiment in principle could be carried out up to pressures of 10^{-3} torr in the diffraction chamber. At this pressure the mean free path of the electrons is still large enough, on the order of the dimension of the apparatus, for detection after scattering from the surface. The limiting factor, however, is not the mean free path of the electrons in the diffraction chamber but the adsorption of impurities on the surface which render the detection and the analysis of the surface structure impossible. Adsorbed gases may form ordered surface structures of their own or may be adsorbed amorphously. In any case, the diffraction features of the substrate may become undetectable upon adsorption of several monolayers or more of gas. Therefore, low-energy electron diffraction experiments have to be carried out in ultra-high vacuum (i.e., at pressures below 10^{-8} torr), at which the rate of incidence of ambient gas atoms onto the single-crystal surface still allows adequate experimental time to detect the diffraction features of the clean surface. Assuming a sticking probability of unity, a surface becomes covered with a monolayer of adsorbed gas in one second at an ambient pressure of 10^{-6} torr. Thus at 10^{-9} torr, experimental times of the order of 1,000 seconds are available to carry out low-energy electron experiments on a clean surface. Continuous pumping of the vacuum chamber and the maintenance of 10^{-9} torr vacuum, permits one to study the properties of a clean surface with minimal interference from undesired gases during the experiment.

When it is necessary to open the diffraction chamber, the apparatus should be brought up to atmospheric pressure with dry nitrogen or some other relatively inert gas rather than air in order to avoid the adsorption of unwanted gases with large surface binding energies such as oxygen or water on the walls of the chamber. In order to maintain oilfree conditions mechanical pumps are not generally used to evacuate the system. Generally, the pressure is reduced to 10^{-3} torr with the use of adsorption pumps that employ large surface area of zeolites cooled to liquid nitrogen temperature. Further reduction of the pressure to about 10^{-7}–10^{-8} torr is accomplished by the use of a vacuum ionization pump and/or an ion sublimation pump. At that point the chamber is baked at 250°C to obtain ultra-high vacuum conditions. Baking is necessary since it facilitates and accelerates the desorption of gases from the surfaces of the stainless steel diffraction chamber. After this baking process, a vacuum of the order of

10^{-9} torr and below can easily be obtained in a leakfree vacuum system. Modern vacuum technology makes the attainment and maintenance of ultra-high vacuum very easy during a low-energy electron diffraction experiment.

VII. LOW-ENERGY ELECTRON DIFFRACTION STUDIES OF DISORDERED SURFACES

A. Effects of Surface Disorder on the Diffracted Beam Intensities

The surfaces of a crystal are far from being perfectly ordered and clean. A real surface is heterogeneous; there are atoms in different crystal positions which are distinguishable by their different numbers of nearest neighbors and thus by variations of their binding energies. There are surface atoms in (1) steps or at ledges. There are also (2) vacancies and impurity sites, (3) dislocations, (4) mosaic structures, low-angle grain boundaries, and (5) liquid-like regions or disorder due to surface preparation, melting, vaporization, or adsorption of foreign substances. The effect of these defects on the scattered intensity is discussed in this section.

If N_T is defined as the total number of scattering centers in an array using the simplest kinematic model, the intensity is given by

$$I=|f_q|^2\left[N_T+\sum_{n\neq m}^{N_T}\sum^{N_T}\cos\{(\mathbf{K}'-\mathbf{K}^\circ)\cdot\mathbf{r}_{nm}\}\right] \tag{100}$$

If the array of atoms forms a perfect three-dimensional arrangement, then the summations collapse to the particularly simple form of $N(N-1)$ for $(\mathbf{K}'-\mathbf{K}^\circ)\cdot r_{nm}$ equal to some integral multiple of 2π. Under these circumstances, $I=|f_q|^2N^2$ and the intensity is proportional to the square of the number of scattering centers.

If, on the other hand, there is a random relationship between $\mathbf{r}_n$ and $\mathbf{r}_m$, and the array of atoms is completely disordered, then the terms in the summations tend to be out of phase and cancel so that $I=|f_q|^2N$; that is, the intensity varies linearly with the number of scattering centers. Therefore, to a first approximation, the intensity for electrons scattered from a completely disordered surface is proportional to the number of scattering atoms and gives rise to a relatively featureless diffuse background, while the intensity from a perfectly ordered surface is proportional to N^2 and is characterized by sharp and discrete diffraction spots where the condition $(\mathbf{K}'-\mathbf{K}^\circ)\cdot\mathbf{r}_{nm}=n2\pi$ is met.

In LEED on real surfaces (assuming the utility of the kinematic models) we are mostly concerned with situations between the two extreme cases of complete order and complete loss of periodicity (disorder). An interesting study of the influence of disorder on LEED intensities is provided by the results of Jona,[58] plotted in Figure 18. The dotted line in Figure 18 corresponds to the expectation that 90% of the coherent intensity is from the top three layers and Equation 100 is accurate. While the fit is not perfect, the results indicate that qualitatively the effect of disorder on LEED diffraction intensities (in the figure the circles, squares, and triangles refer to intensities at diffraction maxima at the indicated beam voltages) can be described by a simple kinematic (single scattering) model.

An interesting effect frequently noted in LEED studies is that random surface irregularities on a macroscopic scale (10^4 Å or larger) do not, in any apparent way, affect the intensity or the size of the diffraction spots. Goodman[60] has shown that sharp high-intensity diffraction features may be obtained from surfaces having a great concentration of pits, ledges, grain boundaries, etc., of about 1–10 μ in size. These results support the basic consideration discussed earlier concerning coherence. That is, if the surface is ordered in patches of perhaps 1,000 atoms, then the intensity is unaffected as long as all the patches are oriented with respect to one another. Macroscopic steps, pits, ledges, dislocations, grain boundaries, and so on, have virtually no effect on either spot size or intensity in LEED, as long as the spacing of the defects is of the order of the coherence length of the electron beam or larger. However, surface imperfections closer to each other than the coherence length will contribute to an intensity decrease in the diffraction features.

Ion bombardment of single-crystal surfaces using high-energy noble gas ions broadens the diffraction spots and simultaneously reduces the spot intensities. Such ion bombardment damage is detectable in all single-crystal surfaces where surface diffusion rates are negligible at the temperature at which the ion bombardment is carried out. Annealing the crystal by heating the surface to higher temperature increases the diffraction spot intensity and reduces the spot size. Clearly, ion bombardment has reduced the size of the ordered domains to below the coherence length of the low-energy electrons. Annealing increases the order and increases the sizes of these domains above that of the limiting size given by the coherence length. The main result of experiments on disordered surfaces is that the presence of disorder in concentrations of up to 10% of the available surface atoms have only limited or no detectable effect on the diffraction pattern. On the other hand, if impurities, vacancies, or other crystal imperfections are arranged in ordered or periodic array on the crystal surface, the LEED patterns are greatly affected, and new diffraction features appear immediately. Linear disorder occurs whenever atomic spacings along one crystallographic direction are disturbed while order is maintained in the others. A good example is provided by the work of Ellis on uranium dioxide,[61] in which deliberately cutting a crystal face off-axis at a small angle introduces very high step densities in the surface and streaking in the diffraction pattern. Such streaking is very frequently observed in low-energy electron diffraction patterns during the formation of ordered surface structures of adsorbed gases or during surface phase transformation from one ordered surface structure to another ordered surface structure. Linear disorder often reveals that the process by which a new periodicity characteristic of the new surface structure appears is surface diffusion. Monitoring intensity changes during a transformation which proceeds via linear disorder should be useful in obtaining information about kinetics of surface phase transformations which are surface diffusion controlled. Impurities often give rise to surface structures which are characterized by rotational disorder. Rotational disorder can be defined as disorder in which the

Fig. 19.

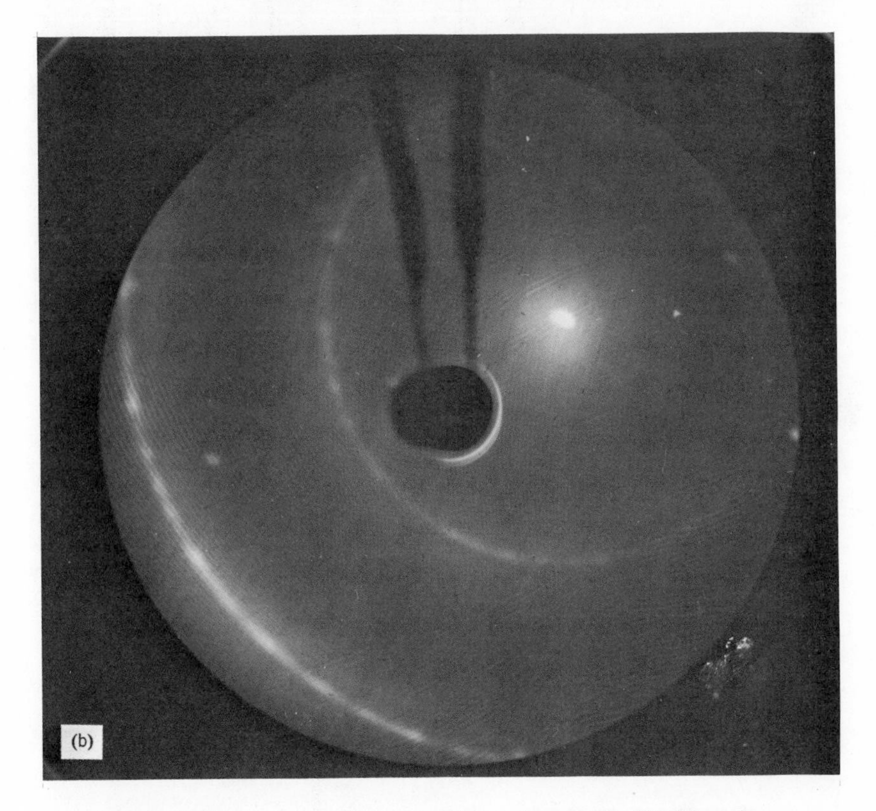

Fig. 19. Diffraction rings due to graphitic carbon on the Pt(100) surface.

surface atoms are ordered within one domain but where there is no preferred orientation of the domains with respect to each other. The result is that the diffraction pattern shows circular symmetry about the specular reflection. Carbon appears to give such a diffraction pattern on several metal surfaces. When amorphous carbon on the surfaces of gold, platinum, silver, and other metals is heated to higher temperature, diffraction rings appear first in segmented form and then finally as a complete circle[62,63,64] (Fig. 19). These ringlike diffraction features indicate the reordering of amorphous carbon on the surface in a graphitic form which is characterized by rotational disorder. Nevertheless, it is ordered parallel to the crystal face. A diffraction pattern of such rings characteristic of carbon on platinum surfaces is shown in Figure 19. Finally, we should consider the complete loss of long-range order; this is referred to as amorphous or liquidlike disorder. Guinier separates disordered structures into two classes: (1) correlated disorder, which refers to disorder in which the atoms are displaced from equilibrium sites a small amount relative to equilibrium interatomic separations, and the average position of all atoms is equivalent to that in the perfect lattice.[65] The most obvious example of this is the disorder introduced by thermal motions. On the other hand, (2) uncorrelated disorder exists when displacement from equilibrium positions may be large and the microscopic atomic density may differ from that of the ordered lattice. A good example is a volume containing a monatomic gas at low pressure. However, most arrangements of atoms in condensed phases do tend to have some of the characteristics of correlated disorder under all conditions of the low-energy electron diffraction experiments.

B. Low-Energy Electron Scattering from Liquid Surfaces

Liquids possess unique characteristics; like all condensed phases they possess some elements of correlated disorder. The distribution of atoms in liquids can be best described by a radial distribution function, $\rho(r)$. Figure 20 shows a typical radial distribution function which was taken from Kaplow's X-ray data on liquid lead.[66] The curve labeled ρ_o represents the average value from the density of the liquid. The main points to observe

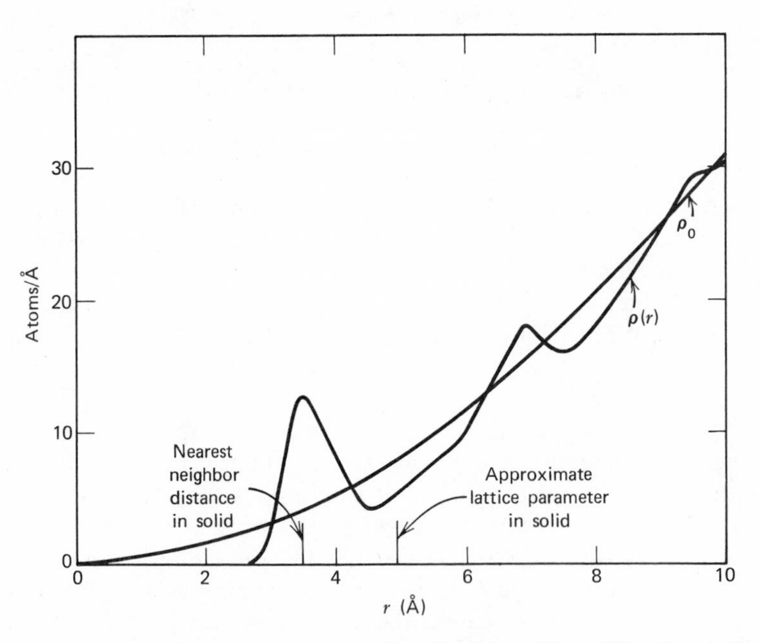

Fig. 20. Radial density function for liquid lead from X-ray diffraction studies.

are that $(\rho_a - \rho_o)$ goes to zero for small distances due to repulsive interactions and that it has a strong maximum near the nearest neighbor distance of the solid, while at large distances $(\rho_a - \rho_o)$ approaches zero again. For diffraction from an array satisfying such a distribution function, Guinier derives the interference function[65]

$$I(s) - 1 = \int_0^{\infty} 4\pi r^2[\rho_a(r) - \rho_0] \frac{\sin(2\pi rs)}{2\pi rs} dr \tag{101}$$

where

$$s = \frac{4\pi \sin \theta}{\lambda}$$

Figure 21 gives the X-ray intensities obtained from liquid lead at 327.4°C as a function of the scattering angle $(\sin \theta/\lambda)$. It should be noted that the

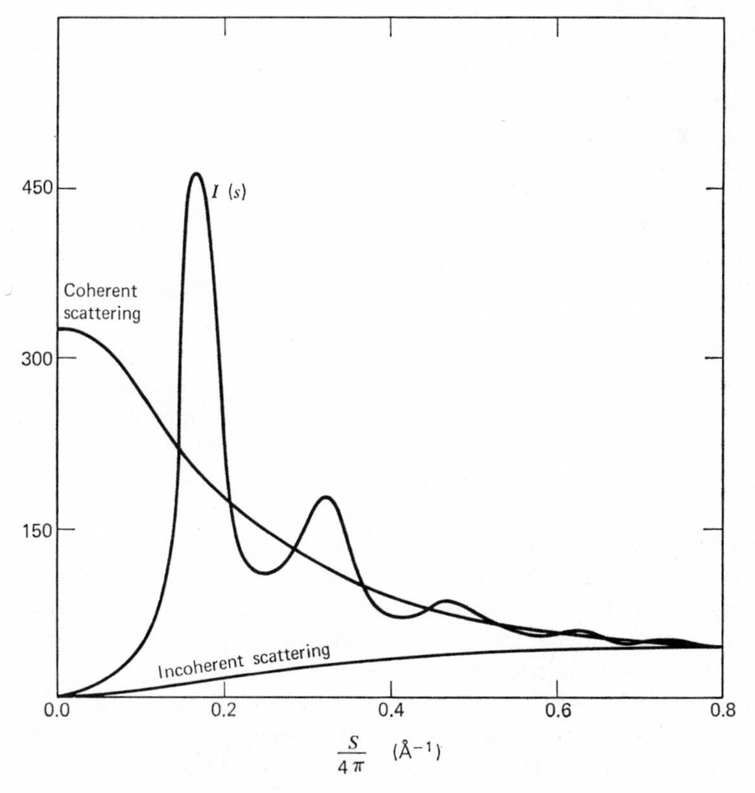

Fig. 21. X-ray intensities obtained from liquid lead at 327.4°C as a function of $s/4\pi = \sin \theta/\lambda$.

peak intensities are, at best, twice the background intensity. Similar results were obtained from liquid metals from high-energy electron diffraction studies.

Elastically scattered electrons or X-rays diffracted from liquid surfaces should show a distribution which reflects two scattering mechanisms: (1) uncorrelated scattering from individual atoms in the liquid, and (2) scattering which is modulated by the density fluctuations in the liquid. The former directly gives the atomic scattering factor, f. This parameter enters into calculations of surface structure from the diffraction beam intensities and its experimental determination is of great value. The latter scattering mechanism gives rise to intensity fluctuations which can be used to obtain the radial distribution function and thus the average interatomic distance and the coordination number at the surface of liquid.

In studies of low-energy electron diffraction, the electrons incident on the single-crystal surface are scattered predominantly by the surface layer or the first few atomic layers. Just as it is useful to determine the structure of solid surfaces, low-energy electron diffraction might also be employed to obtain information about the structure of liquid surfaces. Therefore, the intensities of elastically scattered low-energy electrons from liquid lead, bismuth, and tin surfaces were measured as a function of scattering angle and electron energy.[21] The samples were supported in crucible materials which appeared to show no chemical reaction with the solid or the molten phases.

When the crystal was completely melted and the temperature was about 5° above the bulk melting point the experiment was commenced. Photographic and/or visual observations were made of the diffraction screen; then a telephotometer was focused on the diffraction screen. The three-grid energy analyzer system had been adjusted to minimize the through penetration of the inelastically scattered electrons in order to reduce the inelastic contribution to the total scattered intensity detectable on the fluorescent screen. The output of the telephotometer was plotted on the ordinate of an X-Y recorder while the electron energy was plotted on the abscissa. This way, the plot of screen intensity as a function of beam voltage could be obtained as the voltage was scanned. The visual and photographic evidence indicated that the intensity fluctuation had radial symmetry about the electron gun axis.

The cleanliness of the molten surface was checked by cooling the crystal below its freezing point and observing the resultant diffraction pattern of the crystal surface. The recrystallized surfaces always displayed sharp diffraction patterns, and thus the liquid surfaces were considered to be clean.

The experimental results obtained from the voltage scans at different scattering angles were normalized to constant electron emission in order to eliminate one of the experimental variables from the experimental curves. The most convenient presentation of the measured intensity at different scattering angles and beam voltages from liquid surfaces is in the form of contour maps. The ordinate is taken as the screen angle, ϕ, with respect to the electron gun which is in the center of the screen (surface normal). The abscissa is the beam voltage and the contours are normalized intensities in arbitrary units. Figures 22a, 22b, and 22c show results from molten lead, bismuth, and tin surfaces, respectively. The intensity contours

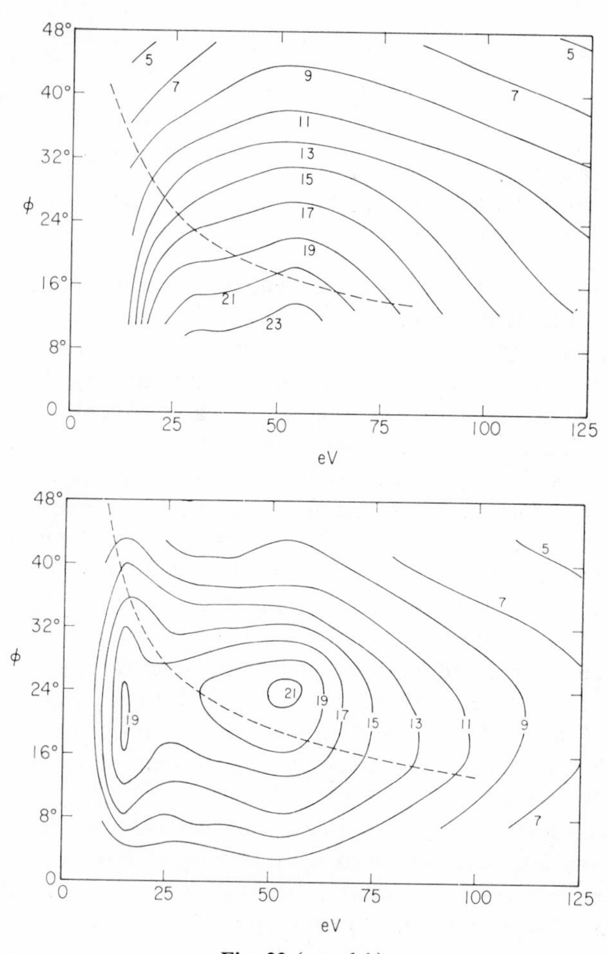

Fig. 22 (a and b)

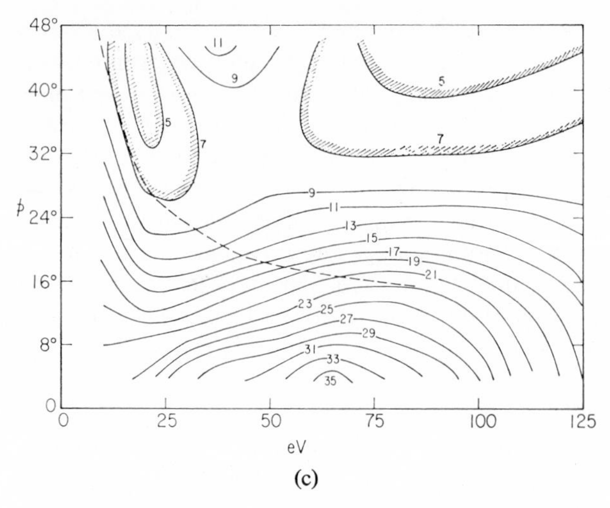

(c)

Fig. 22. Equi-intensity contours as a function of electron energy for scattering of low-energy electrons from lead, bismuth, and tin molten surfaces.

connect those scattering angle and beam voltage values which are characterized by uniform intensity. To convert the screen angle ϕ to a more usual angular variable let the scattering angle, θ, be equal to $\theta = 180° - \phi$, where $\phi = 0$ refers to the exact back scattering or 180° scattering. The intensity units, even though arbitrary, are directly comparable on all three contour maps (Figs. 22a–c). Figure 23 is an intensity map calculated from X-ray data by Kaplow[66] on liquid lead. Similar results were obtained by Richter *et al.*[67] using high-energy electron diffraction. Their data were extrapolated to our region of low energies; the dotted line connects the points where the first maximum should appear. It is readily apparent that none of the experimental curves shows features comparable with this calculated curve. Thus, the intensity fluctuations which we have detected cannot be used to calculate the radial distribution function from liquid surfaces.

Perhaps the most significant result of our LEED studies of liquid surfaces would be if we could associate the observed intensity distribution as being solely due to a single atom scattering mechanism. In that case the data directly gives the atomic scattering factor, *f*.

Let us consider additional scatterings mechanisms which could contribute to the intensities observed by scattering of low-energy electrons from liquid surfaces: (a) inelastic electrons which lose small energy (<2 eV) and may penetrate the repelling grids and contribute uniformly to the background intensity, thereby diminishing the magnitude of the intensity fluctuation: (b) multiple-scattering effects which are due to further

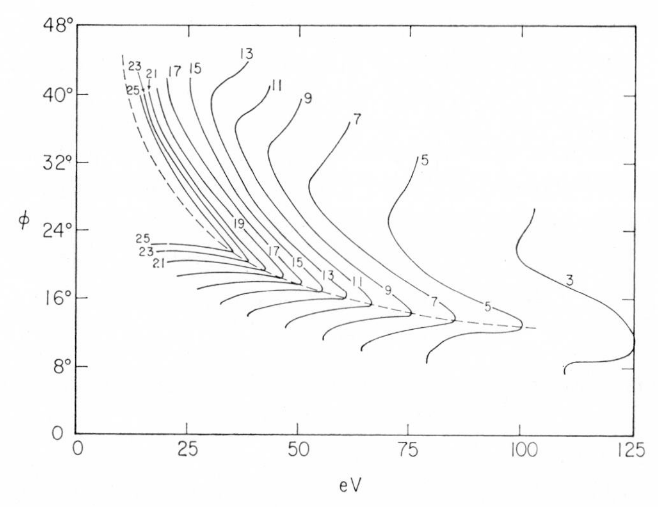

Fig. 23. Intensity contours calculated from X-ray data by Kaplow.[66]

interactions between the scattered electron beams; and (c) thermal diffuse scattering which attenuates the intensities of the diffraction features from liquid surfaces due to increased vibrations of atoms in the disordered surface.

Inelastic contributions to the scattered intensity would uniformly increase the background and may mask intensity fluctuations. Thus, this term would not be expected to change the intensity distributions markedly. The effect of multiple scattering is difficult to assess. It might introduce additional intensity fluctuations which would be superposed on the single-scattering distribution so that single- and multiple-scattering contributions could not be separated. Thermal diffuse scattering has been discussed by Webb et al., in some detail. Since this term is proportional to the atomic scattering factor and to the Debye-Waller factor, it also should not change the intensity distribution markedly.

Therefore, it may be concluded that the intensity distribution of the low-energy electrons scattered by lead, bismuth, and tin liquid surfaces gives us directly the low-energy electron atomic scattering factor for which single- and multiple-scattering contributions are inseparable.

Although it is somewhat disappointing that intensity fluctuations due to the radial distribution function in the liquid surface could not be detected by low-energy electron diffraction (LEED) (in view of the easy detectability of this effect in the bulk liquid by high-energy electron diffraction (HEED), X-ray, and neutron diffraction), the absence of these features can easily be rationalized. All three effects, (a), (b), and (c), mentioned above would be instrumental in reducing the intensity changes by increasing appreciably the background intensity. It should be remembered that the peak intensities are never more than a factor of two higher than the background intensity in all these experiments with bulk liquids near the melting point. There may be an additional reason for masking the scattering due to the liquid "surface structure" (i.e., correlated disorder).[36] Due to the low penetration depth of the electron beam in LEED experiments the number of atoms which contribute to coherent scattering is at least two orders of magnitude smaller than in HEED experiments.

VIII. THE MEAN SQUARE DISPLACEMENT OF SURFACE ATOMS

A. The Mean Square Displacement of Surface Atoms Perpendicular to the Crystal Surface

In one of the preceding sections the theory describing the thermal effects in LEED was discussed. Experimentally, the procedure for determining the surface Debye temperature is quite simple: at room temperature an intensity scan is made and a maximum in the I_{00}(eV) curve is determined. The telephotometer is focused on the 00-spot at the voltage corresponding to the maximum intensity. The crystal is heated (e.g., for Pd to about 650°C) and the power is turned off. The telephotometer output signal (monitoring the spot intensity) is plotted as ordinate; the thermocouple reading as abscissa, producing a curve as shown in Figure 24. In this way there is no interference from fields caused by the heater current. Generally it takes 1–5 min for a crystal to cool to below 100°C. The lower (essentially horizontal) curve in Figure 24 is obtained by rotating the 00-spot into the center of the screen and recording the intensity of the "background" at the same voltage as the previous intensity curve. To obtain the effective Debye temperatures, the intensity of the diffraction spot is read off the curve at different temperatures; the background value is subtracted from this value and the $\log_{10}(I_{00} - I_{\text{BKGRD}})$ calculated. Figure 25 is a plot of $\log_{10}(I_{00} - I_{\text{BKGRD}})$ vs. T(°K) obtained from Figure 24.

The effective Debye temperatures which were obtained at the lowest electron energies were taken as values characteristic of the surface atoms, $\theta_{D,\,\text{surf}}$. In some cases where extrapolation to zero electron volt could be carried out with some confidence the extrapolated value was taken as $\theta_{D,\,\text{surf}}$. Then the root-mean-square displacement of surface atoms perpendicular to the surface plane was calculated from the equation

$$\langle u_\perp{}^2 \rangle_{\text{surf}}^{1/2} = \left[\left(\frac{3Nh^2}{Mk} \right) \frac{T}{\theta_{D,\,\text{surf}}^2} \right]^{1/2} \tag{102}$$

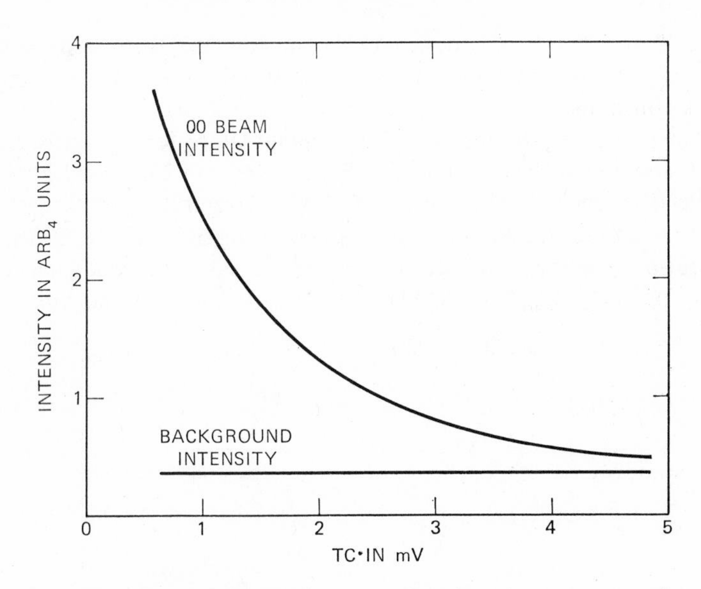

Fig. 24. Intensity of the (00) beam as a function of temperature and the background intensity.

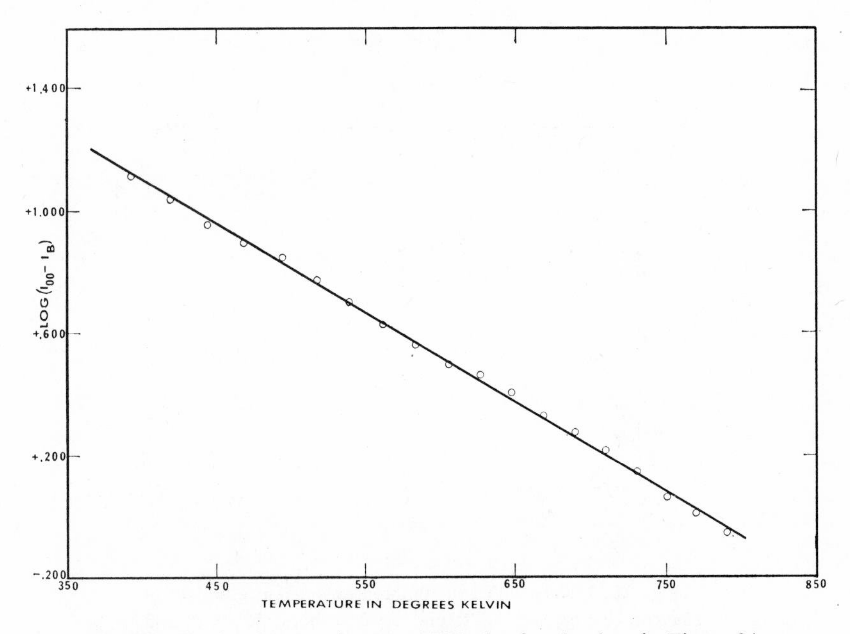

Fig. 25. The log (I_{00}-I_{bkgrd}) versus T°K plot for the data in Figure 24.

where M and T are the atomic weight and the temperature of the solid, respectively, N is Avogadro's number, and k and h are the Boltzmann and Planck constants.

As an example, the effective Debye temperature for two crystal faces of lead and palladium are plotted in Figures 26a and 26b as a function of the electron energy. The surface Debye temperatures are a factor of about 1.8–2 times smaller than the corresponding bulk values. Conversely, the root-mean-square displacement of surface atoms perpendicular to the surface plane $\langle u^2\rangle^{1/2}_{\text{surf}}$, is roughly 140–200% larger than the bulk value.

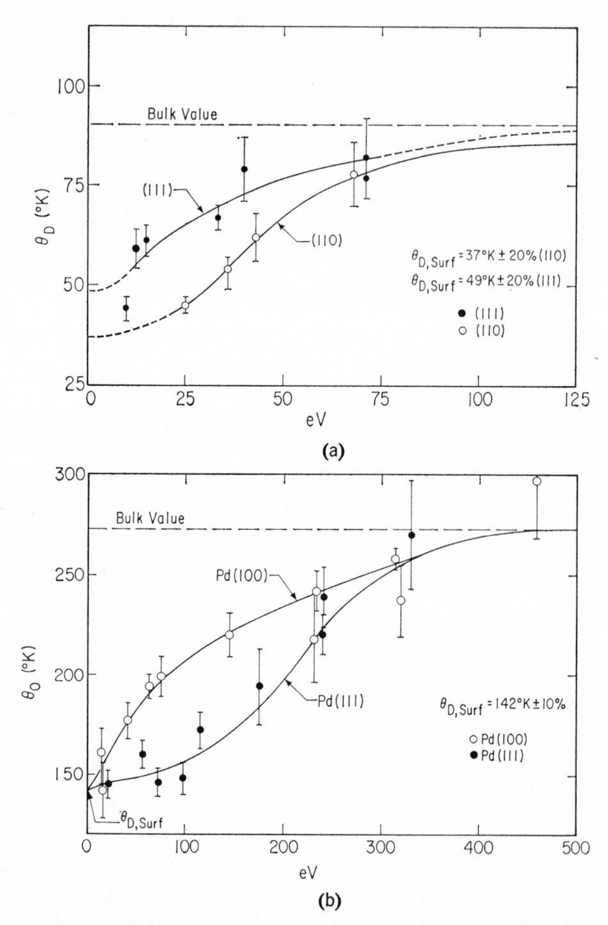

Fig. 26. Effective Debye temperatures as a function of electron energy for different crystal faces of lead and palladium.

TABLE I

The Surface and Bulk Root-Mean-Square Displacement Ratios and Debye Temperatures for Several Metals

	$\frac{\langle u_\perp \rangle \text{ (surface)}}{\langle u \rangle \text{ (bulk)}}$	θ (surface), °K	θ (bulk), °K
Pb (110), (111)[32,60]	2.43 (1.84)	37 (49)	90
Bi (0001), $(01\bar{1}2)$[60]	2.42	48	116
Pd (100), (111)[32]	1.95	142	273
Ag (100),[68] (110),[68] (111)[33,68]	2.16 (1.48)	104 (152)	225
Pt (100), (110), (111)[31]	2.12	110	234
Ni (110)[34]	1.77	220	390
Ir (100)[60]	1.63	175	285

In Table I we list all the data from different crystal surfaces which have been determined so far by experiments. We list the surface and bulk Debye temperatures and the surface and bulk root-mean-square displacement ratios for comparison.

For all the materials which have been studied so far the surface root-mean-square displacement perpendicular to the surface is much larger than the bulk value of the root-mean-square displacement. Conversely, the surface Debye temperature is much smaller than the Debye temperature characteristic of the bulk atoms. There seems to be little difference in the mean square displacements of surface atoms in different low-index planes with respect to the large difference between the bulk and surface values. This is in agreement with previous experimental and theoretical predictions within the experimental accuracies. In calculating the root-mean-square displacements we have not corrected the electron energy for the presence of the inner potential. The attractive potential that the electron experiences at the surface adds an energy increment to the electron energy which is of the order of 5–25 eV. Such a correction would have a little effect on the effective mean displacements which are calculated from the Debye-Waller factor determined at lower electron energies. A root-mean-square displacement can be corrected for the presence of the inner potential by using the formula $\langle u^2 \rangle^{1/2}$ (corrected) $= \langle u^2 \rangle^{1/2}$ (uncorrected) $((ev + ip)/ev).^{1/2}$ For example, at $ev = 50$ eV and for $ip = 20$ eV, $\langle u^2 \rangle^{1/2}$ is $0.86\langle u^2 \rangle^{1/2}$ (uncorrected). Since the inner potential value has not been determined accurately, all the data are given without inner potential correction. An inner potential correction will tend to decrease somewhat the calculated mean surface vibrational amplitudes.

B. The Mean Square Displacement of Surface Atoms Parallel to the Crystal Surfaces

Figure 27 represents the general LEED situation useful for calculating the Debye-Waller factor for the parallel component of the mean displacement of surface atoms. 2ϕ is the angle between the electron gun which is the source of the incident electrons and the (00)-spot (specular reflection) on the fluorescent screen; $2(\theta - \phi)$ is the angle subtended by the diffraction spot (h, k) and the electron gun. From Figure 27 we obtain that

$$|\Delta \mathbf{K}'|^2 = 4|\mathbf{K}^\circ|^2 \cos^2(\theta - \phi) = |\Delta \mathbf{K}'_\perp|^2 + |\Delta K'_\parallel|^2 \tag{103}$$

where $|\Delta \mathbf{K}'_\perp| = |\Delta \mathbf{K}'| \cos\theta$ and $|\Delta \mathbf{K}_\parallel| = |\Delta \mathbf{K}'| \sin\theta$. For any isotropic surface [i.e., (100) or (111), but not (110) for f.c.c. crystals] we can write

$$\langle u^2 \rangle_{\text{surf}} = \langle u_\parallel{}^2 \rangle_{\text{surf}} + \langle u_\perp{}^2 \rangle_{\text{surf}} \tag{104}$$

which can be substituted into Equation 20 to give

$$\exp[-|\Delta K'|^2 \langle u^2 \rangle] = \exp[-|\Delta \mathbf{K}'_\perp|^2 \langle u_\perp{}^2 \rangle - |\Delta \mathbf{K}'_\parallel|^2 \langle u_\parallel{}^2 \rangle] \tag{105}$$

for the Debye-Waller factor. Substituting Equation 103 into 105 and letting $K^\circ = 2\pi/\lambda$ gives

$$\exp[-2W'] = \exp\left[4\left(\frac{2\pi}{\lambda}\right)^2 \cos^2(\theta - \phi)\cos^2\theta \langle u_\perp{}^2 \rangle + 4\left(\frac{2\pi}{\lambda}\right)^2 \cos^2(\theta - \phi)\sin^2\theta \langle u_\parallel{}^2 \rangle\right] \tag{106}$$

Simplifying,

$$\exp[-2W'] = \exp -\left\{\frac{16\pi^2 \cos^2(\theta - \phi)}{{}^2\lambda} [\langle u_\perp{}^2 \rangle \cos^2\theta + \langle u_\parallel{}^2 \rangle \sin^2\theta]\right\} \tag{107}$$

Using Equation 19 and changing to LEED variables as in Equation 21, Equation 109 becomes

$$\exp_{10}[-2W'] = \exp_{10}\left\{-CVT\cos^2(\theta - \phi)\left[\frac{\cos^2\theta}{\theta_{D\perp}^2} + \frac{\sin^2\theta}{\theta_{D\parallel}^2}\right]\right\} \tag{108}$$

where C is the same constant as in Equation 21, $\theta_{D\perp}$ is the effective Debye temperature describing thermal motions normal to the surface, and $\theta_{D\parallel}$ is the effective Debye temperature for thermal motions in the plane of the surface. $\theta_{D\perp}$ is the quantity determined from the previously described measurements on the specular reflection. The extension to nonisotropic

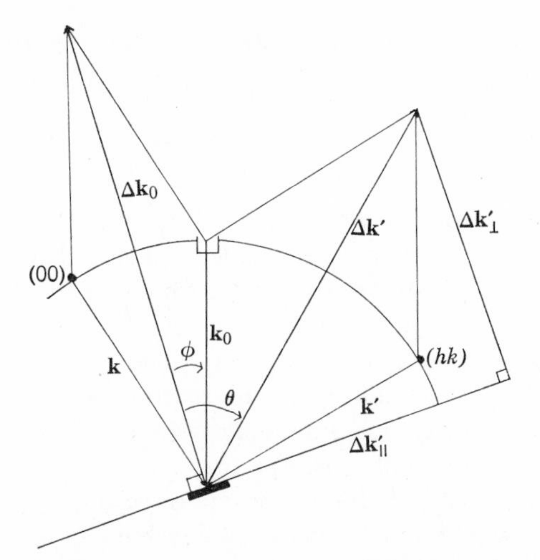

Fig. 27. Determination of scattering vectors for nonspecular (hk) and specular (00) diffraction beams.

$$
\begin{aligned}
|\Delta\mathbf{k}'|^2 &= |\mathbf{k}_0|^2 + |\mathbf{k}_0|^2 - 2\mathbf{k}_0 \cdot \mathbf{k}_0 \\
&= 2|\mathbf{k}_0|^2 - 2|\mathbf{k}_0|^2 \cos\,[\pi - 2(\theta - \phi)] \\
|\Delta\mathbf{k}'|^2 &= 2|\mathbf{k}_0|^2[1 - \cos\,\{\pi - 2(\theta - \phi)\}] \\
|\Delta\mathbf{k}'|^2 &= 2|\mathbf{k}_0|^2[1 + \cos 2(\theta - \phi)] \\
&= 2|\mathbf{k}_0|^2[1 + 2\cos^2(\theta - \phi) - 1] \\
|\Delta\mathbf{k}'|^2 &= 4|\mathbf{k}_0|^2 \cos^2(\theta - \phi)
\end{aligned}
$$

surfaces would require the definition of an azimuthal angle and the splitting of $\langle u_{\|}\rangle$ into components along the main surface coordinates. This will not be discussed here as the extension of the above method to this situation is obvious.

The most obvious characteristic to observe about Equation 108 is that $\theta_{D\|}$ can only be determined by the difference in two experimental determinations of $\log_{10}(I_{hk})$ vs. T. Further, since in conventional back-reflection LEED systems, ϕ cannot exceed 24° or θ exceed 48°, the two experimental slopes $(\Delta \log (I_{hk} - I_B)/\Delta T)$ will be of comparable magnitude. In practice it has been found that an uncertainty in either experimental determination of slope propagates about tenfold in determining a value for $\theta_{D\|}$, so that an uncertainty in $\theta_{D\perp}$ of 5% produces a 50% uncertainty in $\theta_{D\|}$, making extrapolations of $\theta_{D\|}$ (eV) curves difficult. Experiments using grazing angle incidence low-energy electron beams are therefore necessary to determine the parallel components of the mean square displacement with any degree of certainty. Surface Debye-Waller factor measurements under these conditions have been carried out for the Ni(110) face by MacRae.[34]

IX. THE STRUCTURE AND PHASE TRANSFORMATIONS OF CLEAN ORDERED SURFACES

A. Sources of Surface Impurities

We have discussed in some detail how the low-energy electron diffraction pattern is sensitive to the appearance of ordered structures or ordered impurities in very small concentrations. On the other hand it is insensitive in the presence of disordered impurities in concentrations as high as 5–10% of the total number of surface atoms. Therefore, one needs additional experimental techniques which might be used in combination with low-energy electron diffraction to ascertain the cleanliness of the surface. Appropriate techniques are Auger spectroscopy, flash desorption experiments, application of mass spectrometry, and ellipsometry.[68] Although no definitive experimental criterion for cleanliness for each material is possible, if these additional experimental techniques and the low-energy electron diffraction results do not suggest the presence of surface impurities, and the surface structures which have been detected are reproducible in several laboratories, it could then be assumed that the observed surface structure is characteristic of a clean surface. The presence of impurities below the detection limit of any of these presently available techniques cannot be excluded. These impurities may still act catalytically, favoring or inhibiting certain structural transformations though the impurities themselves do not necessarily participate in forming the new surface structure.

The impurity concentration at surfaces could change during an experiment and in many ways may influence the experimental results. Therefore it is essential that we consider the different sources of surface impurities during a low-energy electron diffraction experiment. One of the most probable sources of surface impurities is the adsorption of reactive gases on the surface. In ultra-high vacuum systems that are presently used carbon monoxide and water vapor are those reactive molecules which are present in large concentrations. These molecules could react with the studied crystal surfaces. Adsorption of these molecules could start chemical surface reactions which can alter the nature of the solid surface. In any given experiment after extended periods of experimentation they may induce irreversible chemical or crystallographic changes in the surface. Their interaction with the electron beam can result in the deposition of unwanted surface species such as carbon on the surface. The crystal bulk is another source of impurities. Bulk impurities may migrate to the surface and segregate out while the crystal is being heated to elevated temperatures. Although this is an excellent method to free the crystal bulk from impurities which are removed to the surface, surface contamination may be unavoidable especially in the beginning of the experiment. Once these impurities reach the surface they may be removed by ion bombardment or some chemical reaction such as oxidation. One of the frequently detectable bulk impurities of this type is sulfur, which has commonly been observed to diffuse to the surface of different metals.[69] Carbon is another contaminant which has been found to move from the bulk to the surface and segregate out.[70] Nevertheless, these could be removed from the crystal surface after careful chemical treatments, repeated ion bombardment, and annealing of the single-crystal samples. Perhaps the most tenacious impurities are those which are in the bulk of the single-crystal samples and have diffusion rates similar to the self-diffusion rate of the host lattice itself. These impurities will not diffuse to the surface and will not segregate out easily, and therefore they are permanently embedded in the single-crystal samples. Such an impurity is, for example, tantalum in niobium where the tantalum has similar diffusion rates to that of the self-diffusion rate of niobium single crystals.[71] In these cases several single-crystal samples should be used in the hope that the levels of impurities are different in the different samples. Inherent irreproducibility of surface structural features could be a sign of impurity-controlled surface properties. The third source of unwanted impurities at the surface could be due to the interaction of the electron beam with adsorbed gases or the host lattice itself. It was found that ionic crystals, for example, alkali halides, interact chemically with the electron beam, which leads to the decomposition of the surface.[72] Halogen evolution and the precipitation of alkali metal in an amorphous form at the crystal surface is commonly observed in studies of different alkali halides by low-energy electron diffraction. In order to avoid excessive decomposition which would affect the diffraction spot intensities and other surface parameters, one may heat the crystal to a temperature at which the damage introduced by low-energy electron beam could be removed by the annihilation of defects at the surface. Finally, it is found that certain adsorbed molecules, such as carbon monoxide, may desorb from metal surfaces due to electron impact by low-energy electrons. Although the exact mechanism of the electron beam excitation of the adsorbed species has not been investigated it is particularly interesting that carbon monoxide interacts strongly with the electron beam while other gases with weaker surface binding energies such as xenon or the olefins do not seem to desorb from metal surfaces at a detectable rate.

B. Order-Order Surface Phase Transformations

Most of the surfaces which have been studied by low-energy electron diffraction so far were high-density, low-index crystal faces of monatomic or diatomic solids. Without exception, all these faces exhibited ordered structures on an atomic scale. These ordered surfaces may be divided into two classes: (1) those that have unit cells identical to the projection of the bulk unit cell onto the surface; and (2) those that are characterized by unit cells which are larger than the unit cell dimensions in the bulk. The solids which belong to the first class have diffraction patterns which are characteristic of a (1 × 1) surface structure. The different crystal faces of tungsten [(110), (100)], nickel, and aluminum [(111) and (100)], for example, seem to belong to this class. Most semiconductors and some of the metal surfaces which have been studied so far belong to the second class. These surfaces exhibit diffraction patterns with extra diffraction features which are superposed on the diffraction pattern of the substrate unit mesh (predicted by the bulk unit cell). In Tables II and III we list some of the solid surfaces which exhibit these surface rearrangements.

TABLE II

Surface Structures Found on Clean Semiconductor Surfaces

Material	Surface structure
Si[27,78,82,83]	(100)—(4 × 4), (111)—(7 × 7)
Ge[25,78]	(100)—(4 × 4), (111)—(2 × 8), (110)—(2 × 1)
GaAs[78,79]	(111)—(2 × 2)
GaSb[78,79]	(111)—(2 × 2)
InSb[78,79]	(100)—(2 × 2), (111)—(2 × 2)
CdS[80]	(0001)—(2 × 2)
Te[81]	(0001)—(2 × 1)

TABLE III

Surface Structures Found on Clean Metal Surfaces

Surface Structures	Temperature range of established stability, °C
Pt (100)—(5 × 1)[31,73,74,90]	25–1300
Au (100)—(5 × 1)[62,75]	25–500
Au (110)—(2 × 1)[62]	
Ir (100)—(5 × 1)[76]	25–1500
Pd (100)—*c*(2 × 2)[76]	550–850
Bi ($1\bar{1}20$)—(2 × 10)[77]	25–melting point
Sb ($1\bar{1}20$)—(6 × 3)[77]	25–250

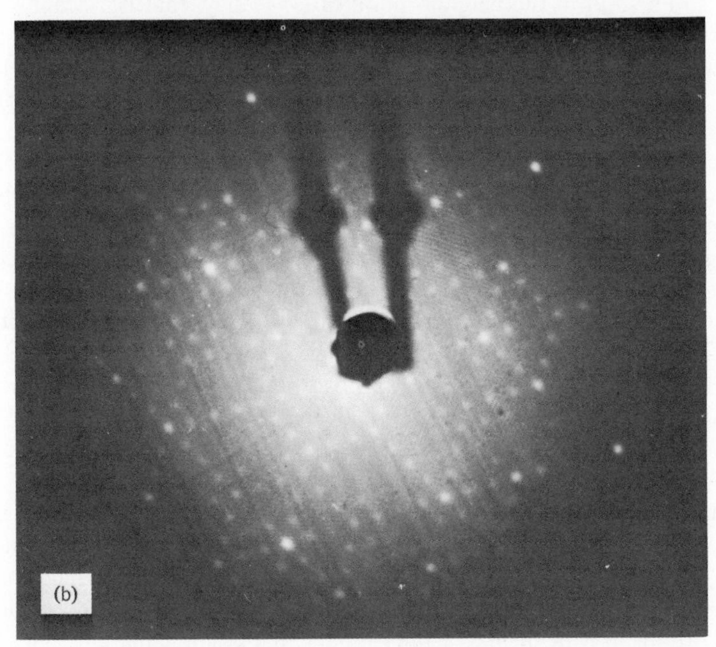

Fig. 28. Diffraction patterns of (a) the Si(111)-(1 × 1) and (b) the Si(111)-(7 × 7) structures.

The surface structures on semiconductor surfaces appear to have well-defined temperature ranges of stability. At temperatures above and below this range the surface undergoes a transformation into another ordered surface structure. For example, the (111) face of silicon has a (7×7) surface structure $[\mathrm{Si}(111) - (7 \times 7)]$ which forms upon heating the crystal to about 700°C.[82,83] The diffraction pattern corresponding to this structure is shown in Figure 28. Above 800°C this surface structure transforms into a (2×2) structure, while below 700°C the (1×1) surface net predominates. The (111) surfaces of other semiconductors with the diamond structure seem to form (2×2) surface structures.

What is the mechanism of surface rearrangements? There are several possible mechanisms which are still under investigation.

1. *Relaxation of Surface Atoms*

The surface structure is formed by a periodic displacement of surface atoms out of the surface plane. The surface thus exhibits a periodic "buckling" which gives rise to new, characteristic diffraction features. This mechanism may best be illustrated by considering what happens with atoms in a solid in the neighborhood of a vacancy, that is, vacant lattice position. If we remove an atom from its equilibrium position in the bulk to the gas phase, the atoms surrounding the now vacant site "relax," that is, are displaced slightly toward the vacancy. They are no longer restrained from larger displacement in the direction of the empty site by the strong, repulsive atomic potential. Therefore, the free energy of removing an atom from its bulk, equilibrium position to the gas phase is partially offset by the energy of lattice "relaxation" about the vacancy. The free energy of vacancy formation from a rigid lattice which is not allowed to relax can be approximated by the cohesive energy, the energy necessary to break a solid into single atoms infinitely separated from each other. We find that the free energy of forming vacancies is always appreciably smaller than the cohesive energy for most solids where these quantities have been measured. For example, for silicon, the cohesive energy is 81.0 kcal/mole, while the free energy of vacancy formation is 32.2 kcal/mol.[84] This leaves over 48 kcal/mole, a very large relaxation energy.

Surface atoms are in an anisotropic environment as though they were surrounded by atoms on one side and by vacancies on the other. The atoms can "relax" out of plane, perpendicular to the surface, which is not allowed for the bulk atoms. Depending on the bonding properties of the solid, atoms may be displaced out of plane in a periodic manner. This way there is an increased overlap of localized electronic orbitals. Calculations indicate[85,86] that the formation of some of these buckled surfaces can be energetically favored over the formation of flat surfaces in temperature ranges below the melting point of the solid. The appearance of any new surface periodicity will be reflected in the characteristics of the LEED diffraction pattern. It is likely that surface structural rearrangements in germanium, silicon, and other semiconductor surfaces occur by this mechanism.

It should be noted that periodic out-of-plane surface relaxation should be very sensitive to the presence of impurities or to certain types of lattice defects emerging at the surface (dislocations, vacancies). These could cause the collapse of surface structures by changing the chemical environment about the surface atoms or, in some cases, could also catalyze their formation.

2. *Surface Phase Transformation*

Some of the metal surfaces were also found to undergo atomic rearrangements (see Table II). For example, the (100) surfaces of gold[76,87] and platinum[88] exhibit a diffraction pattern which is shown in Figure 29. The presence of the $n/5$-order diffraction spots indicate the appearance of a new periodicity which is five times as large along one principal axis and the same as that in the bulk unit cell along the other. The pattern is a result of the superposition of domains of two structures of this type rotated 90° to one another. These surface structures may be designated as Au(100) – (5 × 1), and Pt(100) – (5 × 1). The diffraction patterns can be interpreted as indicating the presence of a hexagonal arrangement of scattering centers superposed on the underlying square (100) substrate. The interatomic spacing in the hexagonal surface layer is $\frac{5}{6}$ that of the substrate along one principal axis and the same along the other. Thus, the atoms in the hexagonal surface are coincident with every fifth substrate atoms and this could generate the observed fivefold surface periodicity. A small compression ($\sim 5\%$) in the hexagonal layer would allow six rows of the surface layer to fit onto the five rows of the square substrate.[87,88]

The chemical properties of this surface structure [its sensitivity to chemisorbed gases $(5 \times 1) \xrightarrow[\text{adsorption}]{\text{gas}} (1 \times 1)$] make it likely that the surface structure is again the result of periodic "buckling" of the surface plane. In this case, however, the surface relaxation resulted in the formation of a hexagonal surface structure, that is, there is a change of rotational multiplicity (from fourfold to sixfold). For semiconductors the surface structure maintained the rotational symmetry of the bulk unit cell even though the surface net became enlarged. Furthermore, the (5 × 1) surface structure and surface structures on other metal surfaces[75,76] are stable from 300°K up to almost the melting point of the solid. It appears that

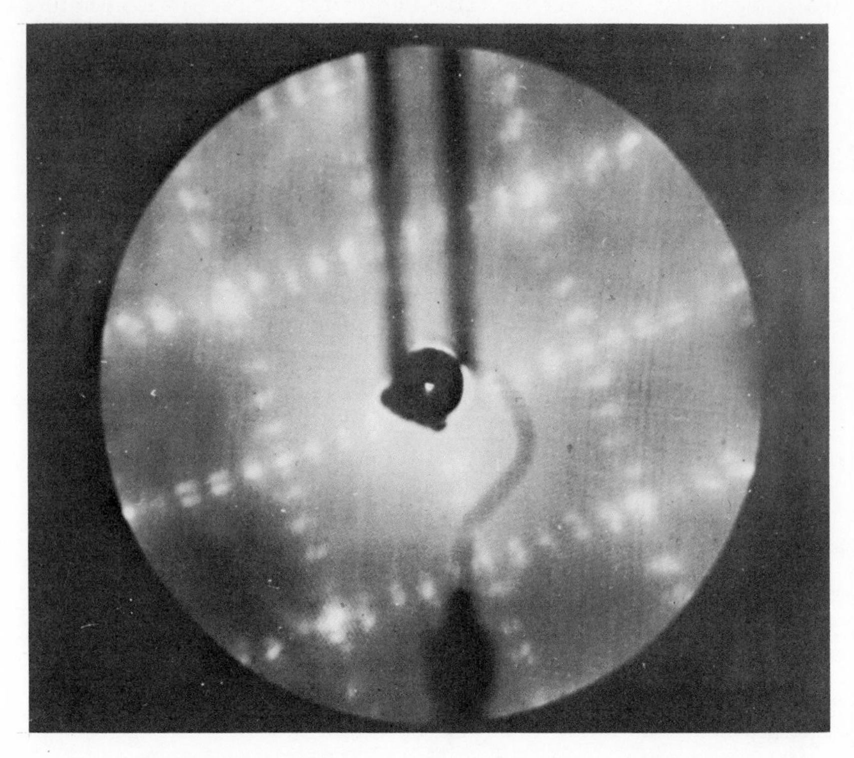

Fig. 29. Diffraction pattern of the Pt(100)-(5 × 1) structure.

these surfaces have undergone a phase transformation from a face-centered cubic to a hexagonal close-packed surface structure while no corresponding transformation has occurred in the bulk of the solid.

The crystal structure which a solid will take up has been shown to depend primarily on the number of unpaired *s* and *p* valence electrons per atom which are available for binding.[89] For example, atoms which have one unpaired *s* or *p* electron have body-centered cubic crystal structure when condensed to solid (like Na, W). Atoms with two unpaired *s* and/or *p* electrons will crystallize in the close-packed hexagonal structure (Zn, Os); three unpaired valence electrons will give face-centered cubic (Pt, Ag), and four unpaired valence electrons will give diamond crystal structures (Ge, C). A theory based on this concept, when extended to include the contribution of unpaired *d* electrons to the binding can explain and predict the structure and stability range of most alloys.[90,91]

Surface atoms, in addition to being in an asymmetric environment, have fewer neighbors than atoms in the bulk of the solid. Therefore, their electron density distribution should be different from that in the bulk; they may have more or fewer valence electrons available for binding than the bulk atoms. Thus, they may undergo phase transformations in the surface plane with respect to their crystal structure in the bulk. It appears that on the (100) face of gold and platinum a face-centered cubic-close-packed hexagonal surface phase transformation has occurred.

It should be noted again, just as in the case of surface relaxation, impurity atoms with different numbers of unpaired valence electrons per atom may cause or accelerate surface phase transformation of this type on transition metal surfaces, or conversely may inhibit it. For example, carbon (four unpaired valence electrons per atom) stabilizes the (1×1) surface structure on the platinum and gold (100) surfaces.[92] On the other hand, there appears to be evidence that oxygen on these noble metal surfaces, which accept electrons from the platinum and/or gold surface atoms, can stabilize the hexagonal surface structure (5×1).[62]

This mechanism would also predict the formation of surface alloys with a variety of structures and other interesting physical chemical properties. These may be prepared by the deposition of other suitable metal atoms with different numbers of unpaired valence electrons. There is evidence that tungsten surfaces undergo structural rearrangements during carbon diffusion which indicates a b.c.c. $\rightarrow$ h.c.p. surface phase transition long before hexagonal W_2C precipitates out at the surface. For example the W(100) surface develops a (5×1) structure first, which is followed by the formation of the carbide structure.

3. *Faceting*

Some of the more open, lower atomic density surface planes appear to be unstable upon heat treatment. At temperatures where surface atoms have enough mobility to diffuse, the surface undergoes rearrangement. New, high-density crystal planes form with the simultaneous disintegration of the more open crystal face. This process is frequently called faceting. LEED studies on the (110) faces of silver, for example, have indicated that faceting begins at low temperatures (<140°C).[68]

4. *Structural Changes Due to Variation of Surface Chemical Composition*

As noted above, the structure of metal surfaces appear to be very sensitive to the electron density at the surface. Furthermore, the presence of impurities have been shown to be very important in initiating or stabilizing surface transformations. There is, therefore, strong evidence that the structure of a surface should be very sensitive to slight variations in chemical composition. These considerations should be particularly important in studies of diatomic and polyatomic crystals where nonstoichiometry may easily be induced in the surface by heating in vacuum. As yet, this area has received relatively little attention.

The following study may be taken as an example of the importance of these considerations.

The structure of alumina surfaces is different from that which is expected by projection of the bulk unit cell to the various crystal surfaces.[93,94,95,96] The (0001) crystal face exhibits its (1×1) bulklike structure up to $\sim 1250°C$ in vacuum. It rearranges above this temperature to give a weak $(\sqrt{3} \times \sqrt{3})$-(rotated 30°) surface structure, and upon further heating to the final rotated $(\sqrt{31} \times \sqrt{31})$ surface structure, which is stable to the highest studied temperature of 1700°C. The diffraction patterns are shown in Figure 30.

It is customary to designate the complex surface structures by the coefficients of its transformation matrix which generate the structures with the unit cell vectors of the bulklike substrate.[97] This is given, for the rotated $(\sqrt{31} \times \sqrt{31})$ pattern, by

$$A = \begin{vmatrix} 11/2 & \sqrt{3}/2 \\ -\sqrt{3}/2 & 11/2 \end{vmatrix} \qquad B = \begin{vmatrix} 11/2 & -\sqrt{3}/2 \\ \sqrt{3}/2 & 11/2 \end{vmatrix}$$

These matrices generate the two domains which must be present on the surface simultaneously in order to generate the observed diffraction pattern. These domains are formed from the original unit mesh by expanding the unit vectors by a factor of $\sqrt{31}$ and rotating them either $+9°$ or $-9°$. We shall show that the alumina surface which exhibits the rotated $(\sqrt{31} \times \sqrt{31})$ surface structure is oxygen deficient.

The other two crystal faces, the (1012) and (1123) orientations, which have been studied, give (2×1) and(4×5) surface structures, respectively, at high temperatures ($>900°C$).[96]

Heating, by radiation, the freshly etched (0001) alumina surface which exhibits the (1×1) surface structure in vacuum, above 1250°C, readily produces the rotated $(\sqrt{31} \times \sqrt{31})$ surface structure (Fig. 30). During its formation oxygen evolution is detectable by mass spectrometer.

In order to establish that the stable high-temperature rotated $(\sqrt{31} \times \sqrt{31})$ surface structure has a chemical composition which is different from that of the low-temperature (1×1) surface structure, and to establish its stoichiometry, the (0001) face was heated in excess oxygen and aluminum vapor.

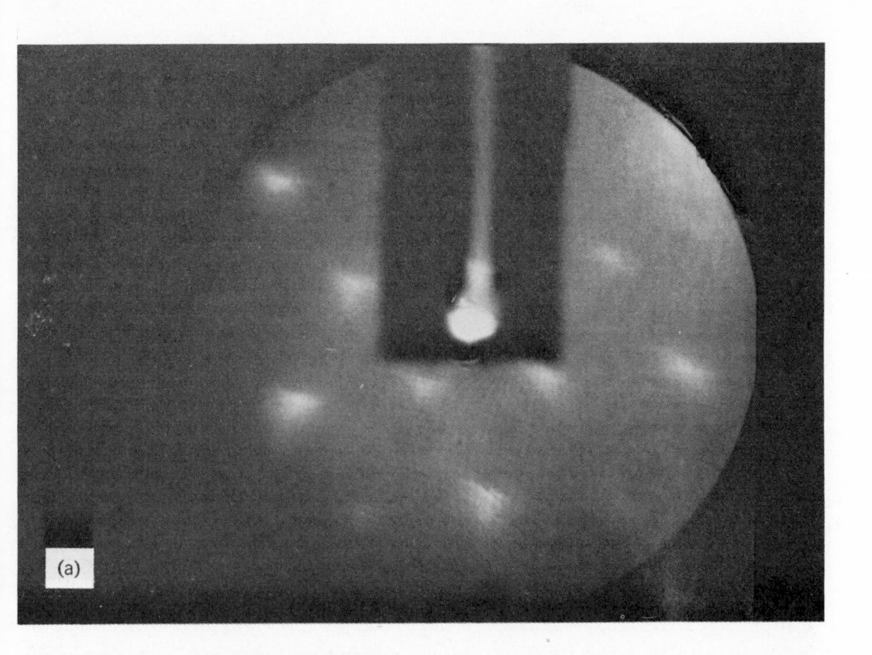

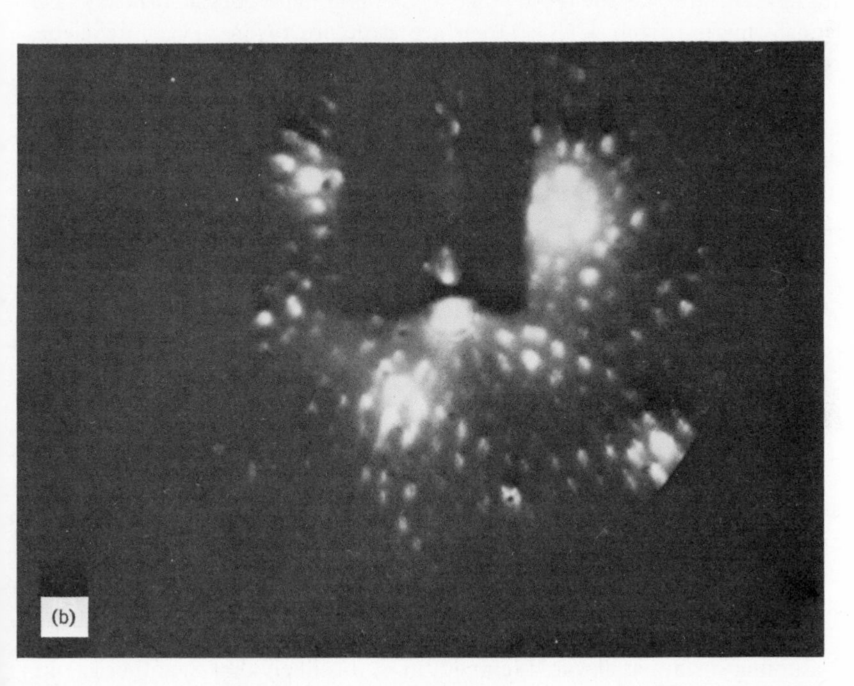

Fig. 30. Diffraction patterns of (a) the $Al_2O_3(0001)$-(1×1) and (b) the $Al_2O_3(0001)$-$(\sqrt{31} \times \sqrt{31})R9^\circ$ structures.

When the rotated ($\sqrt{31} \times \sqrt{31}$) surface structure is heated in oxygen at pressures $>10^{-4}$ torr (these pressures are considered to be high in ultra-high vacuum LEED studies) at 1200°C the (1 × 1) surface structure was obtained. Removal of the oxygen and heating to slightly higher temperature (1250°C or higher) in vacuum caused the reappearance of the rotated ($\sqrt{31} \times \sqrt{31}$) surface structure. This reversible phase transformation could be induced at will upon introduction or removal of oxygen.

When aluminum metal was condensed on the (0001) alumina surface which exhibits the (1 × 1) surface structure, the rotated ($\sqrt{31} \times \sqrt{31}$) surface structure was formed with heating to 800°C. In the absence of excess aluminum on the surface, the (1 × 1) surface structure would have been stable. Thus, the structural changes which occur in vacuum (mass spectrometric detection of oxygen while the rotated ($\sqrt{31} \times \sqrt{31}$) structure forms), in oxygen (the (1 × 1) surface structure is regenerated in a temperature range, ~1200°C, where the rotated ($\sqrt{31} \times \sqrt{31}$) structure is stable), and with aluminum (the rotated ($\sqrt{31} \times \sqrt{31}$) structure is formed in a temperature range, ~800°C, where the (1 × 1) surface structure is stable) indicate that the (0001) face of alumina undergoes a surface phase transformation from a (1 × 1) surface structure to an oxygen-deficient, rotated ($\sqrt{31} \times \sqrt{31}$) surface structure which is stable at high temperatures.

It is difficult to explain the appearance of large surface unit cells which are also rotated with respect to the bulk unit cell without invoking significant chemical rearrangements in the surface layer. The rotated ($\sqrt{31} \times \sqrt{31}$) unit mesh signifies marked mismatch between the newly formed surface structure and the underlying hexagonal substrate.

It appears that if the high-temperature oxygen-deficient rotated ($\sqrt{31} \times \sqrt{31}$) surface structure has a composition which corresponds to Al_2O (or AlO) it is likely to form a cubic overlayer in which the cation is appreciably larger than in the underlying hexagonal (0001) substrate. Strong mismatch due to the differences in structure and ion sizes in the two layers should be expected.

One can generate the rotated ($\sqrt{31} \times \sqrt{31}$) surface structure by placing a cubic overlayer in which the interatomic distance has been increased to adjust for the increased cation radius on top of the (0001) substrate. There are several cubic structures which can generate the rotated ($\sqrt{31} \times \sqrt{31}$) unit mesh by coincidence with (0001) substrate. One of these surface structures is given in Figure 30.

If the reduced oxides of aluminum, Al_2O or AlO, are stable in the α-alumina surface at elevated temperatures, it is likely that the other group III oxides of the M_2O type might also be stable in the surface environment.

Investigation of the surface structures of Ga_2O_3 and In_2O_3 would be of interest. It is also likely that oxides of other metals (MgO or BaO, for example) may have unusual oxidation states which are stabilized in the surface environment. It should be noted that vanadium pentoxide, V_2O_5, has been reported recently[98] to undergo a change of surface composition (accompanied by loss of oxygen) upon heating in vacuum with a corresponding order-order transformation of its surface structure.

C. The Structure of Vaporizing Surfaces

Experiments in which the evaporation rate of metals into vacuum was measured as a function of temperature indicated that the vacuum vaporization rates are equal to the ideal maximum rates of vaporization, and independent of crystallographic orientation. It is apparent from these results that every atom on the surface has equal probability of vaporizing. This is surprising since the surface is heterogeneous. There are several surface sites in which atoms have different numbers of nearest neighbors and which are distinguishable by their different binding energies. It is therefore not to be expected that all these surface sites can participate equally in the vaporization process. In order to explain such a high vacuum vaporization rate it was proposed that the vaporizing surface may be liquidlike. The concentration of disordered atoms was proposed to equal the total number of surface atoms and these atoms, having high surface mobility, can wander around on the surface and vaporize when sufficient energy is imparted to them. In order to study the structure of the vaporizing surfaces, the diffraction pattern of several metal surfaces (silver, chromium, and nickel) have been monitored while these surfaces vaporized into vacuum.[60] These metals were heated to a temperature at which the vapor flux away from the surfaces was appreciable, of the order of 100–1000 Å/sec. It was found that during vaporization the surface remains ordered and it is characterized by sharp diffraction features. Electron microscopic pictures of similar vaporizing surfaces indicate a large degree of heterogeneity and extreme roughness on a scale of about 10,000 Å to 200,000 Å. Nevertheless, on an atomic scale the surface appeared to be ordered during vaporization. However, the low-energy electron diffraction pattern is not particularly sensitive to the presence of disordered surface atoms up to concentrations of 8% of the total number of surface atoms. Therefore it is possible that such a concentration of disordered atoms might be present on the surface during vaporization. Nevertheless, most of the surface atoms appear to be in their ordered, equilibrium position at the vaporizing surface during the vaporization process.

D. Low-Energy Electron Diffraction Studies of Surface Melting of Lead, Bismuth, and Tin Surfaces

Studies of the mean square displacement of surface atoms by measuring the temperature dependence of the low-energy electron diffraction beam intensity (the surface Debye-Waller factor) have shown that for several monatomic face-centered cubic metals the mean square displacement of atoms in the surface is appreciably larger than the mean square displacement of atoms in the bulk. There is at least one model of melting which indicates that the mean square displacement plays an important role in determining the melting temperature.[99,100] Therefore the results would indicate that the surface may disorder, that it loses its long-range order at temperatures below the bulk melting point. In order to explore the importance of surfaces in the melting process and to investigate whether the surfaces premelt (i.e., melt at a temperature below the bulk melting point), low-energy electron diffraction studies have been carried out to monitor the surface structure up to the melting point and the order-disorder phenomena on the surface at the melting point.[21] The surface structures of the (111), (110), and (100) crystal faces of lead, the (0001) and $(01\bar{1}2)$ faces of bismuth, and the (110) face of tin single crystals were monitored up to the melting temperature and during melting. These metals are particularly suitable for low-energy electron diffraction studies, which must be carried out in ultra-high vacuum, since they have very low vapor pressures—less than 10^{-8} torr at their respective melting points. There are, however, important differences in many physicochemical properties of these materials. They have different crystal structures. Lead and tin, like most solids, expand upon melting. Bismuth, however, undergoes a negative volume change. On melting it contracts. Thus we can study the effect, if any, of these properties on the melting and freezing kinetics. The diffraction spot intensities decreased monotonically according to the temperature dependence predicted by the Debye-Waller factor but were always detectable until the bulk melting point was reached. In Figure 31 the observed diffraction patterns below and above the bulk melting point are given. In every experiment the diffraction pattern remained intact until at the bulk melting point the molten interface reached that region of the surface where the electron beam was focused and the diffraction spots disappeared. In one experiment using a large lead disc, a temperature gradient was introduced along the surface such that melting commenced near one edge of the disk and the melting front proceeded across the surface very slowly. By suitable manipulation of the trimming magnets, the electron beam was focused near the hottest part of the crystal and as the pattern from this

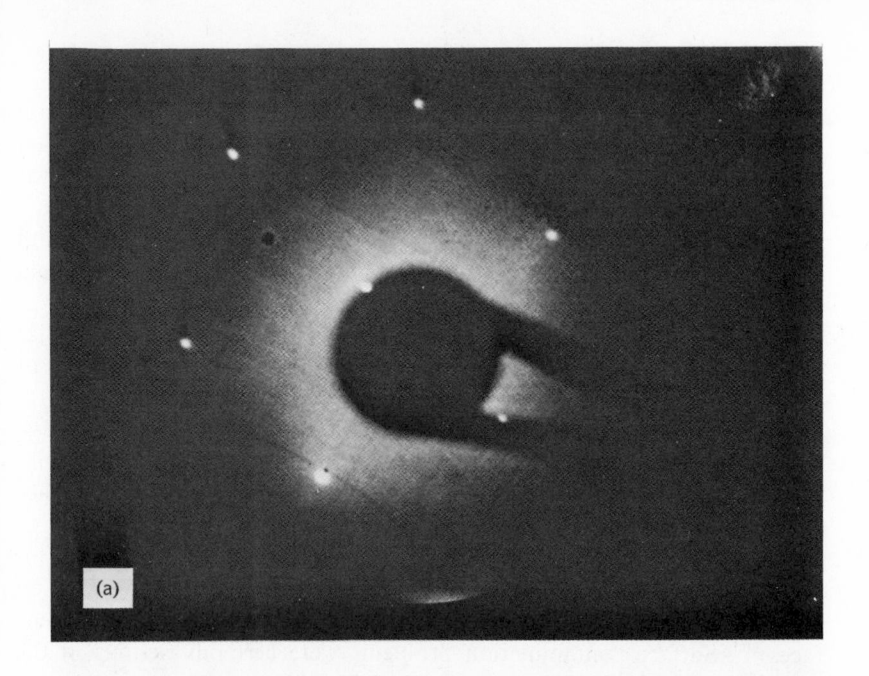

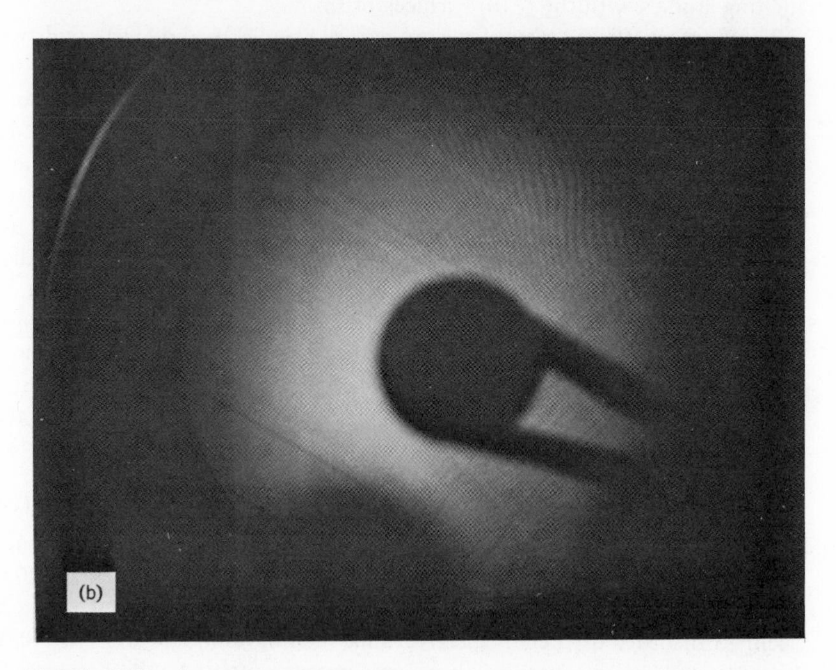

Fig. 31. Diffraction patterns of the (111) face of lead below and above the melting point (327°C).

area disappeared due to melting, the beam was moved to an adjacent, still solid portion; a diffraction pattern was again obtained until that region melted, and so on. In some experiments with bismuth, heating was performed from the bottom. Since the solid is less dense than the liquid the surface solid remained intact and floated on the molten bismuth beneath. As the crystal melted completely the last solid portion would float around on the liquid and the diffraction spot would move correspondingly. The melting of the lead (110) surface was studied with particular care since it is the lowest density and the highest free energy surface of the three lead crystal surfaces studied. In fact, once melted, the (110) orientation never appeared on the recrystallized lead samples. Nevertheless, the (110) surface proved to be ordered and stable to the bulk melting point of lead just observed with the (111) and (100) crystal faces. The surface melting experiment with the tin (110) surfaces were more difficult to perform. In every case a surface structure had formed on this face. This structure agrees with the (3×1) surface structure reported by Jackson and Hooker for slow epitaxial deposition of tin on niobium (110) surfaces.[101] Surface contamination problems were certainly serious in the melting studies with the (110) surfaces of tin.

In summary, three crystal faces of lead, (111), (100), and (110), and two different crystal faces of bismuth, (0001) and $(01\bar{1}2)$, were studied and showed no premelting; they remained stable to the bulk melting point and they melted spontaneously at that point. Contamination of lead and bismuth surfaces could be completely avoided. Formation of a surface structure and contamination problems make the melting studies with tin (110) surfaces difficult to perform. The experimental criterion used to ascertain melting was the loss of the diffraction features that is, the disappearance of the diffraction spots which are due to long-range order in the crystal surfaces (Fig. 31). If the surface remains ordered, the diffraction spots should be visible up to the temperature at which the loss of long-range order occurs. It should be noted again, however, that the concentration of disordered surface atoms could be as high as 5–10% of the total surface concentration before there is an experimentally detectable decrease in the LEED spot intensities.

Low-energy electron diffraction studies of the melting of low-index, lead, bismuth, and tin single-crystal surfaces, in which the disappearance of the diffraction pattern characteristic of long-range order was taken as a sign of melting, indicated no surface premelting. The different surfaces seem to disorder at their respective bulk melting temperature. Although bismuth undergoes negative volume change upon melting and has a crystal structure different from that of lead, the melting behavior of its surfaces are similar to that of lead surfaces. The low-energy electron diffraction pattern is insensitive to the presence of disordered atoms on the surfaces as long as their concentration is only a few percent of the total surface concentration. Thus the presence of a LEED pattern from the different surfaces, which suggests a dominance of long-range order on the surface up to the bulk melting point, does not rule out the presence of disordered atoms in a few atom percent surface concentration.

There are several additional experimental observations accumulated in recent years which give us indications of the mechanism of melting. Turnbull et al. showed that bulk quartz and P_2O_5 crystals could be superheated by 300°C and 50°C, respectively, due to the slow propagation of the viscous molten interface into the solids.[102] Melting was found to nucleate always heterogeneously at emerging dislocations or imperfections and then move into the bulk. In order to avoid nucleating the melt at the surface, Kass and Magun[103] heated the inside of an ice single crystal while keeping the surface below the bulk melting point; in this way they were successful in observing superheating. Similar results were obtained by other investigators using gallium crystals.[104] Several experiments show that in the presence of small temperature gradients, the melting rate varies along different crystallographic orientations. These observations indicate that melting has to be nucleated and that the crystal surfaces appear to provide nucleation centers most efficiently. Thus when melting occurs in the presence of a surface, a condition almost always met in melting experiments, superheating cannot be observed due to the large concentration of surface nucleation sites. Although most of the surface remains ordered up to the bulk melting point, it is likely that the nucleation sites are already present before melting commences. As soon as the liquid phase becomes thermodynamically stable the solid-melt interface may propagate along the surface or into the bulk from these nucleation centers equally well. A melting theory, to be successful, should explain the kinetic, thermodynamic, and statistical properties of the melting phase transformation; these are as follows: (a) low-index surfaces of single monatomic solids remain chiefly ordered up to the bulk melting point; (b) superheating of solids occurs only in the absence of nucleation sites or because of the slow propagation of the melt interface; (c) nucleation of melting occurs most easily at the surface and the melt propagates into the bulk from these selected surface nucleation sites; (d) X-ray, neutron, and high-energy electron diffraction experiments indicate that melting occurs with the loss of long-range order; and (e) melting is a first-order phase transition with well-defined thermodynamic parameters. So far none of the melting models proposed have been able to account for all these properties. It is to be hoped that in the near future a judicious synthesis of the favorable features of some of these proposed models, which will be briefly numerated below, will produce a melting model that allows quantitative prediction of the melting characteristics of different solids.

There are several melting models which explain the kinetic properties of melting demonstrated by recent experiments. Turnbull has proposed a melting model which allows the computation of the propagation velocity of the molten interface in the temperature gradient provided by superheating.[105] Agreement between that theory and experiments could be reached only if it were assumed that melting occurs only at some fraction of the surface sites at the solid melt interface. What is the nature of those surface sites where melting may be nucleated? None of the experimental melting studies so far have been able to identify these centers. They may be vacancies and vacancy aggregates or disordered regions around dislocations which emerge at the surface. Stark[106] has proposed that the vacancy concentration of the surface builds up faster than in the bulk. When a critical concentration of vacancies is reached melting is nucleated at the surface. Stranski has viewed melting as the dissolution of a solid in its own melt.[107] He has observed that certain crystal surfaces (high index) are wetted by their own melt while other faces (low index) remain stable and are not wetted by their own melt even at temperatures very near the melting point. Thus, Stranski postulates that melting is initiated on high-index surfaces, the low-index crystal faces being stable at all temperatures up to the melting point. These melting models recognize the importance of surfaces in nucleating melting. There are several other melting models which describe either the thermodynamic or the statistical properties of melting without consideration of the importance of the crystal surface in nucleating and initiating melting. Perhaps the most notable and successful is that proposed by Lennard-Jones and Devonshire.[108] They adopted the Bragg-Williams model of one-dimensional order-disorder transition in which the first-order transition is generated with the help of a disorder parameter. Born[109] has considered the major difference between the solid and liquid phases to be the lack of resistance of liquids to low-frequency shearing stresses. Using the elastic continuum model he predicts that as melting commences the shear modulus of the crystal, C_{44}, vanishes. Sound velocity measurements in different crystals, however, did not bear out this prediction and the model was later retracted. Kuhlmann-Wilsdorf[110] has proposed a model in which free energy of formation of a dislocation is taken as positive in solids and negative for liquids.

The melting temperature is postulated to be the temperature at which the free energy of formation is zero. The liquids are thus treated as infinitely dislocated solids.

E. Studies of Freezing of Molten Lead and Bismuth by LEED

These investigations were carried out to discover the experimental parameters which influence the surface structure of recrystallized metals and their kinetics of freezing and growth.[21,60] Studies of the surface structure of metal crystals during solidification should provide a great deal of information on the mechanism of crystal growth from the melt. The molten lead and bismuth samples were cooled using cooling rates in the range of 2–0.02°C/sec. It was found that during freezing more than one crystallite formed. These crystallites were nucleated at the holder walls, as expected. Although the size of these crystallites varied, most of them were large enough to show sharp diffraction features, allowing us to monitor their orientation and surface structure. Their orientation was checked by locating the specular (00) spot for each prominent crystallite. In discussing the effect of cooling rates on surface orientation one may take the cooling rate of 0.5°C/sec as the dividing line between rapid and slow freezing rates. Rapid freezing rates (larger than 0.5°C/sec favored the growth of the (100) surfaces of lead, while slow cooling rates (less than 0.5°C/sec) favored the formation of the lead (111) surfaces. For bismuth surfaces, the opposite results were obtained; rapid cooling rates favored the appearance of crystallites oriented with the (0001) or hexagonal axis perpendicular to the crystal surface, and slow freezing rates favored crystallites with the (0112) axis [which is a pseudo-cubic (100) axis] oriented perpendicular to the crystal surfaces. Undercooling of liquid lead of the order of 8° was frequently observed during studies of the recrystallization of lead. However, bismuth did not show undercooling in any of the crystal growth experiments.

It was found that slow freezing rates yield a dominantly (111) surface orientation for lead and the $(01\bar{1}2)$ orientation which is the pseudo-cubic (100) for bismuth crystallites. Conversely, rapid cooling rates produce the (100) orientation for lead and the (0001) or pseudo-cubic (111) surface for bismuth. We might argue that near equilibrium lead, which has to contract upon freezing, should prefer to build its lattice from surfaces which show the densest packing of atoms [(111) face]. Bismuth, which expands upon freezing, should prefer a more open surface which still has low-surface free energy. The result that growth conditions far from equilibrium (fast cooling rates) produce opposite surface orientations of the two solids should have to be taken into account in future theoretical studies of crystal growth kinetics.

F. LEED Studies of Magnetic Surface Structures

In several magnetic transitions the magnetic structures which form are characterized by a unit cell which is not the same as the atomic unit cell. For example, nickel oxide is antiferromagnetic; it has a transition temperature of 525°C. Along the (100) surface it should have a (2 × 1) magnetic unit cell. It was found that upon heating the surface near the Neèl temperature, new diffraction spots appeared which correspond to the appearance of a new surface periodicity.[111] This could be associated with antiferromagnetic ordering in the crystal. Thus it appears that magnetic ordering, can in addition to neutron diffraction, also be studied by low-energy electron diffraction. Some of these studies have been carried out, in addition to nickel oxide surfaces, on chromium surfaces as well.[112] One of the difficulties in these experiments is that the surface Debye-Waller factor, being large, decreases the intensity of the diffraction spots due to magnetic ordering near the magnetic transition temperature, and one has to use low electron beam energies and well-ordered surfaces to be able to detect the onset of magnetic ordering. Nevertheless, low-energy electron diffraction studies promise to be an important tool in the studies of magnetic structure and magnetic phase transitions at surfaces.

X. LEED STUDIES OF THE ADSORPTION OF GASES ON SINGLE-CRYSTAL SURFACES

Adsorption studies may be divided arbitrarily into two classes, physical adsorption and chemisorption. Physical adsorption involves gases which have heats of adsorption less than about 10 kcal/mole. Chemisorption, however, indicates strong, mostly electrostatic interactions between the adsorbed gas and the solid surface with heats of adsorption in excess of 15 kcal/mole.

A. Physical Adsorption

Due to the low heats of adsorption of these gases, physical adsorption studies have to be carried out at low temperatures (below room temperature). Only a few systems have been studied so far. Low-energy electron diffraction studies of the physical adsorption of xenon and bromine on graphite single crystal surfaces showed that well-defined surface structures may form at low temperatures.[113] These studies seem to provide the first evidence that physical adsorption takes place via the formation of ordered surface structures even for noble gas adsorbates, for which the bonding between adsorbed atoms is weak. The two-dimensional condensation of adsorbed bromine to the liquid state could also be monitored by low-energy electron diffraction. Physical adsorption studies of several gases on silver single-crystal surfaces, however, indicate that adsorption takes place without the formation of ordered surface structures.[114] These studies have been carried out in combination with ellipsometry measurements. The adsorption isotherms of several gases have been measured and the heats of adsorption as a function of surface coverage have been computed. It was found that the adsorbate-adsorbate interaction was just as strong at low coverages as the interaction between the adsorbate and the metal surface. Consequently, most of the adsorbed atoms are likely to be situated in patches on the silver surface. At present there are only a few experimental results, making it somewhat difficult to assess the role of ordering in physical adsorption processes. It appears that larger heats of adsorption are necessary to localize the adsorbed atoms on the surface. Otherwise, surface diffusion (which might require only low activation energies) will tend to disorder the adsorbed surface layer. The formation of ordered surface structures should be favored by decreasing temperature and by the adsorption of gases with increasing heats of adsorption.

B. Chemisorption

Chemisorption has been studied extensively in several laboratories in the past few years primarily on monatomic and diatomic solids of different crystal structure. In Table IV we have summarized the experimental information available by listing the solid surfaces which were studied, the adsorbed gas, and the surface structures which were found under a variety

TABLE IV

Structures of Adsorbed Gases

Surface	Adsorbed gas	Surface structure	Refs.
		Face-Centered Cubic Structures	
Ni(111)	O_2	(2 × 2)-0	117, 136, 137
		($\sqrt{3} \times \sqrt{3}R30°$-0	117, 138
	Co	(2 × 2)-CO	136
Dissociates to $C + CO_2$		$(16\sqrt{3} \times 16\sqrt{3})R30°$-C $+(2 \times \sqrt{3})$-CO_2	138
	H_2	(1 × 1)-H	137
		Disordered	139
	N_2	Not adsorbed	171
	CO_2	(2 × 2)-CO_2	138
		$(2 \times \sqrt{3})$-CO_2	

TABLE IV—*Continued*

Surface	Adsorbed gas	Surface structure	Refs.
Face-Centred Cubic Structures			
	Dissociates in electron beam to	$(16\sqrt{3} \times 16\sqrt{3})$-C + $(\sqrt{3} \times \sqrt{3})$R30°-0	138
	C_2H_2	(2×2)-C_2H_4	116
	C_2H_6	(2×2)-C_2H_6	116
	C_3H_6	(2×1)-C_3H_6	116
		$(\sqrt{7} \times \sqrt{7})$R19°-C	116
Ni(100)	O_2	(2×2)-O	117, 141, 142, 143
		$c(2 \times 2)$-O	144, 145, 146, 147
			117, 148, 149, 150
	CO	$c(2 \times 2)$-CO	147, 148, 151
	H_2	Disordered	139
	N_2	Not adsorbed	139
	C		142, 145, 152
Ni(110)	O_2	(2×1)-0	136, 117, 150, 153, 154
		(3×1)-O	117, 139, 143, 155
		(5×2)-0	117
		(5×1)-0	117
	CO	(1×1)-CO	117, 156
	H_2, D_2	(1×2)-H	139, 156, 157, 158
	H_2O	(2×1)-H_2O	157
	C	$c(2 \times 2)$-C	157
Ni(210)	I_2	Facet to Ni (540)	159
Pt(111)	O_2	(2×2)-O	131, 160, 161, 162
	$H_2 + O_2$	$(\sqrt{3} \times \sqrt{3})$R30°	161
	CO	$c(4 \times 2)$-CO	92
	C_2H_2	(2×1)-C_2H_2	92
	C_2H_4	(2×1)-C_2H_4	92
	C_3H_6	(2×1)-C_3H_6	92
	C_4H_8 (2-butene)	(2×2)-C_4H_8	92
	C_4H_6 (butadiene) (*cis* and *trans*)	(2×2)-C_4H_6	92
	C_4H_8 (isobutylene)	$(\sqrt{7} \times \sqrt{3})$R13.9°	92
Pt(100) —(5×1)	O_2	(1×1)-O	160
	H_2	(2×2)-H	88, 163
	CO	$c(4 \times 2)$	88, 92, 162, 164
		$(3\sqrt{2} \times \sqrt{2})$R45°	
		$(\sqrt{2} \times \sqrt{5})$R45°	88, 163
	C_2H_2	$c(2 \times 2)$-C_2H_2	88, 92
	C_2H_4	$c(2 \times 2)$-C_2H_4	88, 92
	C_3H_6	Disordered	92

TABLE IV—*Continued*

Surface	Adsorbed gas	Surface Structures	Refs.
		Face-Centered Cubic Structures	
	C_4H_8 (2-butene) (*cis* and *trans*)	Disordered	92
	C_4H_8 (isobutylene)	Disordered	92
	C_4H_6 (butediene)	Disordered	92
	$CO + H_2$	$c(2\times 2)$-$(CO + H_2)$	88, 163
Pt(110)	O_2	(1×2)	161
		(2×4)	161
Pd(100)	CO	Disordered	165
		$c(4\times 2)$-CO	165
		Compressed	165
Cu(111)	O_2	$(11\times 5)R5°$-O	166
Cu(100)	O_2	$c(2\times 2)$-O	167
		(1×1)-O	167
		(2×1)-O	167
	N_2	(1×1)-N	141
Cu(110)	O_2	(2×1)-O	166, 167
		$c(6\times 2)$-O	166
Cu(035)	O_2	(1×1)-O	167
Cu(014)	O_2	(1×1)-O	167
Al(100)	O_2	Disordered	120, 121, 129
Rh(100)	O_2	(2×8)-O	168
	CO	(4×1)-CO	168
Rh(110)	O_2	$c(2\times 4)$-O	169, 170
		$c(2\times 6)$-O	
		$c(2\times 8)$-O	
		(2×2)-O	
		(2×3)-O	
$UO_2(111)$	O_2	(3×3)-O	140
		$(2\sqrt{3}\times 2\sqrt{3})R30°$-O	140
		Body-Centred Cubic Structures	
W(110)	O_2	(2×1)-O	150, 172, 173
		$c(14\times 7)$-O	
		$c(21\times 7)$-0	173
		$c(48\times 16)$-0	
		$c(2\times 2)$-0	173
		(2×2)–0	
		(1×1)-0	173
	O_2 + CO co-adsorption	$c(11\times 5)$-CO + O_2	119
	CO	Disordered	174
		$c(9\times 5)$-CO	

TABLE IV—*Continued*

Surface	Adsorbed gas	Surface structure	Refs.
Body-Centered Cubic Structures			
	CH_4	(15 × 3)-3	126
		(15 × 12)-C	
W(111)	O_2	To (211) facets	175
	CH_4	(6 × 6)-C	126
W(211)	O_2	(2 × 1)	175
		(4 × 3)	
		(1 × 2)	
		(1 × n)-O, $n = 1, 2, 3, 4$	176, 177
	NH_3 thermal breakup	$c(4 \times 2)$-NH_2 12% stretch	178
	CO	$c(6 \times 4)$-CO	179
		(2 × 1)-CO	
		$c(4 \times 2$-)CO	
W(100)	O_2	(4 × 1)-O	123
		(2 × 1)-O	
	CO	$c(2 \times 2)$-CO	123, 180
	N_2	$c(2 \times 2)$-N	118, 123
	CH_4	(5 × 1)-C	126
	NH_3	Disordered	124
		$c(2 \times 2)$-NH_2	124
		(1 × 1)-NH_2	124
	($CO + N_2$)	(4 × 1)-($CO + N_2$)	118
	H	$c(2 \times 2)$-H	181, 182
		(2 × 5)-H	182
		(4 × 1)-H	182
Ta(110)	O_2	(3 × 1)-O	183
		(3 × 2)R18° 16′-O	122
		Oxides	
	N_2	Not adsorbed	122
	H_2	(1 × 1)-H	122
	CO	Disordered	122
		Decompn. to $C + CO_2$	
Ta(112)	O_2	(3 × 1)-O	122
		Oxides	
	N_2	Nitride forms epitaxially on (113) planes	122
	H_2	(1 × 1)-H	122
	CO	Disordered	122
		Decompn. to C + CO	122
Nb(110)	O_2	(3 × 1)-O	184
		(3 × 2)R18° 16′-O	
	H_2	(1 × 1)-H	185

TABLE IV—*Continued*

Surface	Adsorbed gas	Surface structure	Refs.
		Body-Centered Cubic Structures	
V(100)	O_2	(1 × 1)-O	17
		(2 × 2)-O	
	H_2	Disordered	17
V(110)	CO	Disordered	184
Cr(100)	O_2	(2 × 2)-O	158
	CO	(2 × 2)-CO	
	N_2	(2 × 2)-N	
α-Fe(110)	O_2	*c*(2 × 2)-O	18, 219
		c(3 × 1)-O	
		c(1 × 5)-O	18
		(2 × 8)-O	18
		FeO(111) (cubic)	219
		γ-Fe_2O_3 (spinel)	219
Mo(110)	O_2	(2 × 2)-O	186, 187, 188
		(2 × 1)-O	
		(1 × 1)-O	187, 188
		c(2 × 2)-O	188
	CO	(1 × 1)-CO	186, 188
		c(2 × 2)-CO	156
	H_2	Adsorbed (no structure given)	186
	CO_2	Disordered	156
Mo(100)	H_2	*c*(4 × 2)-H	189
		(1 × 1)-H	
	O_2		
		Disordered	190
		c(2 × 2)-O	187, 188, 190
		(1 × 1)-O	190
		$\sqrt{5}$(1 × 1)R ± 26° 34′-O	188, 190
		(2 × 2)-O	190
	N_2	(1 × 1)-N	188
	CO	(1 × 1)-CO	188
		Diamond Structures	
Si(111)-			
(1 × 1)	O_2	(1 × 1)	191, 192
(7 × 7)		Disordered	
	H_2S	(2 × 2)-S	193
	H_2	Not adsorbed	
	H_2Se	(2 × 2)-Se	193

TABLE IV—*Continued*

Surface	Adsorbed gas	Surface structure	Refs.
	Diamond Structures		
	I_2	(1 × 1)	
	NH_3	(8 × 8)-N	194
	PH_3	$(6\sqrt{3} \times 6\sqrt{3})$-P	195
		(1 × 1)-P	195
		$(2\sqrt{3} \times 2\sqrt{3})$-P	195
Si(100)	O_2	(1 × 1)	191, 192, 196, 197
		(111) facets	
	I_2	(3 × 3)	193
C			
C(diamond)			
(111)	O_2	Ordered	199
	CO_2		
C(graphite)			
(0001)	O_2	Not adsorbed	200
	CO	Not adsorbed	
	H_2O	Not adsorbed	
	I_2	Not adsorbed	
	Br_2	Not adsorbed	
C(diamond)			
(100)	O_2	Disordered	199
		Ordered	
Ge(111)	O_2	(1 × 1)	191, 197
		Disordered	201
	I	(1 × 1)	201
Ge(100)	O_2	(1 × 1)	191, 197
		Disordered	
	I_2	(3 × 3)	201
Ge(110)	O_2	(1 × 1)	191, 197
		Disordered	
	Hexagonal Structures		
Ti(0001)	O_2	(1 × 1)	197
	CO	(1 × 1)	197
Re(0001)	O_2	(2 × 2)-O(CO)	202
		(1 × 1)-O(CO)	
	CO	(2 × 2)-CO	202
Be(0001)	O_2	Disordered	203
	CO	Disordered	203
	N_2	Not adsorbed	203
	H_2	Not adsorbed	203
CdS(0001)	O_2	Disordered	204

of experimental conditions of surface density of adsorbed atoms and of surface temperature. It is apparent from the available experimental data that ordering of adsorbed atoms into surface structures of different kind is an essential part of the chemisorption process. Although in some cases disordered adsorbed structures have been detected, in most experimental situations ordering is preferred over disordered adsorption. It is possible to classify the different chemisorbed structures into a few well-defined types which are already apparent from the experimental data.

1. *Chemisorption "on Top"*

Gases may chemisorb on the surface and arrange themselves in different surface structures. The arrangement of atoms depends on the crystallographic orientation of the substrate, the atomic density of the adsorbate, and the particular temperature at which the experiment is taking place. By adsorption "on top" we mean that the reactants will adsorb on the surface, or if they dissociate, the products of dissociation will stay on the surface and will not subsequently diffuse into the bulk to participate in bulk chemical processes. The structure which forms is a two-dimensional arrangement in which the participating atoms are those of the adsorbed gas, and it does not include substrate atoms to any large extent. The adsorption of olefins on platinum surfaces provides a good example for this type of chemisorption. Ethylene, propylene, butenes, and butadiene can adsorb on both (100) and (111) faces of platinum single crystals. While these gases form ordered structures [(2×2) surface structures] on the (111) crystal face, they appear to adsorb in a disordered manner on the (100) face of platinum (with the exception of C_2H_4 and C_2H_2).[115] One common characteristic of adsorbed gases is clearly apparent from these studies. The adsorbed atoms or molecules, if available in sufficiently large concentration, seem to form structures that give rise to the highest possible surface coverages. If ordered structures form, this means that the unit cell of the surface structure is as small as the closest packing of the molecules allows. Adsorbed atoms of all types may form ordered structures provided that they are far apart. In most cases, however, disordered adsorption is preferred over the formation of large unit cell surface structures. The underlying symmetry and size of the substrate unit cell to a large extent determines the structures of the adsorbed gas layer. In the case of large polyatomic molecules on the surface the substrate structure determines whether they can order on the surface or not. There can be several side reactions which may interfere with chemisorption. Often the adsorbed molecules in the two-dimensional surface structures may undergo chemical decomposition. In the case of olefin adsorption, it was found that upon heating the surface to temperatures in excess of 200°C cracking of the molecules may occur. The deposition of carbon could be monitored from the appearance of new diffraction features. Careful cleaning of the surface prior to adsorption studies must always be carried out. Adsorption of olefins on graphitic surfaces which may cover platinum is different than on pure platinum surfaces. Adsorption of saturated hydrocarbons (ethane for example) have been studied successfully on nickel surfaces.[116] It was found that ordered structures form, and again the structures have the smallest possible unit cell, indicating close packing of these organic molecules on the surface. The adsorption of saturated hydrocarbons on platinum surfaces, however, could not be studied because of the competition for adsorption sites on the surfaces between carbon monoxide, which is one of the major constituents of the ambient, and the hydrocarbon molecules. The carbon monoxide, adsorbing preferentially, has prevented the study of the adsorption of saturated hydrocarbons at the low pressures used in most of the low-energy electron diffraction experiments.

Most of the adsorption studies used ambient pressures between 10^{-9} and 10^{-4} torr. Above these pressures the vacuum pumps may not be able to effectively remove the gases which were introduced into the system. Due to the low pressures which were employed in LEED experiments some of the results of low-energy electron diffraction studies may not be directly correlated with studies of surface adsorption or surface reactions which were carried out at high pressures. When adsorption or reaction of diatomic molecules is accompanied by the dissociation of the molecule, the pressure-dependent dissociation might prevent certain chemical reactions or adsorption from occurring, though these processes are clearly detectable at high pressures. In the future, considerable effort should be made to establish a pressure region where the low- and high-pressure studies would overlap so one could extrapolate with confidence the results of low-pressure low-energy electron diffraction experiments to high pressures as well. Low-energy electron diffraction studies should be extended to as high pressures as experimentally feasible.

2. *Reconstruction*

It has been reported from several studies that a strongly exothermic surface reaction, such as the chemisorption of oxygen on nickel or on other metal surfaces, can dislodge the substrate atoms from their equilibrium positions and cause rearrangement of the surface structure which is commonly called reconstruction.[25,117] The reconstructed surface structures is composed of both metal and chemisorbed atoms in periodic arrays. Although changes in the diffraction pattern during chemisorption can be analyzed in several different ways, complementary experimental evidence seems to indicate that reconstruction is the most likely interpretation of the structural changes observed during the oxidation of many metal surfaces. Reconstruction of the surface may be looked upon as a precursor for oxidation reactions or other chemical reactions which proceed into the bulk (i.e., carbide formation via carbon diffusion or nitridation via nitrogen diffusion into the bulk via a diffusion-controlled mechanism). Since reconstruction displaces and rearranges metal atoms on the surface, these structures may be stable to much higher temperatures than two-dimensional surface structures which are solely due to adsorbed gases. The type of surface structure which forms depends on the structure of the substrate and on the surface density of adsorbed atoms. For example, during the initial stages of chemisorption of oxygen on the nickel (110) surface, (2×1) and (3×1) surface structures are formed.[117] Heating these surface structures in vacuum causes their disappearance, indicating that diffusion of oxygen from these surface structures into the bulk has occurred. Further oxygen dosing of surfaces at high temperature re-forms these surface structures, which appear to be surface intermediates during the dissolution of oxygen in the bulk nickel lattice. The dissolution of oxygen via the oxygen surface structures continues until the solubility limit of oxygen in the metal crystal is reached. At that point the metal oxide may precipitate out as a second phase. The formation of a second phase is accompanied by the appearance of streaking in the surface diffraction patterns and then the gradual appearance of new diffraction features, which can be attributed to the newly formed oxide. Although reconstructed surfaces may persist to higher temperatures than those due to gases adsorbed only on top of the surface, they can often be removed by well-chosen surface chemical reactions. Oxide structures or structures due to chemisorbed oxygen could be removed by heating in hydrogen. Ion bombardment or high-temperature treatment in vacuum, which causes the vaporization of the topmost atomic layers, can also be used to restore the surface to its original unreconstructed state.

Surface reconstruction processes that have been discovered by LEED studies gives us a new view of the mechanism of chemisorption. Reconstructed surfaces may well be the active surface structures in many exothermic catalytic surface reactions.

3. *Co-Adsorbed Structures*

Low-energy electron diffraction studies further showed that the surface structures formed during the simultaneous adsorption (co-adsorption) of two gases would not form in the presence of only one or the other gas component. The formation of these mixed surface structures seems to be a general property of adsorbed gas layers on tungsten surfaces. It was shown that the simultaneous adsorption of nitrogen and carbon monoxide on the (100) surface of tungsten gives a series of surface structures, not all of which can be formed by the individual gases.[118] Similar results were obtained by the co-adsorption of oxygen and carbon monoxide on tungsten (110) faces,[119] or hydrogen and carbon monoxide on the (110) surfaces of platinum.[92] The appearance of such surface structures indicate that within the adsorbed layers there is a strong interaction between the different molecules, which arrange themselves in a mixed structure where both molecules appear to participate in the primitive unit cell. These structures form most frequently when both gases that are being adsorbed have approximately equal probability of adsorption. If one gas adsorbs much more strongly than the other (for example, during the co-adsorption of xenon and carbon monoxide), then the more tenacious species (carbon monoxide) will replace and displace the other species (xenon) adsorbed on the surface. In this case the co-adsorbed structures are unlikely to form. The observation of such co-adsorption phenomena indicates that in many chemical reaction studies it is important how the different reactive gases are introduced into the chemical reaction. When one gas is pre-adsorbed on the surface and the other gas is allowed to react with the adsorbed species, one might find different chemical reaction rates and reaction products, than if the two gases are introduced as a mixture simultaneously onto the surface.

4. *Amorphous Surface Structures*

It has been found during studies of the adsorption of oxygen on some metal surfaces that chemisorption takes place via the formation of a disordered layer. For example, the chemisorption of oxygen on aluminum surfaces takes place in such a manner.[120,121] The adsorption of carbon monoxide on the (100) faces of tantalum is another example of this type of adsorption.[122] With aluminum, when the chemisorbed disordered oxygen layer is heated, oxygen from the aluminum surface diffuses into the bulk and the surface returns to its original clean, ordered metallic state. Further dosing with oxygen at high temperatures increases the concentration of oxygen in the bulk of the metal but the surface structure remains that of clean aluminum. This is in contrast to the behavior by oxygen on nickel or on tungsten surfaces. Once the bulk of the aluminum crystal is saturated with oxygen the surface finally loses its ordered aluminum structure and forms a disordered oxide which can no longer be removed by heat treatment. Under high-temperature treatment, in some cases there is a degree of ordering that may be taking place on the surface. However, the oxide which appears on the surface of aluminum at room temperature is characterized by the lack of ordering. Bedaire, Hoffman, and Smith[121] have observed the partially ordered growth of Al_2O_3 when the disordered oxide on the (100) face of aluminum was heated in oxygen at about 300°C. Although the experimental information presently available is scanty, it appears that those oxide layers that form nonporus resistant surface films form disordered surface structures. The lack of crystallinity of the surface structures may be correlated with the degree of nonporosity of the deposited oxides in future studies. In some cases, heating the adsorbed disordered structure may result in partial or complete ordering. For example, ammonia adsorbs on the (100) surface of tungsten in a disordered manner at room temperature.[123,124] Upon heating to elevated temperatures, a $c(2 \times 2)$ surface structure forms with the evolution of hydrogen, indicating that this structure consists of NH_2 groups adsorbed on the tungsten surface. Upon further heating, the structure is rearranged into a (1×1)–NH_2 surface structure. Carbon monoxide seems to chemisorb at room temperature on several crystal surfaces in disordered manner.[122,125] Heating increases the surface order and aids the formation of ordered surface structures. It seems that the formation of these surface structures requires surface diffusion to occur. Therefore, it is important that in chemisorption studies sufficient attention be given to the thermal history and the thermal treatment carried out after adsorption has taken place.

5. *Three-Dimensional Structures*

We have already discussed that during the chemisorption of gases that may induce exothermic chemical reactions at the surface, reconstruction of the solid surface may occur. This reconstruction may be followed by further chemical reactions that take place in the bulk of the solid. As the surface species diffuse into the bulk the chemical reaction is no longer two dimensional but actually involves the species which are below the surface. In the final stages of oxidation when the second phase (e.g., nickel oxide) is beginning to precipitate, other surface structures may appear which are characteristic of that of the bulk oxide or some mixture of the metal and the oxide structures.[117] Three-dimensional structures also form during the carburization of tungsten.[126] Methane decomposition yields a layer of carbon on tungsten surfaces which subsequently diffuses into the bulk. There are ordered structures at the surface during this process in which the surface unit cells are some integral multiple of the bulk tungsten unit cell. That is, the body-centered cubic tungsten structure appears to be maintained during the carbon diffusion process. The surface structures change from one ordered structure to another during carbon diffusion. Finally a structure indicating the precipitation and formation of tungsten carbide, W_2C, appears at the surface. Although LEED studies give us information about the structure of the surface, or perhaps structures which are a few atomic layers deep at the surface, there is little doubt that these oxide or carbide structures are three dimensional. The condensation of the second phase can conveniently be followed by low-energy electron diffraction due to the streaking of the diffraction pattern by the strain introduced in the phase transformation. Such studies provide us with new information about the formation of bulk phases or bulk phase transformation.

TaO(111) has been observed by Boggio and Farnsworth[127] to grow epitaxially on the (110) face of tantalum when the Ta(110) – (5 × 1) structure was exposed to oxygen at room temperature and higher temperatures. The rate of oxidation was found to increase with temperature, and an activation energy for oxidation of 0.24 eV was obtained. The sixfold symmetry of the oxide diffraction pattern indicated the existence of two types of domains which were rotated through 180° with respect to one another. The $[11\bar{2}]$ direction of the oxide was found to coincide with the $[\bar{1}10]$ direction of the tantalum substrate. There is about an 8% difference in the nearest neighbor spacing between the atoms present in the Ta(110) face and those in the TaO(111) orientation.

Pignocco and Pellisier[18] have studied the epitaxial growth of thin films of iron oxide on a clean Fe(110) surface. When the iron surface was exposed to oxygen at room temperature, several surface structures were formed and then the development of a discrete thin film of FeO(111) was observed. As with TaO(111) on Ta(110), the orientation of the epitaxial film was related to that of the substrate and the hexagonal symmetry of the diffraction pattern indicated that two types of domains were present in the oxide structure.

MacRae has observed the epitaxial growth of NiO when the (100), (110), and (111) faces of nickel were oxidized in 10^{-6} torr of oxygen at about 500°C.[128] In all three cases, the (100) face of the oxide (rock salt structure) was the exposed surface. Particularly on the (110) nickel surface, there were strong indications that the oxide was nucleated at separate sites and that the crystallites then grew until the entire surface was covered. The orientations of the oxide films were related to those of the substrates.

C. Correlation of Properties of Adsorbed Gas Surface Structures

It is apparent from inspection of Table IV that chemisorption yields ordered surface structures of adsorbed gases for most systems that have been studied so far. The structure of adsorbed gases to a very large extent is determined by the symmetry, the unit cell size, and the chemistry of the underlying substrate. It is not surprising that a chemisorbed gas forms the same structures on different solid surfaces which exhibit similar electronic structure, the same crystal structure, and the same surface orientation. In fact, the structural changes which are a function of surface concentration of adsorbed atoms are also similar in many surfaces. Because of the large body of information which has been accumulated in the last several years, several tentative correlations may be established which, if used judiciously, will allow one to predict what types of surface structures might form on different solid surfaces which have not been studied so far by low-energy electron diffraction. It appears that (1) ordering of adsorbed molecules on the surface requires heats of adsorption in excess of 10 RT. The lack of ordering in the few cases where physical adsorption of molecules was studied indicate that heats of adsorption of certain magnitude may be necessary to localize the atoms on the surface. (2) Adsorbed atoms form ordered structures which correspond to their closest packing arrangement on the surface. The chemisorption is exothermic although there might be some activation energy for the adsorption of diatomic molecules. An increase in the surface density of the surface molecules decreases the free energy of the substrate-adsorbate system. Adsorbate-adsorbate interactions may also be attractive until critical packing density is attained. These factors lead to a condition where the adsorbed molecules should prefer a close-packing arrangement on the surface. (3) Two-dimensional ordering is more likely on surfaces of high rotational symmetry. The fact that unsaturated hydrocarbons form ordered surface structures on the (111) face of platinum while adsorbed in a disordered manner on the (100) face of platinum is an indication that the multiplicity of the rotation axis may play an important role in ordering during chemisorption.[115] (4) On surfaces with unequal unit cell vectors (such as the (110) face for the face-centered cubic crystals) chemisorbed gases are likely to form $(n \times m)$ type ordered structures where $n \neq m$. It should be noted that the $(n \times 1)$ type domain surface structures where $n = 2, 3, \ldots$ is frequently observed, presumably because its formation leads to greater packing densities.

D. The Interaction of the Electron Beam with Surfaces

The electron beam used in LEED studies has energies of the order of 5–500 eV. These energies are much larger than the binding energies which hold the adsorbed atoms at the surface or hold the substrate atoms together.[129] Thus it is not unlikely that the electron beam may interact with the substrate or with the adsorbed gas and induce desorption or chemical reactions. Fortunately, the efficiency of the interaction of the electron beam with the surface is very low. In most cases the electron beam desorbs surface atoms with an efficiency of $<10^{-5}$ (i.e., one incident electron out of 10^5 may be effective in desorbing a surface atom). The desorption efficiencies of the electron beam for carbon monoxide, hydrogen, and oxygen on tungsten surfaces have been studied.[130] Carbon monoxide from certain binding states appears to desorb rather rapidly during electron bombardment while oxygen desorbs slowly. Carbon monoxide is rapidly desorbed by the electron beam from other metal surfaces[92,131,132] as well. Ammonia desorbs from the (100) face of tungsten by electron impact.[133] The electron beam appears to excite the adsorbed atoms and the atoms then desorb from this repulsive excited states. De-excitation processes are effective in removing the excitation energy in most cases before desorption can take place. It was found that the electron beam may cause rearrangement of the surface structure into new structures or it may convert atoms adsorbed in one binding state to atoms adsorbed in a different state. These studies have been carried out using tungsten surfaces where the conversion of carbon monoxide and nitrogen from one adsorption state to another was found.[134]

There is one group of materials, the alkali halides, which appear to interact chemically with the electron beam.[135] Electron bombardment seems to dissociate the alkali halide surface and leads to halogen evolution and/or the precipitation of the alkali metal atoms. Heat treatment removes the alkali metal atoms either by vaporization or by diffusion via a vacancy mechanism into the bulk of the crystals. Such an interaction makes intensity measurements on alkali halide surfaces difficult to perform since the surface structure deteriorates as a function of time in a broad temperature range during low-energy electron diffraction studies. In most of the low-energy electron diffraction experiments, the electron beam density is low enough that heating of the surface by the electron beam can be neglected. However, under conditions of electron bombardment heating, where high energies and high electron densities are used, chemical changes can occur.

XI. LEED STUDIES OF THE STRUCTURE OF CONDENSABLE VAPORS (EPITAXY)

Thin films grown on a single-crystal substrate are frequently crystallographically oriented relative to the substrate. This orderly growth is known as epitaxy. The sensitivity of low-energy electron diffraction to ordering makes it an ideal tool for studying the epitaxial development of such films. Many different systems have been studied. They may be arbitrarily categorized as metal on metal, metal on insulator, insulator on insulator, and insulator on metal, depending upon the nature of the substrate material and that of the thin film.

One of the earliest LEED studies of epitaxy was that by Farnsworth of silver on the (100) face of gold.[205] An epitaxial film of silver was grown in register with the (100) face of the gold substrate. The intensities of the diffracted beams from the film were similar to those from bulk silver, indicating that the silver film on the gold had the same structure as the top layers of a pure bulk silver crystal. The mismatch in lattice parameters between gold and silver is less than one percent. More recently, Farnsworth and Haque have studied the carefully controlled growth of a nickel monolayer on copper.[206] Gradman has concluded that this nickel monolayer must be pseudomorphic with the copper substrate.[207] That is, the first nickel layer must be constrained to take on the interatomic spacing of the copper, while subsequent layers will relax back to the nickel interatomic distances. The mismatch between the lattice parameters for nickel and copper is 2.5%.

Several metal-metal systems with larger mismatches have been studied. Taylor has investigated the epitaxial deposition of copper[208] onto a single crystal (110) face of tungsten under ultra-high vacuum conditions. He concluded that evaporation onto a clean tungsten surface at room temperature resulted in partial alloying and then the formation of well-oriented, uniformly thin copper (111) surface. The copper (111) plane is parallel to the W(110) with the copper $(\bar{1}\bar{1}2)$ direction parallel to the $(\bar{1}10)$ tungsten direction. The lattice mismatch in the W(110) direction is 1% while that in the (001) direction is 19%. Thus, the primary diffraction spots for the tungsten and the copper essentially coincide in the W(110) direction but not in the (001) direction, leading to a fairly complicated diffraction pattern. Moss and Blott have also studied the epitaxial growth of copper on a W(110) face.[209] Their observations are similar to those of Taylor. However, their conclusions are slightly different. In their interpretation, alloy formation is not involved; rather they concluded that the first monolayer is deposited in a strained configuration and further deposition leads to the growth of a film with a periodicity characteristic of bulk Cu(111). Heating above 600°K led to the formation of large three-dimensional copper islands on the surface. Further heating to above 1050°K resulted in the evaporation of these islands leaving only the first, strained monolayer of copper.

Taylor has also studied the effect of oxygen on the epitaxial growth of copper on W(110).[208] He found that even half a monolayer of chemisorbed oxygen severely inhibited epitaxy even after the deposition of 20 layers of copper. However, if some physisorbed oxygen was present in addition to the chemisorbed, there was a marked improvement in the epitaxy even though it was still considerably worse than that on clean tungsten.

Alloying has been observed when Nb_3Sn was grown on niobium. Jackson and Hooker have evaporated tin onto clean Nb(110) under a variety of conditions.[101] Amorphous films of tin were deposited at high evaporation rates while slow deposition rates resulted in a diffraction pattern; this was interpreted as being due in part to the presence of Sn(110). Both the ordered and the disordered tin films on niobium produced $NbSn_3$(110) when heated between 500°C and 950°C, depending upon the history of the film. At temperatures below the formation temperature of Nb_3Sn, a hexagonal pattern was observed that was interpreted as resulting from an attempt by the tin to match the substrate. Heating to temperatures above the Nb_3Sn formation temperature regenerated the Nb(110) diffraction pattern.

Pollard and Danforth have studied the deposition of thorium onto the (100) face of a tantalum substrate held at 950°C.[16] At coverages below a monolayer, they observed the formation of $c(2 \times 2)$ surface structure which reverted to a (1×1) upon further coverage. This (1×1) pattern was still observed even after 15 monolayers had been deposited. A careful investigation of the intensity of the specularly reflected beams as a function of coverage, and auxiliary studies with an optical microscope and electron microscope, led them to conclude that thorium clusters into islands about 500 Å high and several thousand angstroms in length, whose principal crystallographic directions are aligned with those of the substrate. It appears possible for thorium to form a monolayer in registry with the tantalum because of the partially ionic nature of the thorium-tantalum bond. However, further thorium cannot go into registry and therefore migrates to nucleation centers where three-dimensional thorium crystals are formed. The Th-Ta(100) system may be contrasted with the Th-W(100) system where, after the formation of the $c(2 \times 2)$ surface structure, a hexagonal structure that is similar to the (111) plane of bulk thorium is formed.[210]

The (110) face of tantalum has also been used as a substrate for the epitaxial growth of aluminum thin films. Jackson, Hooker, and Haas have found that Al(111) forms on the Ta(110) face in two orientations with the proper substrate temperature.[211] The observation of two or more orientations in an epitaxially grown film is quite common. Individual Al(111) and Ta(110) planes have hexagonal symmetry. However, there are two possible ways, for example, of superposing a second Al(111) layer on the first. The resulting ensembles have trigonal symmetry. *A priori*, both orientations are equally probable and the development of an epitaxial growth containing two types of domains frequently results. An Al(111) $c(2 \times 2)$ structure and an Al(100) – $c(2 \times 2)$ structure have also been observed on a Ta(110) substrate. Further, all these aluminum films have been observed to be somewhat unreactive to oxygen and carbon monoxide. This behavior may be contrasted with that of copper on tungsten, where the presence of oxygen inhibited epitaxial growth.

The high degree of order found in these metal-metal systems is not always observed. Gerlach and Rhodin have studied alkali metal adsorption on several single-crystal nickel surfaces.[212] They have found that at coverages less than one monolayer, the alkali metal atoms appear to repel each other with the result that these atoms are uniformly spaced over the substrate surface. The diffraction patterns indicate that there was a definite anisotropy in the adatom distribution on the (110) face but not on the (100) or the (111) faces. At higher coverages, the adatoms form incoherent hexagonal structures, presumably to maximize the packing. The deposition of several monolayers led to the disappearance of the nickel diffraction spots, implying that at least the outer layers of the film are disordered.

Weber and Peria have used LEED to study the alkali metals sodium, potassium, and cesium on the (100) and the (111) faces of silicon and germanium.[213] The diffraction pattern observed for deposition on the (111) faces were not characterized by well-ordered surface structures. However, those for the potassium- and cesium-covered (100) surfaces were characteristic of a well-ordered overlayer. Supplementary measurements of the retarding field characteristics indicated that the alkali atoms were not nucleated in clusters but were uniformly distributed over the surface. As with the nickel substrate, it is possible that the repulsive adatom forces due to the partial ionization of the electropositive alkali metal atoms is more important in determining the epitaxial geometry than the adatom-substrate forces for the (111) germanium and silicon faces.

Jona has investigated the "amorphous" deposition of silicon onto the (111) face of a silicon substrate at low temperatures.[58] However, ordered epitaxial growth can occur at higher temperatures. Joyce, Neave, and Watts have also studied the autoepitaxy of silicon by decomposing a molecular beam of silane on a heated silicon substrate in an ultra-high vacuum system.[214] They found that a fraction of a monolayer of carbon or a carbon compound changes the growth mechanism from an apparent step movement process to one of discrete three-dimensional nucleation. The amount of impurity involved was too small to effect the Si(111)–(7×7) LEED pattern, but could be detected by Auger spectroscopy.

Silicon has been a very popular substrate material for many studies of metal-insulator systems. Among others, aluminum, lead, tin, calcium, barium, cesium, indium, and gold have been deposited on silicon surfaces and studied with LEED. At least five aluminum phases have been observed on the (111) face of silicon at coverages between $\frac{1}{3}$ and a full monolayer. When aluminum is evaporated onto the β-Si(111)$-\sqrt{3}-$Al structure, a nearly perfectly oriented Al(111) epitaxial film is formed after the deposition of about 5–10 monolayers.[215] The mismatch in the unit mesh is about 25 %. As in many cases, the rotational symmetry has been preserved even when the translational symmetry between the film and the substrate has been discarded. Ultra-high vacuum conditions were necessary for the development of this epitaxial film. When deposited on a Si(111) face, indium also exhibited a complicated set of surface structures that in part resembled those formed by aluminum. However, the development of an epitaxial film was not observed. Estrup and Morrison have studied the deposition of lead and tin on the (111) face of silicon.[216] They found that epitaxial lead films could be grown, but that heating resulted in the clustering of the lead atoms into islands separated by areas of the silicon surface covered with ordered fractional monolayer lead structures. Tin could not be epitaxied, possibly due to clustering or to inhibition by contaminants. The reader is referred to an excellent review by Lander on the usages of silicon as a substrate material.[25]

Epitaxial studies where silicon has been employed as the condensate rather than the substrate have been performed by Chang.[96] Several different faces of α-Al_2O_3 were used as substrates for the deposition of silicon films. The objective of the study was to determine whether any properties of the epitaxial films could be related to the superstructures, or surface structures present on the α-Al_2O_3 faces. An interesting correlation was found for (111) silicon films grown on the α-$Al_2O_3-(\sqrt{31} \times \sqrt{31})$ surface structure. This $\sqrt{31}$ surface structure has double domains rotated by $\pm 9°$ relative to the principal crystallographic directions of α-Al_2O_3. When this substrate was held in a very narrow temperature range near 850°C, the epitaxial Si(111) films also grew in double domains rotated by about $\pm 9°$, indicating that the orientation of the film was determined by that of the surface structure rather than that of the substrate. The sensitivity of this process to substrate temperatures was shown by the observation that silicon films grown on an α-$Al_2O_3-(\sqrt{31} \times \sqrt{31})$ substrate at lower temperatures were indistinguishable from those grown on an α-$Al_2O_3-(1 \times 1)$ substrate and were in register with the [112] silicon direction parallel to the [1010] direction of the substrate. As in other studies, these films had hexagonal symmetry, indicating the existence of considerable twinning. Several other metal-insulator systems that have been studied are silver on mica and silver and gold on potassium chloride. Seah prepared clean substrate by cleaving mica in ultra-high vacuum.[10] The mica was then heated to 300°C and silver was deposited at a rate of several monolayers per minute in an average ambient pressure of less than 1.5×10^{-10} torr. The resulting films showed a high degree of perfection and some evidence of twinning. The epitaxial growth of tellurium on the (111) face of copper has been studied by Andersson, Marklund, and Martinson.[217] As observed in other systems, the growth of the epitaxial film was preceded by several well-characterized surface structures at partial monolayer coverages. A Cu(111)$-(2\sqrt{3} \times 2\sqrt{3})$30°-Te surface structure was observed at room temperature after about a twelfth monolayer coverage. A third monolayer coverage gave a Cu(111)$-(\sqrt{3} \times \sqrt{3})$130°-Te structure when heated to about 300°C. Further deposition of tellurium then resulted in the epitaxial growth of a Te(0001) film.

Another semiconductor, CdSe, has been epitaxied on the $YMnO_3$ (0001) face by Aberdam, Bouchet, and Ducros.[218] The diffraction patterns indicated that the degree of orientation of the films was greatest for low deposition rates, about 3 Å/sec at 210°C. They observed that the CdSe[1010] direction was parallel to the [1120] direction in the $YMnO_3$. Here, the mismatch in lattice parameters is about 17%.

The preceding enumeration of epitaxial systems that have been studied by low-energy electron diffraction is by no means an exhaustive compilation but is designed solely as an illustration of the unique applicability of LEED in studying such systems. The very nature of low-energy electron diffraction, particularly the low penetration depths involved and the sensitivity to order and disorder, makes it an ideal tool for investigating such important questions as whether the condensed film is ordered or amorphous, what the crystallographic orientation of such a film is, and what the relationship is between the film and substrate orientation. Frequently, these questions can be answered simply by the observation of the geometry of the diffraction pattern. In more complicated cases, an analysis of the intensities of the diffraction features may be helpful if and when such analysis can be performed on a routine basis. Auxiliary techniques such as Auger spectroscopy, electron microscopy, conductivity, and work-function measurements are frequently very useful in supplying complementary information, such as that about the presence of impurities and macroscopic structuring, that is not readily extracted from low-energy electron diffraction data.

From the existing studies, a number of generalizations may be made, though it should be borne in mind that these may be frequently violated. It has often been observed that where more than one physically equivalent orientation is possible, the diffraction pattern may have a higher symmetry than that of the film, indicating that twinning has occurred. The orientation of many, if not most, ordered films bears some relation to that of the substrate, and the rotational orientation is usually preserved even when the translational symmetry is violated. The pseudomorphic growth of a film may necessitate a small lattice mismatch between the film and the substrate while clustering may occur when the mismatch is large. The state of the surface, such as the presence of surface structures, the presence of contaminants, and so forth, may frequently affect the nature of the film growth. Substrate temperatures, deposition rates, and ambient pressures also have been shown to be very important in many systems.

References

1. C. Kittel, *Introduction to Solid State Physics*, 2nd ed., John Wiley and Sons, Inc., New York, 1956.
2. E. A. Wood, *J. Appl. Phys.*, **35,** 1306 (1964).
3. R. L. Park and H. H. Madden, Jr., *Surface Sci.*, **11,** 188 (1968).
4. D. S. Boudreaux and V. Heine, *Surface Sci.*, **8,** 426 (1967).
5. P. M. Marcus and D. W. Jepsen, *Phys. Rev. Letters*, **20,** 925 (1968).
6. H. H. Farrell and G. A. Somorjai, *Phys. Revs.* **182**[3], 751 (1969).
7. J. B. Pendry, 1969, to be published.
8. H. Eyring, J. Walter, and G. E. Kemble, *Quantum Chemistry*, John Wiley and Sons, Inc., New York, 1963.
9. R. M. Stern and A. Gervais, *Surface Sci.*, **17,** 273 (1969).
10. M. P. Seah, *Surface Sci.*, **17,** 132 (1969).
11. B. Segall, *Phys. Revs.*, **125,** 109 (1962).
12. J. J. Lander and J. Morrison, *J. Appl. Phys.*, **35**[12], 3593 (1964).
13. L. H. Germer and A. U. MacRae, *Ann. N.Y. Acad. Sci*, **101,** 605 (1963).
14. R. L. Gerlach and T. N. Rhodin, *Surface Sci.*, **8,** 1 (1967).
15. E. G. McRae and C. W. Caldwell, Jr., *Surface Sci.*, **2,** 509 (1964).
16. J. H. Pollard and W. E. Danforth, in *The Structure and Chemistry of Solid Surfaces*, G. A. Somorjai, Ed., John Wiley and Sons, Inc., New York, 1969.
17. K. K. Vijai and P. F. Packman, *J. Chem. Phys.*, **50**[3], 1343 (1969).
18. A. J. Pignocco and G. E. Pellisier, *Surface Sci.*, **7,** 261 (1967).
19. J. M. Ziman, *Principles of the Theory of Solids*, Cambridge University Press, London, 1964.
20. A. O. E. Animalu and V. Heine, *Phil. Mag.*, **12,** 1269 (1965).
21. R. M. Goodman and G. A. Somorjai, *J. Chem. Phys.*, **52**, 6331 (1970); **52**, 6325 (1970).
22. J. J. Lander and J. Morrison, *J. Appl. Phys.*, **34,** 3517 (1963).
23. R. O. Jones and J. A. Strozier, Jr., *Phys. Rev. Letters*, **22**[22], 1186 (1969).
24. R. A. Armstrong, *Can. J. Phys.*, **44,** 1753 (1966).
25. J. J. Lander, in *Advances in Solid State Chemistry*, Vol. II, MacMillan Co., New York, 1965.
26. H. Raether, *Surface Sci.*, **8,** 233 (1967).
27. J. J. Quinn, *Phys. Revs.* **126,** 1453 (1962).
28. C. B. Duke and C. W. Tucker, Jr., *Surface Sci.*, **15,** 231 (1969).

29. T. W. Haas, J. T. Grant, and G. J. Dooley, *Phys. Revs.*, to be published.
30. R. W. James, *The Optical Principles of the Diffraction of X-Rays*, Cornell University Press, Ithaca, New York, 1965.
31. H. B. Lyon, Jr., and G. A. Somorjai, *J. Chem. Phys.*, **46,** 2539 (1967).
32. R. M. Goodman, H. H. Farrell, and G. A. Somorjai, *J. Chem. Phys.*, **48**[3], 1046 (1968).
33. E. R. Jones, J. T. McKinney, and M. B. Webb, *Phys. Revs.*, **151**[2], 476 (1966).
34. A. U. MacRae, *Surface Sci.*, **2,** 522 (1964).
35. A. A. Maradudin and P. A. Flinn, *Phys. Revs.*, **129**[3], 523 (1967).
36. J. T. McKinney, E. R. Jones, and M. B. Webb, *Phys. Revs.*, **160**[3], 523 (1967).
37. Hirabayashi and Takeishi, *Surface Sci.*, **4,** 150 (1966).
38. F. Hoffman and H. P. Smith, *Phys. Rev. Letters.* **19,** 1472 (1967).
39. Y. H. Ohtsuki, *J. Phys. Soc. Japan*, **24,** 5 (1968).
40. H. Bethe, *Ann. Phys.*, **87,** 55 (1928).
41. M. Von Laue, *Phys. Revs.*, **37,** 53 (1941).
42. E. Merzbacher, *Quantum Mechanics*, John Wiley and Sons, Inc., 1961.
43. (a) E. G. McRae, *J. Chem. Phys.*, **45,** 3258 (1968); (b) E. G. McRae, *Surface Sci.*, **8,** 14 (1967); (c) E. G. McRae, *Fundamentals of Gas Surface Interactions*, Academic Press, New York, 1967.
44. K. Kambe, *Z. Naturforsch.*, **22a,** 22 (1967).
45. (a) K. Kambe, *Z. Naturforsch.*, **22a,** 322 (1967); (b) K. Kambe, *Z. Naturforsch.*, **22a,** 422 (1967); (c) K. Kambe, *Z. Naturforsch.*, **23a,** 1280 (1968).
46. J. L. Beeby, *J. Phys., C.* (*Proc. Phys. Soc.*), **1**[2], 82 (1968).
47. C. G. Darwin, *Phil. Mag.*, **27,** 315 (1914).
48. H. Morgenau and G. M. Murphy, *The Mathematics of Physics and Chemistry*, Van Nostrand Co., Inc., Princeton, N.J., 1957.
49. G. Carpart, *Surface Sci.*, **13,** 361 (1969).
50. R. M. Stern, J. J. Perry, and D. S. Boudreaux, *Rev. Mod. Phys.*, **41**[2], 275 (1969).
51. R. M. Goodman, H. H. Farrell, and G. A. Somorjai, *J. Chem. Phys.*, **49,** 692 (1968).
52. E. G. McRae, *Surface Sci.*, **11,** 479 (1968).
53. E. G. McRae, *Surface Sci.*, **11,** 492 (1968).
54. E. Bauer, *Colloque Intern. CNRS 1965 No. 152*, p. 19.
55. E. G. McRae and L. Winkler, *Surface Sci.*, **14,** 407 (1969).
56. G. Gafner, in *The Structure and Chemistry of Solid Surfaces*, G. A. Somorjai, Ed., John Wiley and Sons, Inc., New York, 1969.
57. H. B. Lyon Jr., Ph.D. Dissertation, University of California, Berkeley, 1967.
58. F. Jona, *Surface Sci.*, **8,** 478 (1967).
59. H. D. Heidenreich, *Fundamentals of Transmission Electron Microscopy*, Interscience Publishers, New York, 1961.
60. R. M. Goodman, Ph.D. Dissertation, University of California, Berkeley, 1969.
61. W. P. Ellis, in *Fundamentals of Gas Surface Interactions*, Academic Press, New York, 1967.
62. D. G. Fedak and N. A. Gjostein, *Acta Met.*, **15,** 827 (1967).
63. A. E. Morgan and G. A. Somorjai, *Surface Sci.*, **12,** 405 (1968).
64. A. M. Mattera, R. M. Goodman, and G. A. Somorjai, *Surface Sci.*, **7,** 26 (1967).
65. A. Guinier, *X-Ray Diffraction*, W. H. Freeman and Co., San Francisco, 1963.
66. R. Kaplow, S. L. Strong, and B. C. Averbach, *Phys. Revs.*, **138A,** 1336 (1965).
67. R. Leonhardt, H. Richter, and W. Rossteutscher, *Z. Physik*, **165,** 121 (1961).
68. J. M. Morabito, Jr., R. F. Steiger, and G. A. Somorjai, *Phys. Revs.*, **179,** 638 (1969).

69. D. G. Fedak, J. V. Florio, and W. D. Robertson, in *The Structure and Chemistry of Solid Surfaces*, G. A. Somorjai, Ed., John Wiley and Sons, Inc., New York, 1969.
70. E. J. Scheibner and L. N. Tharp, *Surface Sci.*, **8,** 247 (1967).
71. Ch. A. Wert and R. M. Thompson, *Physics of Solids*, McGraw-Hill, New York, 1964.
72. T. M. French, Ph.D. Dissertation, University of California, Berkeley, 1970.
73. S. Hagstrom, H. B. Lyon, and G. A. Somorjai, *Phys. Rev. Letters*, **15,** 491 (1965).
74. G. A. Somorjai, *J. de Physique*, **31,** C1–139 (1970).
75. P. W. Palmberg and T. N. Rhodin, *Phys. Revs.*, **161,** 586 (1967).
76. J. T. Grant, *Surface Sci.*, **18,** 228 (1969).
77. F. Jona, *Surface Sci.*, **8,** 57 (1967).
78. J. W. May, *Ind. Eng. Chem.*, **57,** 13 (1965).
79. A. U. MacRae and G. W. Gobeli, *J. Appl. Phys.*, **35,** 1629 (1964).
80. B. D. Campbell, G. A. Haque, and H. F. Farnsworth, in *The Structure amd Chemistry of Solid Surfaces*, G. A. Somorjai, Ed., John Wiley and Sons, Inc., New York, 1969, 33–2.
81. S. Andersson, I. Marklund, and D. Anderson, in *The Structure and Chemistry of Solid Surfaces*, G. A. Somorjai, Ed., John Wiley and Sons, Inc., New York, 1969, 72–1.
82. R. E. Schlier and H. E. Farnsworth, *J. Chem. Phys.*, **30,** 917 (1959).
83. J. J. Lander and J. Morrison, *J. Chem. Phys.*, **33,** 729 (1962).
84. N. B. Hannay, *Semiconductors*, Reinhold, New York, 1960.
85. J. J. Burton and G. Jura, in *The Structure and Chemistry of Solid Surfaces*, G. A. Somorjai, Ed., John Wiley and Sons, Inc., New York, 1969.
86. J. J. Burton and G. Jura, *J. Phys. Chem.*, **71,** 1937 (1967).
87. P. W. Palmberg, T. N. Rhodin, and C. J. Todd, *Appl. Phys. Letters*, **10,** 122 (1967).
88. A. E. Morgan and G. A. Somorjai, *Surface Sci.*, **12,** 405 (1968).
89. L. Brewer, in *Electronic Structure and Alloy Chemistry*, P. A. Beck, Ed., Interscience, New York, 1963.
90. L. Brewer, in *High Strength Materials*, V. F. Zackay, Ed., John Wiley and Sons, Inc., New York, 1965.
91. S. L. Altman, C. A. Coulson, and W. Hume-Ruthery, *Proc. Roy. Soc.* (*London*), **A240,** 145 (1957).
92. A. E. Morgan and G. A. Somorjai, *J. Chem. Phys.*, **51,** 3309 (1969).
93. J. M. Charig, *Appl. Phys. Letters*, **10,** 139 (1967).
94. C. C. Chang, *J. Appl. Phys.*, **39,** 5570 (1968).
95. J. M. Charig and D. K. Skinner, in *The Structure and Chemistry of Solid Surfaces*, G. A. Somorjai, Ed., John Wiley and Sons, Inc., New York, 1969.
96. C. C. Chang, *ibid.*
97. T. M. French and G. A. Somorjai, *J. Phys. Chem.*, **74,** 2489 (1970).
98. L. Fiermans and J. Vennik, *Surface Sci.*, **9,** 187 (1968).
99. F. A. Lindemann, *Z. Physik*, **14,** 609 (1910).
100. J. J. Gilvarry, *Phys. Revs.*, **102,** 308 (1956).
101. A. G. Jackson and M. P. Hooker, in *The Structure and Chemistry of Solid Surfaces*, G. A. Somorjai, Ed., John Wiley and Sons, Inc., New York, 1969.
102. R. L. Cornia, J. D. MacKenzie, and D. Turnbull, *J. Appl. Phys.*, **34,** 2239 (1963).
103. M. Kass and S. Magun, *Z. Kristall.*, **116,** 354 (1961).
104. P. R. Pennington, Ph.D. Dissertation, University of California, Berkeley, 1966.
105. W. B. Hillig and D. Turnbull, *J. Chem. Phys.*, **24,** 914 (1956).

106. J. P. Stark, *Acta Met.*, **13,** 1181 (1965).
107. J. N. Stranski, W. Gans, and H. Rau, *Ber. Bunsingessell.*, **67,** 965 (1963).
108. J. E. Lennard-Jones and A. F. Devonshire, *Proc. Roy. Soc. (London)*, **A170,** 464 (1939).
109. M. Born, *J. Chem. Phys.*, **7,** 591 (1939).
110. D. Kuhlmann-Wilsdorf, *Phys. Revs.*, **140**[5A], 1599 (1965).
111. P. W. Palmberg, R. E. DeWames, and L. A. Vredevoe, *Phys. Rev. Letters*, **21,** 682 (1968).
112. R. Kaplan and G. A. Somorjai, to be published.
113. (a) J. J. Lander and J. Morrison, *Surface Sci.*, **6,** 1 (1967); (b) J. J. Lander, *Fundamental of Gas-Surface Interactions*, Academic Press, New York, 1967.
114. R. F. Steiger, J. M. Morabito, Jr., G. A. Somorjai, and R. H. Muller, *Surface Sci.*, **14,** 279 (1969).
115. A. E. Morgan and G. A. Somorjai, *J. Chem. Phys.*, **51,** 3309 (1969).
116. J. C. Bertolini and G. Dalmai-Imelik, *Rept. Inst. de Rech. sur la Catalyse—Villeurbonne*, 1969.
117. A. U. MacRae, *Surface Sci.*, **1,** 319 (1964).
118. P. J. Estrup and J. Anderson, *J. Chem. Phys.*, **46,** 567 (1967).
119. J. W. May, L. H. Germer, and C. C. Chang, *J. Chem. Phys.*, **45,** 2383 (1966).
120. F. Jona, *J. Phys. Chem. Solids*, **28,** 2155 (1967).
121. S. M. Bedair, F. Hoffman, and H. P. Smith, Jr., *J. Appl. Phys.*, **39,** 4026 (1968).
122. T. W. Haas, in *The Structure and Chemistry of Solid Surfaces*, G. A. Somorjai, Ed., John Wiley and Sons, Inc., New York, 1969.
123. P. J. Estrup, *ibid.*
124. P. J. Estrup and J. Anderson, *J. Chem. Phys.*, **49,** 523 (1968).
125. T. W. May and L. H. Germer, *J. Chem. Phys.*, **44,** 2895 (1966).
126. M. Boudart and D. F. Ollis, in *The Structure and Chemistry of Solid Surfaces*, G. A. Somorjai, Ed., John Wiley and Sons, Inc., New York, 1969.
127. J. E. Boggio and H. E. Farnsworth, *Surface Sci.*, **3,** 62 (1964).
128. A. U. MacRae, *Science*, **139,** 379 (1963).
129. H. H. Farrell, Ph.D. Dissertation, University of California, Berkeley, 1969.
130. D. Menzel and R. Gomer, *J. Chem. Phys.*, **41,** 3311 (1964).
131. C. W. Tucker, Jr., *Surface Sci.*, **2,** 516 (1964).
132. R. A. Armstrong, in *The Structure and Chemistry of Solid Surfaces*, G. A. Somorjai, Ed., John Wiley and Sons, Inc., New York, 1969.
133. J. Anderson and P. J. Estrup, *Surface Sci.*, **9,** 463 (1968).
134. J. T. Yates and T. E. Madey, in *The Structure and Chemistry of Surfaces*, G. A. Somorjai, Ed., John Wiley and Sons, Inc., New York, 1969.
135. P. W. Palmberg, C. J. Todd, and T. N. Rhodin, *J. Appl. Phys.*, **39,** 4650 (1968).
136. L. H. Germer, E. J. Scheibner, and C. D. Hartman, *Phil. Mag.*, **5,** 222 (1960).
137. R. L. Park and H. E. Farnsworth, *Appl. Phys. Letters*, **3,** 167 (1963).
138. T. Edmonds and R. C. Pitkethly, *Surface Sci.*, **15,** 137 (1969).
139. J. W. May and L. H. Germer, in *The Structure and Chemistry of Surfaces*, G. A. Somorjai, Ed., John Wiley and Sons, Inc., New York, 1969.
140. W. P. Ellis, *J. Chem. Phys.*, **48,** 5695 (1968).
141. R. E. Schlier and H. E. Farnsworth, *J. Appl. Phys.*, **25,** 1333 (1954).
142. H. E. Farnsworth and J. Tuul, *J. Phys. Chem. Solids*, **9,** 48 (1958).
143. J. W. May and L. H. Germer, *Surface Sci.*, **11,** 443 (1968).
144. H. E. Farnsworth, *Appl. Phys. Letters*, **2,** 199 (1963).

145. R. E. Schlier and H. E. Farnsworth, *Advances in Catalysis*, **9,** 434 (1957).
146. L. H. Germer and C. D. Hartman, *J. Appl. Phys.*, **31,** 2085 (1960).
147. H. E. Farnsworth and H. H. Madden, Jr., *J. Appl. Phys.*, **32,** 1933 (1961).
148. R. L. Park and H. E. Farnsworth, *J. Chem. Phys.*, **43,** 2351 (1965).
149. L. H. Germer, *Advances in Catalysis*, **13,** 191 (1962).
150. L. H. Germer, R. Stern, and A. U. MacRae, in *Metal Surfaces*, ASM, Metals Park, Ohio, 1963, p. 287.
151. M. Onchi and H. E. Farnsworth, *Surface Sci.*, **11,** 203 (1968).
152. H. E. Farnsworth, R. E. Schlier, T. H. George, and R. M. Buerger, *J. Appl. Phys.*, **29,** 1150 (1958).
153. L. H. Germer and A. U. MacRae, *Robert Welch Foundation Research Bull. No. 11*, 1961, p. 5.
154. R. L. Park and H. E. Farnsworth, *J. Chem. Phys.*, **40,** 2354 (1964).
155. L. H. Germer, J. W. May, and R. J. Szostak, *Surface Sci.*, **7,** 430 (1967).
156. A. G. Jackson and M. P. Hooker, *Surface Sci.*, **6,** 297 (1967).
157. L. H. Germer and A. U. MacRae, *Proc. Natl. Acad. Sci. U.S.*, **48,** 997 (1962).
158. C. A. Haque and H. E. Farnsworth, *Surface Sci.*, **1,** 378 (1964).
159. C. W. Tucker, Jr., in *The Structure and Chemistry of Solid Surfaces*, G. A. Somorjai, Ed., John Wiley and Sons, Inc., New York, 1969.
160. C. W. Tucker, Jr., *Appl. Phys. Letters*, **3,** 98 (1963).
161. C. W. Tucker, Jr., *J. Appl. Phys.*, **35,** 1897 (1964).
162. J. M. Charlot and R. Deleight, *Compt. Rend.*, **259,** 2977 (1964).
163. A. E. Morgan and G. A. Somorjai, *Trans. Am. Cryst. Assoc.*, **4,** 59 (1968).
164. C. Burggraf and Sime Mosser, *C.R. Acad. Sci.*, **268,** 1167 (1969).
165. J. C. Tracy and P. W. Palmberg, *J. Chem. Phys.*, **51,** 4852 (1969).
166. G. W. Simmons, D. F. Mitchell, and K. R. Lawless, *Surface Sci.*, **8,** 130 (1967).
167. L. Trepte, C. Menzel-kopp, and E. Menzel, *Surface Sci.*, **8,** 223 (1967).
168. C. W. Tucker, Jr., *J. Appl. Phys.*, **37,** 3013 (1966).
169. C. W. Tucker, Jr., *J. Appl. Phys.*, **38,** 2696 (1967).
170. C. W. Tucker, Jr., *J. Appl. Phys.*, **37,** 4147 (1966).
171. L. H. Germer and A. U. MacRae, *J. Chem. Phys.*, **36,** 1555 (1962).
172. L. H. Germer, *Physics Today*, 19 (July 1964).
173. L. H. Germer and J. W. May, *Surface Sci.*, **4,** 452 (1966).
174. J. W. May and L. H. Germer, *J. Chem. Phys.*, **44,** 2895 (1966).
175. N. J. Taylor, *Surface Sci.*, **2,** 544 (1964).
176. C. C. Chang and L. H. Germer, *Surface Sci.*, **8,** 115 (1967).
177. T. C. Tracy and J. M. Blakely, in *The Structure and Chemistry of Solid Surfaces*, G. A. Somorjai, Ed., John Wiley and Sons, Inc., New York, 1969.
178. J. W. May, R. J. Szostak, and L. H. Germer, *Surface Sci.*, **15,** 37 (1969).
179. C. C. Chang, *J. Electrochem. Soc.*, **115,** 354 (1968).
180. J. Anderson and P. J. Estrup, *J. Chem. Phys.*, **46,** 563 (1967).
181. P. W. Tamm and L. D. Schmidt, *J. Chem. Phys.*, **51,** 5352 (1969).
182. P. J. Estrup and J. Anderson, *J. Chem. Phys.*, **45,** 2254 (1966).
183. J. E. Boggio and H. E. Farnsworth, *Surface Sci.*, **1,** 399 (1964).
184. T. W. Haas, A. G. Jackson, and M. P. Hooker, *J. Chem. Phys.*, **46,** 3025 (1967).
185. H. H. Madden and H. E. Farnsworth, *J. Chem. Phys.*, **34,** 1186 (1961).
186. T. W. Haas and A. G. Jackson, *J. Chem. Phys.*, **44,** 2921 (1966).
187. H. E. Farnsworth and K. Hayek, *Nuovo Cimento Suppl.*, **5,** 2 (1967).
188. K. Hayek and H. E. Farnsworth, *Surface Sci.*, **10,** 429 (1968).

189. G. J. Dooley and T. W. Haas, *J. Chem. Phys.*, **52,** 993 (1970).
190. H. K. A. Kann and S. Feuerstein, *J. Chem. Phys.*, **50,** 3618 (1969).
191. R. E. Schlier and H. E. Farnsworth, *J. Chem. Phys.*, **30,** 917 (1959).
192. J. J. Lander and J. Morrison, *J. Appl. Phys.*, **33,** 2089 (1962).
193. A. J. Van Bommel and F. Meyer, *Surface Sci.*, **6,** 39 (1967).
194. R. Heckingbottom, in *The Structure and Chemistry of Solid Surfaces*, G. A. Somorjai, Ed., John Wiley and Sons, Inc., New York, 1969.
195. A. J. Van Bommel and F. Meyer, *Surface Sci.*, **8,** 381 (1967).
196. L. H. Germer and A. U. MacRae, *J. Appl. Phys.*, **33,** 2923 (1963).
197. H. E. Farnsworth, R. E. Schlier, T. H. George, and R. M. Buerger, *J. Appl. Phys.* **29,** 1150 (1958).
198. J. J. Lander and J. Morrison, *J. Chem. Phys.*, **37,** 729 (1962).
199. J. B. Marsh and H. E. Farnsworth, *Surface Sci.*, **1,** 3 (1964).
200. D. Haneman, *Phys. Revs.*, **119,** 567 (1960).
201. J. J. Lander and J. Morrison, *J. Appl. Phys.*, **34,** 1411 (1963).
202. H. E. Farnsworth and D. M. Zehner, *Surface Sci.*, **17,** 7 (1969).
203. R. O. Adams, in *The Structure and Chemistry of Solid Surfaces*, G. A. Somorjai, Ed., John Wiley and Sons, Inc., New York, 1969.
204. B. D. Campbell, C. A. Haque, and H. E. Farnsworth, *ibid.*
205. H. E. Farnsworth, *Phys. Revs.*, **43,** 900 (1933).
206. H. E. Farnsworth and C. A. Haque, *Surface Sci.*, **4,** 195 (1966).
207. U. Gradmann, *Surface Sci.*, **13**[2], 498 (1969).
208. N. J. Taylor, *Surface Sci.*, **4,** 161 (1966).
209. A. R. L. Moss and B. H. Blott, *Surface Sci.*, **17,** 240 (1969).
210. P. J. Estrup, J. Anderson, and W. E. Danforth, *Surface Sci.*, **4,** 286 (1966).
211. A. G. Jackson, M. P. Hooker, and T. W. Hass, *Surface Sci.*, **10,** 308 (1968).
212. R. L. Gerlach and T. N. Rhodin, *Surface Sci.*, **17,** 32 (1969).
213. R. E. Weber and W. T. Peria, *Surface Sci.*, **14,** 13 (1969).
214. B. A. Joyce, J. H. Neave, and B. E. Watts, *Surface Sci.*, **15,** 1 (1969).
215. J. J. Lander and J. Morrison, *Surface Sci.*, **2,** 553 (1964).
216. P. J. Estrup and J. Morrison, *Surface Sci.*, **2,** 465 (1964).
217. S. Andersson, I. Marklund, and J. Martinson, *Surface Sci.*, **12,** 269 (1968).
218. D. Aberdam, G. Bouchet, and P. Ducros, *Surface Sci.*, **14,** 121 (1969).
219. K. Molière and F. Portele, in *The Structure and Chemistry of Solid Surfaces*, G. A. Somorjai, Ed., John Wiley and Sons, Inc., New York, 1969.

HIGH-RESOLUTION ELECTRONIC SPECTRA OF LARGE POLYATOMIC MOLECULES

IAN G. ROSS

Chemistry Department, School of General Studies, Australian National University, Canberra, Australia

CONTENTS

I. INTRODUCTION

In electronic spectroscopy, as in many other branches of chemical physics, small molecules tend to be studied most heavily by physicists, who have the advanced instrumentation, and large ones by chemists, who have the compounds. This sociological circumstance is reinforced by differences in various physical properties, including the quality of the spectra themselves, which have caused work on the excited states of small molecules and work on large molecules to develop markedly different styles—different to the extent that few spectroscopists habitually operate in both areas.

Lately, however, there has been something of a convergence of techniques. From the large-molecule field, for example, the techniques and theory of crystal spectroscopy are beginning to flow into small-molecule work. And in the contrary direction there has come, to studies of large molecules, an appreciation of the role that high resolution can play. This last is the subject of this review.

By *high resolution* we shall mean, mostly, resolving powers ($d\lambda/\lambda$) in excess of 250,000. Grating spectrographs with this kind of performance are now available commercially at costs comparable with conventional low-resolution recording spectrophotometers. The highest resolution used in published work exceeds 600,000. By *large molecules* we shall mean almost exclusively aromatic hydrocarbons and related compounds, not because of any specialized concern with aromatics per se, but because they are so far the only molecules containing more than about six atoms in which high-resolution studies have made headway. The likely reason for this is the rigidity imposed by ring closure, rather than any special connection with aromaticity.

We should perhaps begin by considering why high-resolution visible-ultraviolet spectroscopy has come to these much-studied compounds so late: the measurements, at least, if not the methods of interpretation, have been technically possible for over half a century. The prime cause seems to be the nature of large-molecule spectra themselves. A spectroscopist looks where he has reason to think he will find. The structure of any well-defined electronic absorption spectrum, of a vapor-phase sample, consists of a coarse vibrational pattern each band of which is elaborated by rotational fine structure. The vibrational structure is readily seen with instruments of modest resolution. It contains much information bearing on the symmetry of the electronic wave functions and, ever since Garforth and Ingold's pioneering demonstration on benzene, the analysis of vibrational structure has been the preferred avenue of attack on the problem of characterizing excited states. Rotational fine structure, however, rapidly becomes harder to see within any family of molecules of increasing size, because individual rotational transitions become crowded to the point that they are more closely spaced than their own Doppler-determined line widths (about 0.02 cm^{-1}). Thus, in a series of homologous compounds the number of rotational lines per unit interval in the spectrum increases by something like the fourth power of the longest molecular dimension, l. The reasoning is simple: the density of packing of rotational levels is proportional to the moments of inertia (i.e., proportional roughly to l^2), and in addition to this the number of levels significantly populated at a given temperature increases with decreasing spacing, introducing a further crowding, again roughly dependent on l^2. Finally, nearly all large molecules are asymmetric tops (i.e., have three unequal principal moments of inertia) and the energy levels of the asymmetric rotor are often thought to lack the simple regularities which, even in a crowded spectrum, might produce recognizable patterns of intensity. As a result of these considerations it is likely that most molecular spectroscopists, had their views been sought, say, fifteen years ago, would have surmised that the application of really high-resolution instruments to the spectra of large molecules would be profitless, that all that could reasonably be expected of any band would be an unresolved envelope, or contour, such as had already been seen well enough at medium resolution. Such contours contained typically little numerical information: commonly a single peak, at best two or three peaks and a recognizable shading. Higher resolution would see no more.

To support this view there was both confirmatory and contrary evidence. If the complexities of asymmetric tops could be avoided by examining a large symmetric top, analyzable structure might be seen. As early as 1942 Turkevich and Fred[1] had published a high-resolution spectrum of a band in the first ultraviolet system of benzene. Obviously, this did contain structure finer than just a broad envelope, but there was no obvious analysis and the spectrum was exhibited with minimal comment. If then, highly symmetric benzene showed so little promise, what could be hoped of other molecules? On the other hand, finer structure had been seen, for example, in each of the three xylenes, by Cooper and Sponer,[2] but the problems of analyzing this structure, requiring massive computation, led them not to pursue the point.

Developments in 1959–1960 set the stage for a change of expectations. The special difficulty about band analysis in electronic spectra, as opposed to the analysis of infrared bands, comes of course from the different moments of inertia in the initial and final states. The greater the difference, the more drastically the bands are distorted away from the fairly well prescribed patterns expected for the infrared bands. In small molecules, moments of inertia can change considerably on electronic excitation: in aromatics there was only the most sketchy guidance about what to expect. Now, however, it has become apparent that the changes of shape and dimensions that accompany electronic excitation in aromatics are in general not drastic, and there is fair understanding of why this is so. This recognition began with the discovery of finely structured rotational band contours in the azines.

The azines are the nitrogen heterocycles derived from benzene; pyridine is the first of the series. In 1959 Mason[3] published a reproduction of the 0-0 band of the visible absorption spectrum of *s*-tetrazine (1,2,4,5-tetra-azabenzene); the band had a strong central peak about 2 cm^{-1} wide, flanked on both sides by several dozen uniformly spaced lines about 0.4 cm^{-1} apart. This structure is immediately reminiscent of the infrared bands of symmetric rotors when the polarization is parallel to the top axis (e.g., such bands in C_2D_6),[4] and was so analyzed. The analysis showed that the polarization of the electronic transition was perpendicular to the molecular plane, as required theoretically for an allowed $\pi^* \leftarrow n$ transition but not previously demonstrated, and provided first estimates (since considerably revised on the basis of other bands in the same spectrum[5]) of the change in moment of inertia that accompanies excitation. Almost simultaneously, Innes and co-workers[6,7] reported similar finely structured bands in the first singlet ← singlet transitions of the three isomeric diazines; their analyses again established allowed $\pi^* \leftarrow n$ character, and gave estimates of effective moments of inertia which included vibrationally excited states as well as the zero-point levels of the two electronic states.

These azine studies were all based on the analysis of fine structure, in the good approximation that these molecules could be approximated as nearly symmetric oblate tops. Dunn[8] shortly afterwards reported (without details; see Table I for recent results) that bands in the ultraviolet spectrum of fluorobenzene could be likewise treated as arising from a quasi-symmetric prolate top, polarized in-plane and perpendicular to the F ··· F direction. This conclusion was notable as being perhaps the least ambiguous experimental demonstration that the first excited singlet state of benzene itself has symmetry B_{2u} and not B_{1u}; the decision is almost impossible to make from evidence contained in the vibrational-rotational structure of benzene's own spectrum. Assuming that fluorine substituents merely act as small perturbations, a $B_{1u} \leftarrow A_{1g}$ benzene transition would correlate with a transition in C_6H_5F polarized *along* the C—F direction.

At about the same time as these investigations on monocyclic compounds, Craig, Hollas, Redies, and Wait[9] used inferences drawn from rotational contours, not containing fine structure, to settle the long-standing problem of the electronic symmetry of the first excited state of naphthalene. They recognized the existence of two kinds of bands. The inconspicuous origin of the transition, and progressions built on it, have single-peaked contours. A strong band which is a false origin and progressions built on it have double-peaked bands. The principal moments of inertia in naphthalene coincide with symmetry axes, as do possible directions of the transition moment. In the infrared spectra of asymmetric tops, bands polarized along the greatest and least intertial axes tend to show a single peak, while bands polarized parallel to the intermediate axes are double peaked. In naphthalene the excitation is certainly $\pi^* \leftarrow \pi$, that is, it has an in-plane transition moment. The origin and the other single-peaked bands were thus identified as polarized along the long molecular axis, and the double-peaked bands as polarized along the short axis, their intensity coming from vibronic coupling with a higher, more intense, short-axis polarized transition. The identifications were supported by the first attempts to calculate band contours for a large molecule. They were necessarily, given the computing resources available at the time, exceedingly approximate calculations, and the authors had no real way of knowing what the geometry of excited naphthalene would be like. They made assumptions, compatible with rough Franck-Condon considerations, used these to find likely upper estimates of the changes in the inertial constants, and then computed contours which bore out their thesis. There is no doubt that their conclusions were correct, though it has since been recognized[10] that not all short-axis electronic transitions in naphthalene need have double-peaked contours, and as described below, the particular contour they observed comes about in a different way from the double-peaked contour of the short-axis polarized infrared bands. The key point, however, was the recognition of the presence of the two band types and of the polarization information they necessarily contain. The success of this approach in settling the first significant large-molecule assignment since benzene 13 years earlier (electronic spectroscopy moves at the stately gait that befits an elderly subject) promptly led to the development of improved computer programs for the accurate calculation of band contours in electronic spectra. The power of these programs has developed in step with the improved speed, storage, and output facilities of computers themselves, and this development has been as important as improved instrumentation in bringing these studies to their present productive state.

What follows is an account of developments in this area since 1960, commencing with brief outlines of rotational theory, and of the computational problem. Results obtained for individual molecules are then described, mostly in summary form (Table I), and there is then a discussion of applications. An attempt is made too to delimit the range of applicability of this kind of study. All the work referred to concerns absorption spectra. For completeness there is finally a brief discussion of emission spectra, mainly, however, with the object of pointing out that the emission spectroscopy of these molecules is far from being an automatic extension of absorption studies.

Applications of the methods of band contour analysis to vibrational spectra are not mentioned, for little work of this kind has yet been done, no doubt because high resolution is easier to realize in the visible-ultraviolet than in the infrared. Attention is drawn, however, to a recent study of the six infrared active fundamentals of cyclopropane,[11] an investigation very much in the spirit of the work reviewed here.

Finally, it will become apparent that the main emphasis here is upon vapor phase spectra. Nowadays, however, there is a constant interplay between vapor spectra and crystal spectra, and in the latter area also advantage is now being recognized in working at increasingly high resolution. For this reason, several references are made to crystal spectroscopy, in an attempt to set the relation between the two areas of study in perspective.

II. ROTATIONAL STRUCTURE OF VIBRONIC BANDS

A. Energy Levels, Intensities, and Line Coincidences

The well-known theory of band structure, described in standard texts,[4,12] is here very briefly summarized with some emphasis on aspects of particular consequence in large molecules.

The rotational energies of a rigid molecule are determined by the principal moments of inertia, conventionally designated $I_a \leqslant I_b \leqslant I_c$. Define $A \geqslant B \geqslant C$ (the rotational constants) in cm^{-1} units by $A = h/8\pi^2 cI_a$, and so on. The magnitudes of A and C for small polyatomics may be exemplified by formaldehyde, H_2CO, where they are 9.4 and 1.1 cm^{-1}, respectively. For aromatics they are nearer to 0.1 to 0.01 cm^{-1}, and the problems of resolving individual transitions correspondingly greater.

If $A \neq B \neq C$ (asymmetric rotor) the general rotational motion is not soluble in closed form either in classical or quantum mechanics. The usual quantum-mechanical treatment uses a basis of symmetric rotor wave functions. The relevant energy expressions for symmetric rotors are

$$E(J, K_a) = BJ(J+1) + (A - B)K_a^2 \text{ (prolate top:} \quad A > B = C) \quad \text{(1a)}$$

$$E(J, K_c) = BJ(J+1) - (B - C)K_c^2 \text{ (oblate top:} \quad A = B > C) \quad \text{(1c)}$$

J quantizes the total angular momentum, and the K's ($=0, \pm 1, \ldots \pm J$) its component along the figure axis a or c. Any asymmetric top A, B, C may be regarded as intermediate between prolate and oblate tops with the same A and C. As B runs from A to C, the asymmetry parameter κ defined by

$$\kappa = (2B - A - C)/(A - C)$$

runs from -1 (prolate) to $+1$ (oblate). The asymmetric top energies are still diagonal in J, but not in K. The symmetric top levels are doubly degenerate for $K \neq 0$. In the asymmetric top the K degeneracy is lifted (asymmetry splitting), and the split levels then correlate with adjacent K levels of the other kind of symmetric top. Each level can then be labeled by J and by K_a, K_c (also called K_{-1}, K_{+1}), the levels with which it correlates in the prolate and oblate limits. The energy levels are then expressed as

$$E(J_{K_a,K_c}) = \tfrac{1}{2}(A + C)J(J+1) + \tfrac{1}{2}(A - C)E'(J_{K_a,K_c}; \kappa) \qquad (2)$$

Each level is still $(2J + 1)$-fold degenerate, corresponding to the possible projections of J on a laboratory-fixed direction. The most populated J states have $J \sim [kT/(A + C)]^{1/2}$, that is, in the range 40–60 for the molecules considered here. Figure 1 shows the reduced-energy functions $E'(\kappa)$ for the case $J = 40$. To reduce the scale to cm^{-1}, values of $E'(\kappa)$ are to be multiplied by $\frac{1}{2}(A - C)$, that is, typically here by $0.1\ \mathrm{cm}^{-1}$. It is thus evident that for any given value of κ the asymmetry splitting, which is most evident only along the upward diagonal, is observationally significant (i.e., exceeds $0.01\ \mathrm{cm}^{-1}$, say) only for about half a dozen pairs of levels, and may well be of small consequence in practical spectra.

Various approximations to $E(J_{K_a,\,K_c})$ have been developed. The best known are the symmetric rotor approximations, in which

$$\bar{B} = \tfrac{1}{2}(B + C)\ \text{(prolate)} \quad \text{or} \quad \tfrac{1}{2}(A + B)\ \text{(oblate)}$$

are substituted for B in the symmetric rotor equations (1). Perturbation theory provides more precise, though cumbrous, continuations of these formulas. Alternatively, and much more recently, use has been made[13] of an asymptotic expansion due to Gora[14] which commences

$$\begin{aligned} E(J_{K_a,\,K_c}) = {} & CJ(J + 1) - \tfrac{1}{4}(2m^2 + 2m + 1)(A + B - 2C) \\ & + \tfrac{1}{2}(2J + 1)(2m + 1)[(B - C)(A - C)]^{1/2} \\ & \{1 - (2m^2 + 2m + 1)(1 - \delta^2)(32\,\delta J(J + 1))^{-1}\} \end{aligned} \tag{3}$$

where $m = J - K_c$ (and thus $= K_a$ or $K_a + 1$), and $\delta = \frac{1}{2}(\kappa + 1) = (B - C)/(A - C)$. K_a does not appear because at this level of approximation the asymmetry splitting is neglected. This formula applies to oblate tops, and especially for high J and low K_a, or more particularly to the domain to the right of the upward diagonal in Figure 1. The comparable formula for prolate tops, applicable to the upper left domain of Figure 1, was obtained by Brown[15] by the same methods, and simply has A and C interchanged, with m redefined as $J - K_a$ and δ as $\frac{1}{2}(1 - \kappa) = (A - B)/(A - C)$. For near-oblate and near-prolate tops, in which $(1 - \delta^2)$ is small, these formulas reduce to the symmetric top approximations (Eqs. (1) with $\bar{B}$ in place of B), but their accuracy persists to considerably greater asymmetries. As Brown points out, these formulas have the considerable advantage that for intermediate values of κ the oblate approximation (3) describes well the lower levels of given J, and the prolate approximation the higher levels. Of these, the oblate approximation (3) is particularly useful, since it describes the lower levels, which are the more readily accessible thermally and therefore carry most of the intensity. If transitions from low K_a levels are strong, one may thus have the curious circumstance that the important transitions in a distinctly prolate asymmetric top

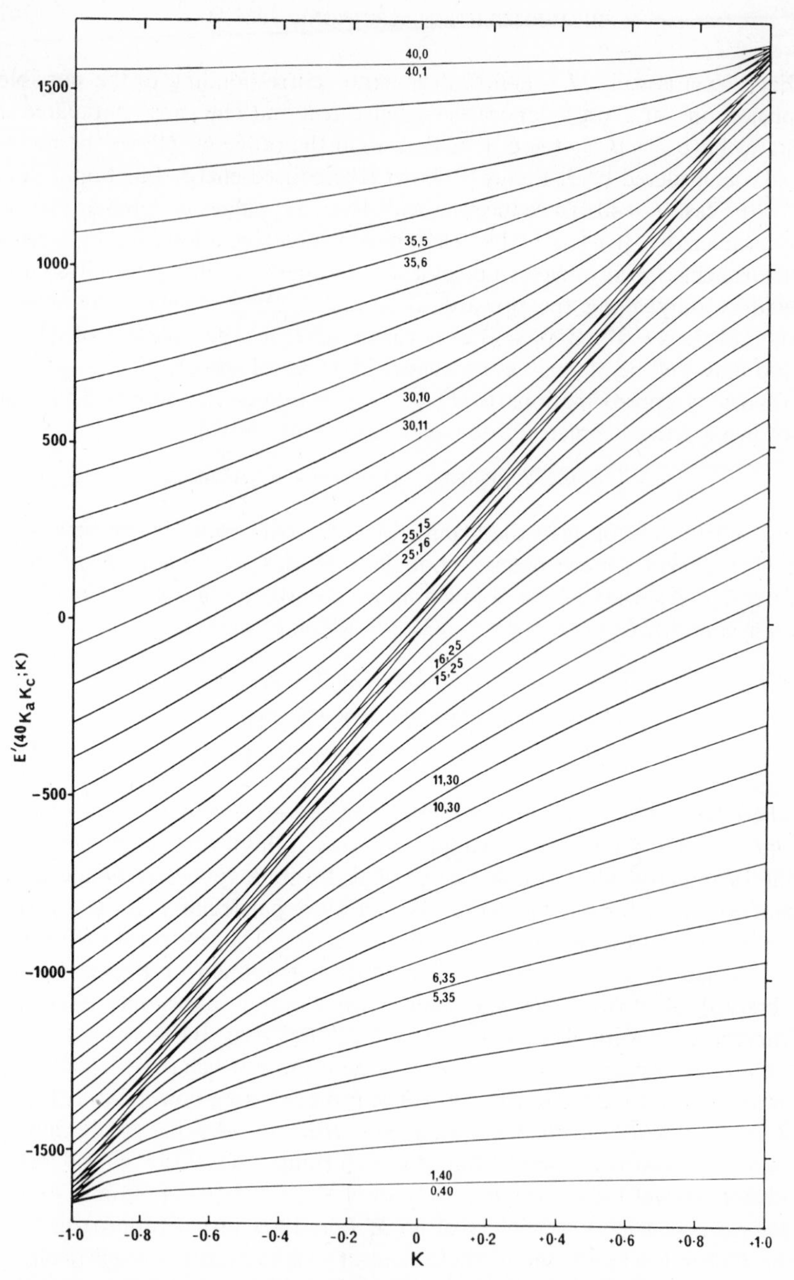

Fig. 1. Reduced asymmetric rotor energies, for the case $J = 40$, and values of κ between the prolate and oblate limits. Every fifth pair of levels, the splitting between which is mostly imperceptible on the scale of the diagram, is labeled by K_a, K_c.

(say $\kappa = -0.7$) can be adequately described by an oblate top approximation.

Using symmetric top functions, the exact calculations of $E(J_{K_a, K_c})$ require diagonalization of a matrix of dimension $2J + 1$ whose only nonzero elements have K indexes differing by ± 2. A simple transformation which exploits the point symmetry (D_2) of a spinning ellipsoid with respect to molecule-fixed axes splits the matrices into four, of dimensions $\sim\frac{1}{2}J$, and imposes on the resulting eigenfunctions and eigenvalues four different symmetry classifications. These, when taken in conjunction with the symmetry of the transition moment vector, determine distinctive selection rules for transitions polarized parallel to the a, b, and c inertial axes. The resulting bands are referred to as types A, B, and C. If the transition moment is inclined to the inertial axes, each component contributes independently to the resulting "hybrid" band.

The selection rule on J is $\Delta J = \pm 1$ or 0. While there is no strict selection rule on K, the symmetric top selection rules $\Delta K = 0$ (for parallel bands) and $\Delta K = \pm 1$ (for perpendicular bands) still determine the strongest lines for states with zero electron spin. In the large-molecule spectra of concern here, the "second-order" lines $\Delta K = \pm 2, \pm 3$ contribute only about 5% of the total intensity and no resolved features due to such lines have yet been identified. In transitions to triplet states, for which Hougen[16] gives the theory, the selection rules permit $\Delta N = \pm 2$ transitions, where N quantizes the total angular momentum excluding spin (in singlet states $N = J$), and also in first order $\Delta K = \pm 2$.

The conventional notation for transitions employs the code ..., P, Q, R, ... for a change in either J or K of ..., $-1, 0, +1$, ... and transitions are designated by the coded symbol ${}^{\Delta K_a \Delta K_c}\Delta J$. If the asymmetry splitting is inconsequential, ΔK_c may be omitted in the case of near-prolate tops, and ΔK_a with near-oblate tops. Thus ${}^{pr}Q$ denotes $\Delta K_a = -1$, $\Delta K_c = +1$, $\Delta J = 0$, and in a prolate top might be written ${}^{p}Q$, in an oblate top ${}^{r}Q$. All lines with the same ΔJ constitute a branch, lines with the same ΔJ and fixed K_a or K_c a sub-branch.

The first-order selection rules permit the following transitions. For type A bands: ${}^{qp}P$ and ${}^{qr}R$, which are strong for low values of K_a, and ${}^{qp}Q$, ${}^{qr}Q$, strong for high values of K_a. For type C bands, ${}^{pq}P$ and ${}^{rq}R$, strong for high K_a and ${}^{pq}Q$, ${}^{rq}Q$ strong for low K_a. The selection rules $\Delta K_a = 0$ (type A), $\Delta K_c = 0$ (type C), just exemplified, come directly from the limiting symmetric top cases for parallel bands. Type B bands represent the perpendicular cases in both kinds of rotor, and both K's must change. Specifically, the possible transitions are ${}^{pr}Q$, ${}^{rp}Q$, strong for intermediate values of K_a; ${}^{pr}P$, ${}^{rp}R$, strong for K_a high; ${}^{rp}P$, ${}^{pr}R$ strong for K_a low; and ${}^{pp}P$, ${}^{rr}R$.

The strengths of these last two transitions are sensitive to κ. The remarks made here about line strengths exclude Boltzmann factors, and come from explicit expressions for the transition-moment matrix elements.

The occurrence of the fine structure which characterizes the majority of the spectra so far analyzed has been the subject of two valuable developmental reviews, and for this reason need not be dealt with in great detail here. In the first, Hollas[17] adopted the quasi-symmetric top approximation and proposed half a dozen ways in which strong lines could happen to coincide in such a way as to produce readily resolvable narrow features: linelike Q sub-branches, numbered by K, occurring if $\Delta\bar{B} = \bar{B}' - \bar{B}'' \sim 0$ (single and double primes identify upper- and lower-state constants respectively); P and R branch heads, recording the reversing point of sub-branches and prominent if the lines are strong where the branch turns back, also numbered by K; and three ways in which lines with the same J and different K can coincide (K here means K_a or K_c, as most nearly appropriate for the molecule in question). An important distinction between these cases is that K-determined structure will show an intensity alternation depending on the nuclear statistical weights of the levels concerned, and J structure will not. In large molecules the statistical weight ratios are often close to unity and the alternation barely visible, yet this property has been successfully exploited in several cases. It is especially useful when the K numbering, determined from fragmentary structure extrapolated to the band origin, is in doubt by ± 1. This is a common circumstance. The weak: strong alternation distinguishes between K even or odd, and can then be decisive.

Brown[15] has updated the discussion of fine structure in terms of Gora's approximations, and the use of the effective quantum numbers $J - K_c$ (oblate domain), $J - K_a$ (prolate). In addition, aromatic molecules are formally planar, and many of their substituted derivatives are planar also, or are effectively so. In this case it was recently shown that Gora's approximations collapse[13] to a particularly concise form, since planarity implies

$$I_c = I_a + I_b \tag{4}$$

whence

$$C = [(B - C)(A - C)]^{1/2}$$

which is a factor in the last term of the oblate top formula (3). Introducing a new number

$$n = J + m = 2J - K_c$$

permits the energy levels in the oblate-top domain to be written as

$$E(J_{K_a,\,K_c}) = (n + 1)^2 C + \tfrac{1}{4}(2m^2 + 2m + 1)(A + B) \tag{5}$$

where once again K_a is not explicitly included. The selection rules can also be rewritten[13] in terms of m and n. For example, for ^{qp}P lines, $\Delta m = 0$ and $\Delta n = -1$, in which case

$$\Delta E = \Delta C n^2 - (2n + 1)C'' - \tfrac{1}{4}(2m^2 + 2m + 1)(\Delta A + \Delta B)$$

For small ΔA, ΔB and low values of m (i.e., low K_a, high K_c) lines with the same n and various m almost coincide, for example, the lines $(J, K_c = J)$, $(J - 1, K_c = J - 2), \ldots$. This should be, and is, a common cause of fine-structure formation in large planar molecules, and since the quadratic coefficient ΔC is independent of the point chosen to start the numbering n, the second differences of the line spacings lead at once to ΔC, rather than the ambivalent $\Delta \bar{B}$ of the symmetric top approximation. Moreover, since the values of K_c in any set of coincident lines are either even or odd, intensity alternation is still observable.

Brown pointed out that this kind of simplicity does not apply in the prolate top domain because the condition (4) is not symmetric in A and C. Nevertheless the formula (5) has proved applicable both to prolate tops, its original use,[13] and to oblate tops as well.[15,18] Brown also re-identified the fine structure in a number of previously analyzed spectra as of this nature, but notes that the conclusions of the original analyses are not seriously affected by use of the more accurate approximation. However, the generality of this fine-structure mechanism should not be allowed to eclipse the relevance of Hollas' earlier analysis, and more than one mechanism can be operative in the same band. A particularly spectacular case[19] is provided by the small-ring molecule F_2CN_2, the bands in which have three maxima and four distinct regions of fine structure with average spacings 0.3, 0.2, 0.8 in the R and 0.4 cm^{-1} in the P branches. The strategy of analysis of any band should include consideration of all possibilities, and the papers of Innes and of Hollas provide admirable case histories.

B. Nonrigid Rotors

The moments of inertia of rotational spectroscopy are in fact effective quantities, embodying various approximations introduced in the separation of rotational and vibrational motion. They reflect the mean-square amplitudes of the vibrational motion, and are potentially sensitive, especially to centrifugal distortion and Coriolis forces. In fact the recognized consequences of nonrigidity to date are not pervasive, though in part this may be due to a concentration of concern with 0-0 bands. Certainly, the detailed studies of Innes and co-workers on the azines have shown that $\bar{B}''$ and $\bar{B}'$ may differ individually, as between vibrational states, by amounts which in this group of compounds are of the same order of magnitude as the observed values of $\Delta\bar{B}$ (see Table I). Inferences about geometrical changes drawn from constants which refer to different vibrational levels of the two electronic states must accordingly be treated with reserve. No attempt has yet been made to rationalize this vibrational dependence. Centrifugal distortion has been deemed negligible by several authors. Recent detailed treatments of the microwave spectra of pyridine[20] and azulene[21] in which all the centrifugal distortion constants are extracted, provide strong quantitative support for this view, at least in respect of molecules which do not have flexible substituents.

Nonrigidity also induces a nonzero value of the inertial defect

$$\Delta = I_c - I_a - I_b \tag{6}$$

which takes small positive values in monocyclic compounds, and negative values in larger molecules and compounds with flexible substituents.[21] The manifestations of nonzero values of Δ in formally planar molecules are of marginal consequence at the present precision of analysis. Table I includes Δ values in cases where this problem has been reckoned with, and for molecules which are definitely not entirely planar.

Coriolis coupling is a different matter. One of the more subtle points of the analysis of naphthalene's spectrum, referred to earlier, was the recognition that the band contours of two key bands were atypical, and that they were close enough and of the right symmetries to be coupled by the Coriolis force.[9] The indicated calculation is within the power of present computational resources, but has yet to be attempted. Meanwhile, Christoffersen and Hollas[22] propose that Coriolis forces are responsible for significant variations in the A rotational constants of benzoquinone, to an extent which severely limits the information obtainable about the shape of the excited states. Coriolis coupling is also largely responsible for the nonzero inertial defects just referred to; semi-quantitative estimates of Δ can be made through the use of appropriate sum-rules on the ζ's and a modified uniform-coupling approximation.[21]

The classic case of Coriolis coupling, however, is provided by benzene, whose spectrum was earlier alluded to as being less than simple. The bands of the celebrated 2500 Å system of benzene show sharp edges on the short wavelength side, and then shade away to longer wavelengths. Earlier estimates of the change of size on excitation, based on the changes in frequency of the ring breathing and symmetric CH stretching modes,[23] and from progression lengths interpreted through the Franck-Condon principle,[24] had set narrow bounds to the probable value of $\Delta\bar{B}$ (the carbon ring expands on excitation by about 3%, and the CH bonds contract by about 1%). Rigid symmetric rotor calculations based on Equation (1c) cannot reproduce the characteristic sharp band edge. Callomon, Dunn, and Mills,[25] in a major study, investigated the consequences of Coriolis coupling, noting that all bands involve at least one doubly degenerate vibrational state. ($^1B_{2u} \leftarrow {}^1A_g$ is a forbidden transition; the strong bands all have one quantum of an e_{2g} vibration in one or the other state.) The effect of the Coriolis coupling is to modify the frequency at which the red-graded R sub-branches turn back, and the sharp edge (or terminus) occurs when the first lines of the returning R branches, which set in at successively higher values of $J \geqslant K$, are carried more rapidly to the red than the sub-branch origins are carried to the violet by the K degradation. In this way the Coriolis coupling constants for two vibrational modes were determined, and the appearance of different types of contour in the same spectrum explained: they involve different vibrations with different Coriolis constants.

C. Computation of Band Contours

Up to 1960, the limit of published tables of asymmetric rotor energies and line strengths was $J = 40$. The band contours of benzene, naphthalene, and anthracene derive considerable intensity from rotational levels extending at least to $J = 80$, 130, and 200, respectively. The weight of numbers involved in any calculation of the energy levels is immediately obvious, and impressive. A calculation on naphthalene, for example, requires in principle the calculation of the eigenvalues and eigenvectors of over 1000 matrices, of dimensions up to 60. It is the eigenvectors, required to calculate line strengths, that particularly tax the computer. For naphthalene, the eigenvalues for the two states collectively have some 750,000 elements. It is thus easy to appreciate that in driving the analysis of rotation envelopes in the direction of larger and larger molecules, it has until recently been easy to state a problem beyond the capacity of reasonably economic computation.

Perhaps now, however, computers and improved methods have matched the needs of the problem. At least, methods have passed through the stage when interpolation of energies and line strengths[26] had to be used to contend with the embarrassment of numbers. A typical large-molecule contour may be the resultant of some half-million significant individual lines. It would seem reasonable that in assembling all these lines into a contour, numerically characterized by just a few numbers, interpolation or some systematic selection of lines might be admissible. The recent experience of several authors is quite to the contrary. On the other hand, the use of symmetric rotor approximations for high K values appears to be satisfactory, and the Gora approximation, apparently not yet incorporated in any program, could be even more effective.

Some six separately written band contour programs appear to be in active use at present, but extensive details have been published only by Parkin,[26] who developed the first comprehensive asymmetric rotor program, and lately by Pierce.[27] All have many features in common. Eigenvalues and eigenvectors are calculated by methods of demonstrated speed and stability for tridiagonal matrices. Transition frequencies are calculated with chosen selection rules for increasing values of J'', until the Boltzmann-weighted intensities fall below a preset value. The intensities are collected in storage locations corresponding to suitable small steps in frequency. If desired, the intensity of any given line may be partitioned among adjacent storages according to a suitable line shape function (e.g., triangular or Gaussian). Intensities are then output directly on an X-Y plotter or, if collected in frequency intervals finer than the instrumental resolution, are convoluted with a slit function prior to output. The method described calculates intensity as absorbance. If the experimental record is a microphotometered photographic plate, conversion to a transmittance scale is desirable. Figure 2 is an example of the excellent matching of observed and computed spectra obtained in the most recent work.

For the calculation of eigenvalues, the quotient-difference algorithm[28] is commonly used, but conservative origin shifting designed to avoid division by zero is unnecessary[29] and the process can be much accelerated. Alternatively, the traditional continued-fraction method can also be much accelerated by judicious first approximations to each successive eigenvalue.[27] In neither case need the number of iterations exceed about two per eigenvalue, but which method is the faster has yet to be determined. Eigenvectors can be found following Wilkinson.[30] It is possible, and permissible, to identify the location of the significant elements among the eigenvectors and to calculate only them, with a useful saving of time and economy of storage.[27] It is necessary to store at any one time the eigenvectors for four values of J (e.g., one in the lower state, three in the upper), and for a calculation extending up to $J = 130$, complete liberation from constraints of number storage requires a core store of about 80 K words.

The actual analysis may be facilitated by prior estimates, from fine structure, of changes in one or more of the inertial constants, but the final steps are inevitably a pedestrian search, a matter of trial and frequent error. Fortunately, provided the ground-state constants can be tolerably well estimated, the analysis can concentrate solely on ΔA, ΔB, ΔC and if the molecule is planar, the condition $\Delta = 0$ (Eq. 6) reduces the effective unknowns to just two. In one of the more elaborate programs[29] (which also

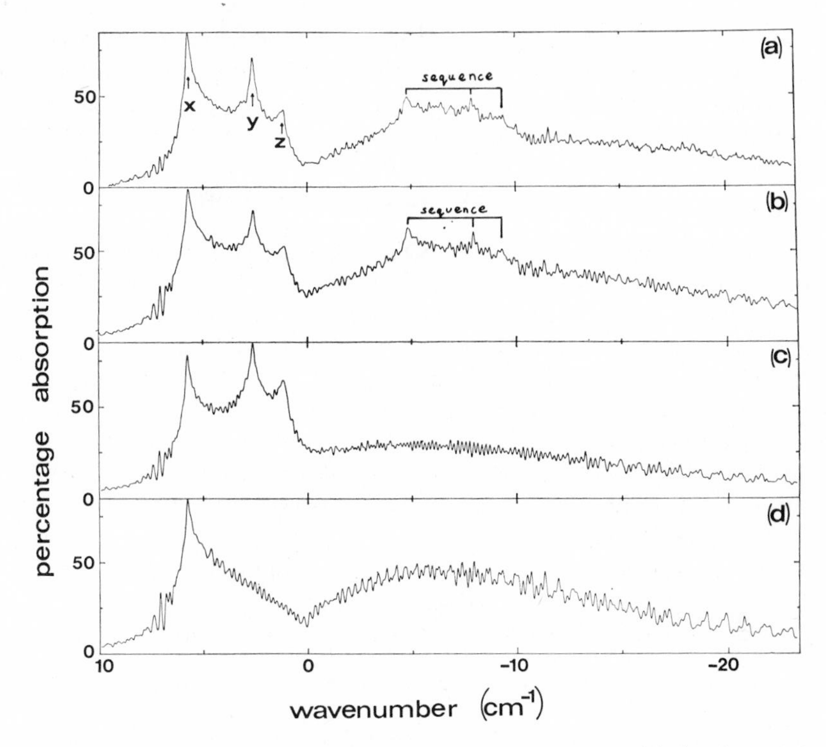

Fig. 2. Observed (a) and computed (b) band profiles for the origin band, an *A-B* hybrid, in 2-aminopyridine; (c) and (d) are computed profiles for pure type *B* and type *A* bands, respectively. In (b), a weak sequence band has been added near -5 cm^{-1}. The figure is reproduced from J. M. Hollas, G. H. Kirby, and R. A. Wright, *Molecular Physics* **18**, 327 (1970) by permission of the authors and publishers.

includes centrifugal distortion handled as a perturbation) the search procedure is improved by using the eigenvectors to obtain derivatives of upper state energies, with respect to changes in A', B', C'. These are then used, without recalculating intensities, to determine the derivatives with respect to A', B', C' of a suitable figure of merit extracted from a comparison of observed and computed contours; these derivatives provide the means for a systematic improvement of the fit. Innes et al.[31] have carried the process a step further by analyzing an individual sharp-feature "line"; that is, the constants are adjusted until the coincidence of a dozen strong lines is just precise enough for the envelope to reproduce the observed feature. The refinements which can be achieved in this way depend on the accuracy with which the profiles of fine-structure lines can be measured (there is little doubt, looking at plates, that their profiles do vary). The next phase of the technique may well involve the use of resolving powers as high as improved gratings and interferometric methods will permit in order to refine the experimental data on composite-line profiles to the eventual Doppler limit.

III. ANALYZED BAND CONTOURS

It is instructive, as a preliminary to a listing of actual analyses, to dissect a pair of related contours. Figure 3 shows two type B contours for azulene[10] (which closely resembles naphthalene). The first contour has $\Delta A = \Delta\bar{B} = 0$, the second has equal, small decreases in ΔA and $\Delta\bar{B}$. The first case corresponds to infrared bands, and the diagram shows that the two peaks come mainly from the oppositely running ^pQ and ^rQ branches. Also seen is the asymmetry splitting which at $K_a = 10$ is effective in the spectrum only for $J = 30$ to 40. In the second case the decrease in both inertial constants directs all sub-branches towards lower wave numbers, flattening out the P and ^pQ sub-branches to the extent that the low-frequency peak has disappeared, and causing the R and ^rQ sub-branches to form heads. The strong ^rQ branches (low K_a) now form a set of closely spaced heads which combine to form a sharp peak, but the ^rR branch heads are formed

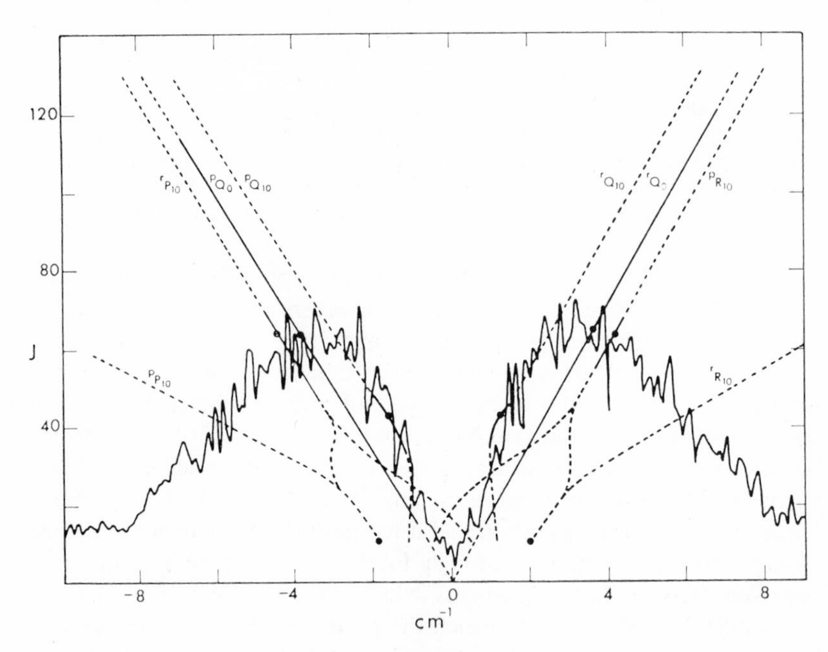

Fig. 3

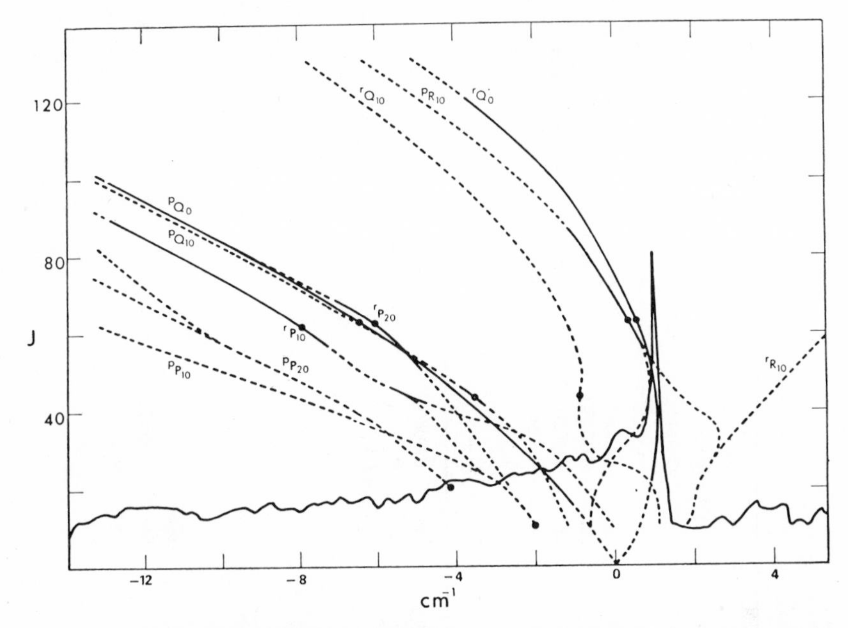

Fig. 3. Calculated type B bands, and some component sub-bands in an asymmetric top for azulene ($\kappa = 0.61$, 300°K). First figure: $\Delta A = \Delta \bar{B} = 0$; second figure $\Delta A = \Delta \bar{B} = -0.0010\ \text{cm}^{-1}$. The curves plot line positions for various K_a'' against $J''(K_a'')$. A dot marks the strongest line in each sub-branch, and curves are shown as solid lines where the line intensities exceed half the intensity of the strongest line in the entire band. Reproduced from A. J. McHugh and I. G. Ross, *Spectrochimica Acta* **26A**, 441 (1970) by permission of the publishers.

at values of J too high to generate any prominent feature. If ΔA is further decreased the ^rQ heads remain much as before but the ^rR sub-branches reverse at lower values of J and their heads are consolidated sufficiently to produce a recognizable second peak; this is the origin of the second peak in the observed type B bands of naphthalene, referred to in Section I.

Analyses published so far are listed in Table I, where they are arranged roughly in order of increasing molecular complexity. The first, three-membered-ring compound is of course not aromatic, but in every other sense it is a member of the series. The table includes reproductions of the contours, redrawn from original data with some loss of detail of the fine structure, on a common scale of frequency (which increases to the right). Band origins, where marked by the authors, are shown by a vertical line. The bands analyzed are specified by the frequencies of their rotational origins to an accuracy which reflects the precision with which these origins have been determined. Also given are their vibrational assignments:

TABLE I

Assigned Rotational Band Contours in Cyclic Molecules[a]

Compound	Band contour	Frequency assignment and type	Ground-state constants	Change in excited state	Fine structure and remarks	Refs.
Difluorodiazirine F_2CN_2	10cm⁻¹ *	28389 0-0 type B	A (0.2970) B (0.2193) C (0.1529) κ −0.07 μ ~0	ΔA +0.00242 ΔB −0.00250 ΔC −0.00267 $\Delta\kappa$ −0.029 $\|\Delta\mu\|$ 1.5 ± 0.2	${}^{p}P$, ${}^{r}P$ coincidences; ${}^{rp}Q$ branches; ${}^{pr}R$ bands; ${}^{r}R$ sub-branches. Geometry: ΔNN +0.05 Å, ΔFF −0.03 Å	19
Benzene C_6H_6	*	37478.1 38608.5 6_1^0, 6_0^1 perp. (1st type)	B 0.1896 κ +1.0 ζ_6 +0.62	ΔB −0.0086 $\Delta\kappa$ 0 $\Delta\zeta_6$ +0.02	Geometry: ΔCC +0.038 Å, ΔCH −0.01 Å	25
	*	41163.3 7_0^1 perp. (2nd type)		$\Delta\zeta_7$ ~0		
-d_6		37708.9 38787.1 6_1^0, 6_0^1 perp. (1st type)	B 0.1568 κ +1.0 ζ_6 +0.403	ΔB −0.0064 $\Delta\kappa$ 0 $\Delta\zeta_6$ ~0		

Pyridine-4-*d*	34800.33	A	0.201444	ΔA	−0.0029		60
C_5H_5N	0-0 type C	B	0.193624	ΔB	−0.0139		
		C	0.098710	ΔC	−0.0044		
		κ	+0.8478	$\Delta\kappa$	−0.209		
Pyridazine *	26648.75	A	0.20820	ΔA	−0.0037	Coincident ${}^qP$, ${}^qR$ lines, with	31
$o\text{-}C_4H_4N_2$	0-0 type C	B	0.19880	ΔB	+0.0029	various high K_c and same J	6
		C	0.10170	ΔC	−0.0004	($J \sim$ 1–60), due to	90
		B	0.2034	$\Delta\bar{B}$	−0.0004	$\Delta\bar{B}=\Delta C$. Probable	
		κ	+0.8235	$\Delta\kappa$	+0.1222	geometry: CNN increases,	
	27021.45	A, B, C as above		ΔA	−0.0011	NCC decreases, NN and	
	$6a_0^1$ type C			ΔB	−0.0019	CC increase (~0.2 Å)	
				ΔC	−0.0007		
		$\bar{B}$	0.2031	$\Delta\bar{B}$	−0.0013		
		κ	+0.8235	$\Delta\kappa$	−0.0158		
	25983.90	$\bar{B}$	0.2032	$\Delta\bar{B}$	−0.0003		
	$6a_0^1$ type C						
	25356.89	$\bar{B}$	0.2030	$\Delta\bar{B}$	−0.0004		
	$6a_1^1$ type C						
-3,6-d_2	$6a_0^1$ type C	$\bar{B}$	0.1902	$\Delta\bar{B}$	−0.0008		
-d_4	27116.91	$\bar{B}$	0.1778	$\Delta\bar{B}$	−0.0006	a, b axes interchanged	
	$6a_0^1$ type C	κ	(~ +0.8)	$\Delta\kappa <$	−0.05	relative to $-d_0$. Calc.	
	27238.21	$\bar{B}$	0.1777	$\Delta\bar{B}$	0.0000	$\Delta\kappa = -0.2$, whence no J	
	$6b_0^2$ type C					structure in 0-0	
Pyridazine	22487.1			ΔC	+ve	Obs. $\Delta K_c = 0$	31
(triplet system)	0-0					implies ${}^3B_1$ or ${}^3A_2$	

TABLE I—*Continued*

Compound	Band contour	Frequency assignment and type	Ground-state constants	Change in excited state	Fine structure and remarks	Refs.
Pyrimidine	*	31072.6	A 0.20940		${}^{q}Q$, $2J–K_c$ structure	64
m-$C_4H_4N_2$		0-0 type C	B 0.20250			
			C 0.10300	ΔC −0.0075	Probable geometry:	
			κ +0.870		$\Delta C_6N_1 \sim -0.09$ Å,	
					$\Delta CC \sim \Delta C_2N_1 = +0.07$ Å;	
					angle changes small	
Pyrazine	*	30875.80	$\bar{B}$ 0.2049	$\Delta\bar{B}$ −0.0012	${}^{q}P$, ${}^{q}R$, various low K_c,	7
p-$C_4H_4N_2$	SCHEMATIC (from print)	0-0 type C			$J \sim 10–40$ ($\Delta\kappa$ small).	63
					Probable geometry:	91
					$\Delta N \cdots N \sim -0.14$ Å	
		$6a_0^1$ type C		$\Delta\bar{B}$ −0.0013		
		$6b_0^1$ type C	$\bar{B}$ as above	$\Delta\bar{B}$ −0.0009		
		11_0^1 type C		$\Delta\bar{B}$ −0.0021		
		19_0^1 type C		$\Delta\bar{B}$ −0.0013		
		$6a_1^0$ type C	$\bar{B}$ 0.2055	$\Delta\bar{B}$ −0.0019		
	*	31258	A (0.2128)	ΔA +0.0015	${}^{p}P$ and ${}^{r}R$, $2J–K_c$ structure	36
		5_0^1 type A	B (0.1975)	ΔB −0.0023		
			C (0.1024)	ΔC −0.0008		
			κ (+0.7)	$\Delta\kappa$ −0.062		
-${}^{15}N$		30879.96	$\bar{B}$ 0.2022	$\Delta\bar{B}$ −0.00105		
		0-0 type C	C 0.1009			33
		31455.81	$\bar{B}$ 0.2023	$\Delta\bar{B}$ −0.00085		
		$6a_0^1$ type C				
		31262.41		ΔC −0.00073		
		5_0^1 type A				

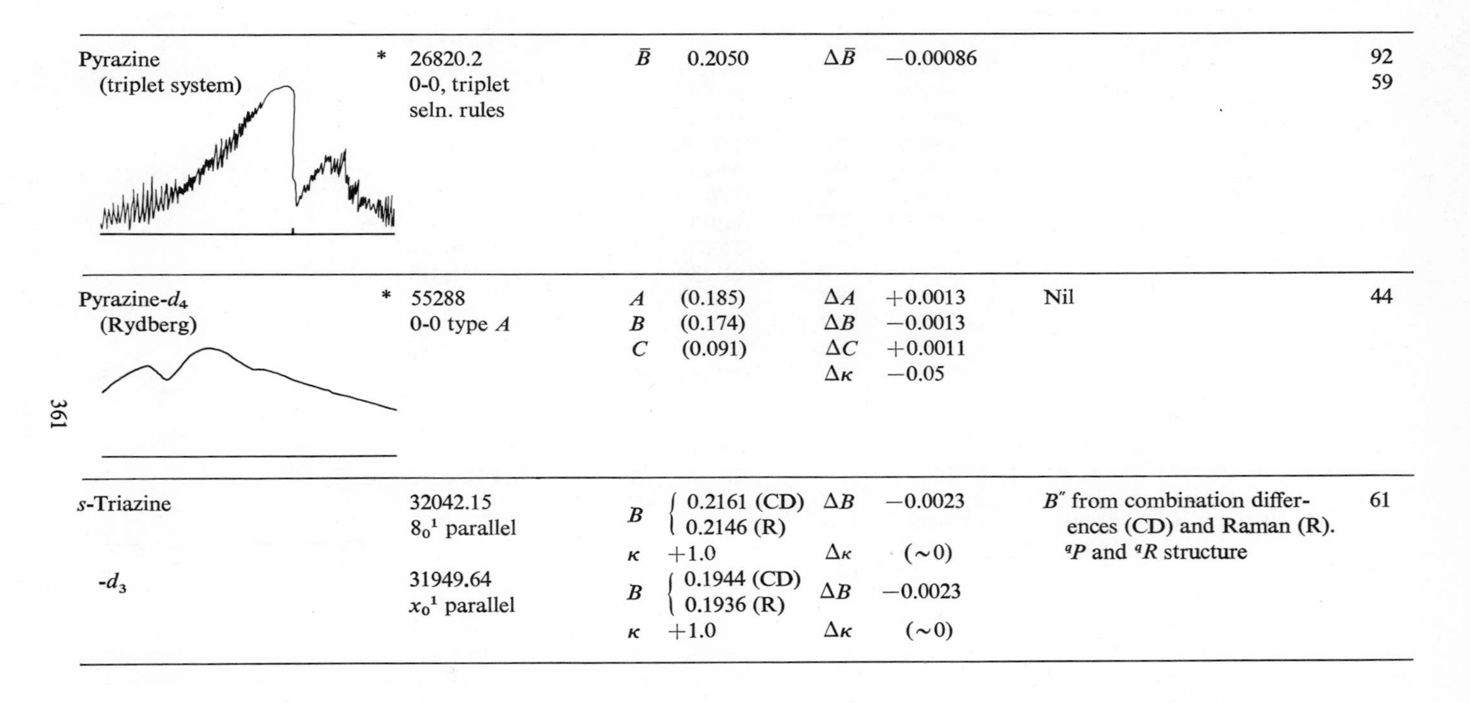

Molecule		Band	Constant	Value	Change	Value	Remarks	Ref.
Pyrazine (triplet system)	*	26820.2 0-0, triplet seln. rules	$\bar{B}$	0.2050	$\Delta\bar{B}$	−0.00086		92 59
Pyrazine-d_4 (Rydberg)	*	55288 0-0 type A	A	(0.185)	ΔA	+0.0013	Nil	44
			B	(0.174)	ΔB	−0.0013		
			C	(0.091)	ΔC	+0.0011		
					$\Delta\kappa$	−0.05		
s-Triazine		32042.15 8_0^1 parallel	B	0.2161 (CD) 0.2146 (R)	ΔB	−0.0023	B'' from combination differences (CD) and Raman (R). ^qP and ^qR structure	61
			κ	+1.0	$\Delta\kappa$	(~0)		
-d_3		31949.64 x_0^1 parallel	B	0.1944 (CD) 0.1936 (R)	ΔB	−0.0023		
			κ	+1.0	$\Delta\kappa$	(~0)		

TABLE I—*Continued*

Compound	Band contour	Frequency assignment and type	Ground-state constants	Change in excited state	Fine structure and remarks	Refs.
s-Tetrazine	*	18912.61	$\bar{B}$ 0.21845	$\Delta\bar{B}$ +0.00033	${}^{q}P$, ${}^{q}R$ various high K_c; J up	5
p-$C_2H_2N_4$		$x_0{}^1$ type C	κ +0.65		to 36. κ from infrared	93
					contour	
				ΔC +0.000589	From $2J$–K_c structure in ${}^{q}Q$.	18
-d		18135.52	A (0.2144)	ΔA +0.0004	Geometry: ΔNCN −11.9°,	
		0-0 type C	B (0.2076)	ΔB +0.0008	ΔCN 0.08 Å, ΔNN −0.17 Å,	
			C (0.1055)	ΔC +0.00056	ΔCH −0.03 Å. These are	
			κ $(+0.8_7)$	$\Delta\kappa$ +0.013	revised figures from	
					Ref. 5b. See text	
-d_2		18143.01	A (0.2144)	ΔA 0.0000		
		0-0 type C	B (0.1941)	ΔB +0.0005		
			C (0.1019)	ΔC +0.00027		
			κ $(+0.6_4)$	$\Delta\kappa$ +0.007		
-p-${}^{15}N_2$-d		18141.87	A (0.2103)	ΔA +0.0003		
		0-0 type C	B (0.2022)	ΔB +0.0012		
			C (0.1031)	ΔC +0.00067		
			κ $(+0.8_5)$	$\Delta\kappa$ +0.017		
-p-${}^{15}N_2$-d_2		18149.37	A (0.2080)	ΔA 0.0000		
		0-0 type C	B (0.1913)	ΔB +0.0008		
			C (0.0996)	ΔC +0.00050		
			κ $(+0.6_9)$	$\Delta\kappa$ +0.013		
-p-${}^{15}N_2$		18906.38	$\bar{B}$ 0.21338	$\Delta\bar{B}$ +0.00041		
		$x_0{}^1$ type C				

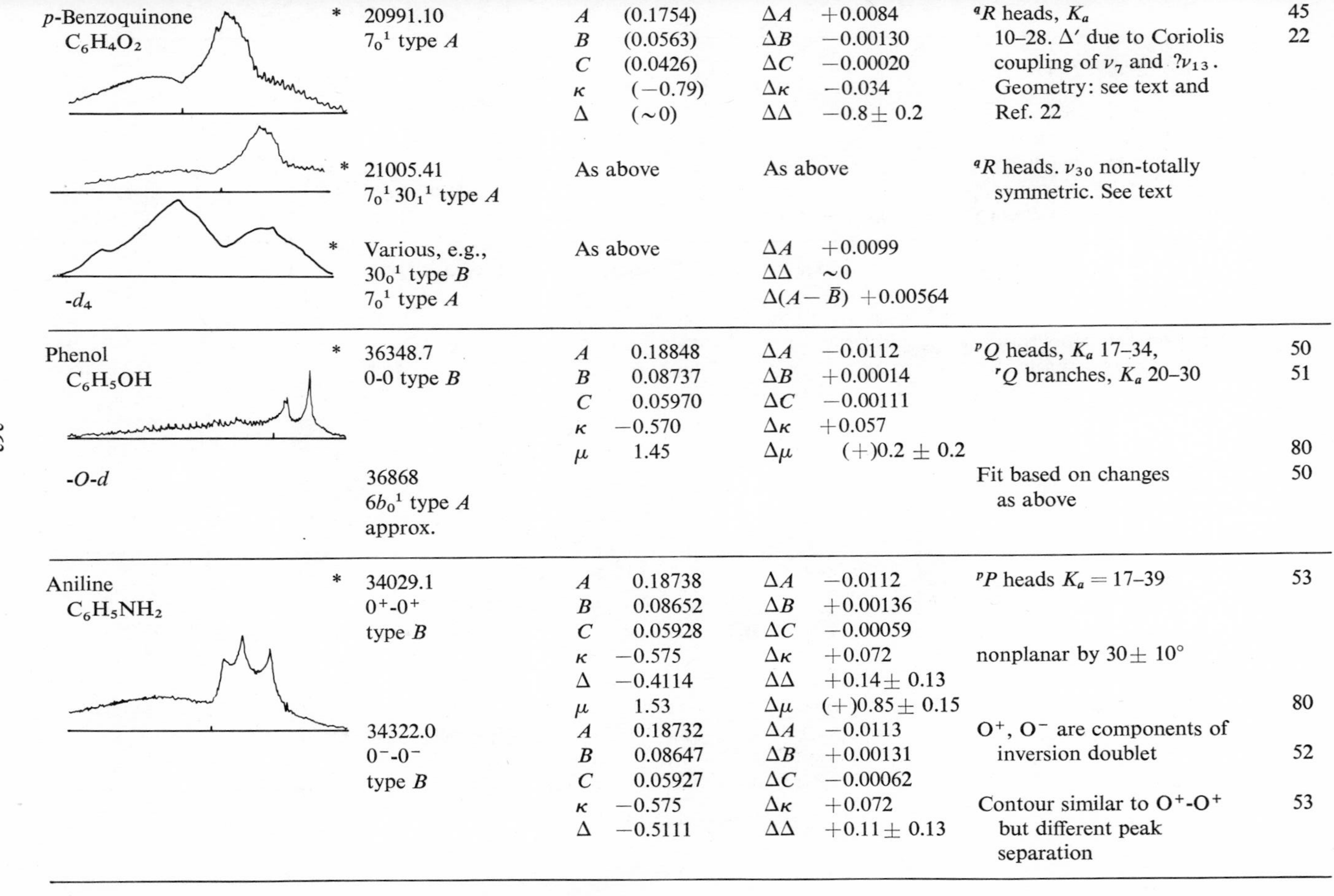

Molecule	Band	Constants		Changes		Remarks	Ref.
p-Benzoquinone $C_6H_4O_2$	* 20991.10	A	(0.1754)	ΔA	+0.0084	^qR heads, K_a	45
	7_0^1 type A	B	(0.0563)	ΔB	−0.00130	10–28. Δ' due to Coriolis	22
		C	(0.0426)	ΔC	−0.00020	coupling of ν_7 and $?\nu_{13}$.	
		κ	(−0.79)	$\Delta\kappa$	−0.034	Geometry: see text and	
		Δ	(~0)	$\Delta\Delta$	−0.8 ± 0.2	Ref. 22	
	* 21005.41 $7_0^1 30_1^1$ type A	As above		As above		^qR heads. ν_{30} non-totally symmetric. See text	
-d_4	* Various, e.g., 30_0^1 type B, 7_0^1 type A	As above		ΔA	+0.0099		
				$\Delta\Delta$	~0		
				$\Delta(A-\bar{B})$	+0.00564		
Phenol C_6H_5OH	* 36348.7	A	0.18848	ΔA	−0.0112	^pQ heads, K_a 17–34,	50
	0-0 type B	B	0.08737	ΔB	+0.00014	^rQ branches, K_a 20–30	51
		C	0.05970	ΔC	−0.00111		
		κ	−0.570	$\Delta\kappa$	+0.057		
		μ	1.45	$\Delta\mu$	(+)0.2 ± 0.2		80
-*O*-*d*	36868 $6b_0^1$ type A approx.					Fit based on changes as above	50
Aniline $C_6H_5NH_2$	* 34029.1	A	0.18738	ΔA	−0.0112	^pP heads K_a = 17–39	53
	0^+-0^+	B	0.08652	ΔB	+0.00136		
	type B	C	0.05928	ΔC	−0.00059		
		κ	−0.575	$\Delta\kappa$	+0.072	nonplanar by 30 ± 10°	
		Δ	−0.4114	$\Delta\Delta$	+0.14 ± 0.13		
		μ	1.53	$\Delta\mu$	(+)0.85 ± 0.15		80
	34322.0	A	0.18732	ΔA	−0.0113	O^+, O^- are components of	
	0^--0^-	B	0.08647	ΔB	+0.00131	inversion doublet	52
	type B	C	0.05927	ΔC	−0.00062		
		κ	−0.575	$\Delta\kappa$	+0.072	Contour similar to O^+-O^+	53
		Δ	−0.5111	$\Delta\Delta$	+0.11 ± 0.13	but different peak separation	

TABLE I—*Continued*

Compound	Band contour	Frequency assignment and type	Ground-state constants	Change in excited state	Fine structure and remarks	Refs.
Benzaldehyde C_6H_5CHO	*	~26916 0-0 type C			Nil	94
	*	Various $A \gg B$ hybrid			Nil	
Phenylacetylene C_6H_5CCH		35897.0 0-0 type B	A 0.18871 B 0.05101 C 0.04015 κ −0.8538	ΔA −0.00710 ΔB +0.00002 ΔC −0.00089 $\Delta\kappa$ +0.019	Two heads 3.4 cm^{-1} apart. Fine structure near heads	95
		Various type A	As above	As above	Single heads	
Benzonitrile C_6H_5CN		36512.9 0-0 type B	A 0.18863 B 0.051581 C 0.040496 κ −0.8504	ΔA −0.0061 ΔB −0.0012 ΔC −0.00103 $\Delta\kappa$ +0.003	Closely similar results obtained also for -4-d compound. $2J$–K_c structure in ^pP. A pair of sequence bands shows pronounced effects due to Coriolis coupling	96

Styrene	34758.8	A	(0.1694)	ΔA	−0.0061	^{qr}R sub-branches, K_a 28–36.	97
$C_6H_5CHCH_2$	0-0 type A	B	(0.0538)	ΔB	−0.00062	Planar	
		C	(0.0408)	ΔC	−0.00072		
		κ	(−0.80)	$\Delta\kappa$	+0.010		
Fluorobenzene	37813.9	A	0.18892	ΔA	−0.0112	$2J$–K_c structure in ^pP	98
C_6H_5F	0-0 type B	B	0.08575	ΔB	−0.0010		
		C	0.05897	ΔC	−0.00159		
		κ	−0.588	$\Delta\kappa$	+0.043		
		μ	1.66	$\Delta\mu$	+0.30 ± 0.07		99
Chlorobenzene	37048.2	A	0.18923	ΔA	−0.0095	$2J$–K_c structure in ^pP;	100
C_6H_5Cl	0-0 type B	B	0.52595	ΔB	−0.00030	^{35}Cl, ^{37}Cl isotope splitting	
	36443.4	C	0.041151	ΔC	−0.00064	negligible	
	29_1^0 type A	κ	−0.8454	$\Delta\kappa$	+0.0015		
Bromobenzene	36991.5	A	0.18905	ΔA	−0.0096	^pQ sub-branch heads,	101
C_6H_5Cl	0-0 type B	B	0.033186	ΔB	−0.00002	K_a 5–16. Isotope splitting	
		C	0.028228	ΔC	−0.00023	negligible	
		κ	−0.9383	$\Delta\kappa$	+0.007		
p-Difluorobenzene	36837.9	A	(0.1881)	ΔA	−0.0119	Very well resolved fine	102
$C_6H_4F_2$	0-0 type B	B	(0.0476)	ΔB	+0.00023	structure spaced up to	
		C	(0.0380)	ΔC	−0.00036	0.9 cm^{-1}, with clear	
		κ	(−0.87)	$\Delta\kappa$	+0.020	intensity alternation. ^rQ	
						and ^pQ heads, $K_a \sim$ 10–40	
p-Chlorofluorobenzene	36275.1	A	(0.1884)	ΔA	−0.0112	^pQ heads, K_a 3–30	103
C_6H_4ClF	0-0 type B	B	(0.0319)	ΔB	+0.00025		
		C	(0.0273)	ΔC	−0.00007		
		κ	(−0.94)	$\Delta\kappa$	+0.009		

TABLE I—*Continued*

Compound	Band contour	Frequency assignment and type	Ground-state constants		Change in excited state		Fine structure and remarks	Refs.
p-Dichlorobenzene		35740.5	A	(0.1887)	ΔA	-0.0105	Poorly defined and irregular	104
$C_6H_4Cl_2$		0-0 type B	B	(0.0224)	ΔB	$+0.0017$		
		36278	C	(0.0200)	ΔC	$+0.00001$		
		x_0^1 type A	κ	(-0.97)	$\Delta\kappa$	$+0.004$		
p-Fluorophenol		35115.9	A	(0.1877)	ΔA	-0.0125	^pP heads, K_a 12–32	105
$C_6H_5(OH)F$		0-0 type B	B	(0.0486)	ΔB	$+0.00073$		
			C	(0.0386)	ΔC	-0.00011		
			κ	(-0.87)	$\Delta\kappa$	$+0.025$		
			μ	2.17	$\Delta\mu$	$+0.4 \pm 0.1$		81
p-Fluoroaniline		32653.3	A	0.18660	ΔA	-0.0111	Well resolved ^pP heads,	105
$C_6H_5(NH_2)F$		0-0 type B	B	0.04834	ΔB	$+0.00099$	K_a 11–43	
			C	0.03843	ΔC	$+0.00008$		
			κ	-0.866	$\Delta\kappa$	$+0.024$		
			μ	2.48	$\Delta\mu$	$+0.8 \pm 0.1$		81
2-Aminopyridine	*	33471.1	A	(0.1923)	ΔA	-0.0120	Weak ^pP heads, and R	43
$C_5H_4N \cdot NH_2$		0^+-0^+, A-B	B	(0.0901)	ΔB	$+0.00135$	structure indexed by	
		hybrid (55% A)	C	(0.0614)	ΔC	-0.00065	J–K_c	
			κ	(-0.56)	$\Delta\kappa$	$+0.075$		
			Δ	(-0.40)	$\Delta\Delta$	-0.1 ± 0.3		

Molecule		Band					Remarks	Ref.
Cyclopentanone C_5H_8O		Various, types A and C	A	0.22063	ΔA	−0.0058	Nil. -α-d_4 and -d_8 compounds also analyzed. ΔCO +0.11 Å, CO angle to plane ~32° (inversion doublets)	106
			B	0.11169	ΔB	+0.0061		
			C	0.08033	ΔC	+0.00022		
			κ	−0.5530	$\Delta \kappa$	+0.108		
Naphthalene $C_{10}H_8$	*	32454.0 and others, B	A	(0.1044)	ΔA	−0.0026	Nil	9
			B	(0.0415)	ΔB	−0.0009	Probable geometry:	58
	*	32019.2 0-0, type A	C	(0.0297)	ΔC	−0.0007	see Ref. 58	
			κ	(−0.68)	$\Delta \kappa$	+0.02		
Azulene $C_{10}H_8$	*	28756.8 0-0 type A	A	0.094797	ΔA	0.0000	^qR, $2J-K_c$ structure	13
			B	0.041857	ΔB	−0.0008		32
			C	0.029044	ΔC	−0.00040		21
			κ	−0.6102	$\Delta \kappa$	−0.010	Geometry: see text	
			Δ	−0.1524	$\Delta\Delta$	−0.15 ± 0.3		
-d_8		28865.3 0-0 type A	A	(0.0893)	ΔA	0.0000		
			B	(0.0373)	ΔB	−0.0007		
			C	(0.0256)	ΔC	−0.00033		
			κ	(−0.58)	$\Delta \kappa$	−0.016		
Indene C_9H_8	*	34723.0 0-0 A-B hybrid (10% B)	A	(0.1256)	ΔA	−0.0035	^qP, $2J-K_c$ structure	36
			B	(0.0526)	ΔB	−0.0004		
			C	(0.0373)	ΔC	−0.00064		
			κ	(−0.66)	$\Delta \kappa$	+0.017		
			Δ	(3.2)	$\Delta\Delta$ ~	−1.5		

TABLE I—*Continued*

Compound	Band contour	Frequency assignment and type	Ground-state constants	Change in excited state	Fine structure and remarks	Refs.
Indole C_8H_6NH	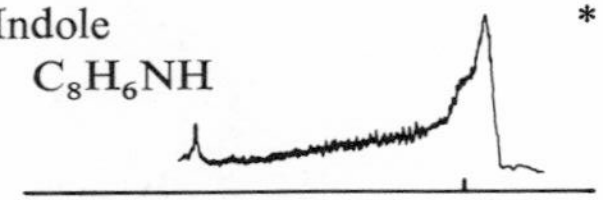 *	35232.1 0-0, *A-B* hybrid (20% *B*)	A (0.1266) B (0.0538) C (0.0378) κ (−0.64)	ΔA −0.0052 ΔB −0.00076 ΔC −0.00084 $\Delta\kappa$ +0.019	${}^{qp}Q$ sub-bands, K_a 21–36; ${}^{qp}P$ sub-bands, K_c 38–68	107
Benzofuran C_8H_6O	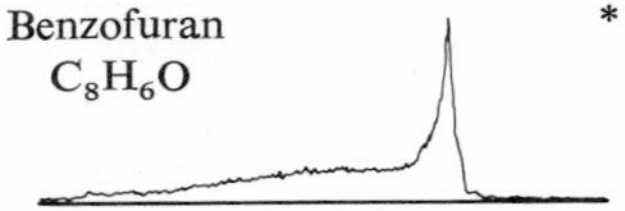 *	35920.0 0-0 *A-B* hybrid (10% *B*)	A (0.1274) B (0.0529) C (0.0374) κ (−0.66)	ΔA −0.0056 ΔB −0.00065 ΔC −0.00084 $\Delta\kappa$ +0.025	${}^{q}Q$ sub-bands, K_a 26–36; ${}^{qp}P$ sub-bands, K_c 41–49 (closer spacing)	42
Thionaphthen C_8H_6S	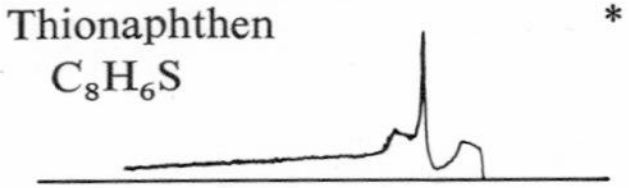 *	34055.7 0-0 *A-B* hybrid (10% *A*)	A (0.1043) B (0.0428) C (0.0303) κ (−0.66)	ΔA −0.0028 ΔB −0.00045 ΔC −0.00046 $\Delta\kappa$ +0.014	K_c sub-bands in ${}^{p}P$	42
2,1,3-Benzo-thiadiazole $C_6H_4N_2S$	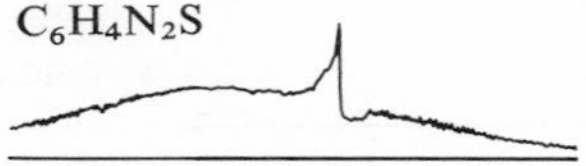 *	30410.5 0-0 type *B*	A (0.1301) B (0.0418) C (0.0317) κ (−0.795)	ΔA +0.0008 ΔB −0.0013 ΔC −0.0008 $\Delta\kappa$ −0.006	${}^{r}R$ heads with $K_a \sim$ 9–34 and groups of ${}^{p}P$ lines (type *B* only)	108
	*	 30937.9 x_0^1 type *A*	As above	As above		

2,1,3-Benzo-selenadiazole $C_6H_4N_2Se$	*	28060.5 0-0 type *B*	A	(0.1217)	ΔA	+00008	Nil	109
			B	(0.0281)	ΔB	−0.0012		
			C	(0.0228)	ΔC	−0.0007		
			κ	(−0.89)	$\Delta \kappa$	−0.012		

[a] See text, Section III, for definitions and units. Band contour illustrations refer to bands marked *. These contours are on a common frequency scale (see first diagram); frequency increases to the *right*. Intensity scales vary; most simply record plate blackening.

thus 0-0 denotes the transition between zero-point levels; a symbol such as $6a_0{}^1$ denotes a $(v' = 1) \leftarrow (v'' = 0)$ transition in a vibration ν_{6a} where $6a$ is the authors' serial number of the vibration (this particular case is chosen as an example because of its prominence in azine spectra: it is the angle deformation mode which flattens the hexagonal ring); $x_0{}^1$ denotes a $1 \leftarrow 0$ transition in an unclassified vibration x. The symmetries of vibrations can be inferred from the statements on band types, compared with the band type of the 0-0 band.

Ground-state constants are given in the next column. Values in parentheses are calculated from observed X-ray structures, or structures postulated from general molecular structural data, and the four-place precision is probably grossly optimistic. Thus, in the case of azulene[32] the X-ray structure yields a value for A in error (relative to the microwave data) by $+0.0006\ \text{cm}^{-1}$ (1.5%), and the electron-diffraction structure a value in error by $-0.0006\ \text{cm}^{-1}$. Or to give another example, in gaseous pyrazine the rotational evidence is that the nitrogen atoms are 0.25 Å further apart than indicated by the X-ray structure of the crystal.[33] For this reason, values of κ not directly observed are quoted only to two decimal places or even to only one in the case of near-oblate tops where κ is an exceedingly sensitive function of the ring dimensions. Ground-state constants quoted to five or six places are microwave results. Ground-state constants quoted to four figures come from the electronic spectra, through the use of combination differences, for example, in the symmetric top approximation.

$$R_{J-1,K} - P_{J+1,K} = 4\bar{B}''(J + \tfrac{1}{2})$$

where the subscripts on R and P are lower-state quantum numbers of the corresponding R and P lines (or clumps of lines), respectively. This particular relation gives $\bar{B}''$.

In a few cases there are given data also for the inertial defect Δ, the Coriolis coupling coefficients ζ_i for vibration i, and the dipole moment μ, obtained for excited states from Stark effect measurements mentioned later. Units are cm^{-1} for A, B, C; amu-Å^2 for Δ; 10^{-18} esu-cm for μ; and κ and ζ are dimensionless.

Excited-state constants are given as changes from the ground state, the quantities directly measured. Ordinarily, no figures have been quoted to more than five places, and even the fifth figure after the decimal should probably be treated with reserve unless it has been very specifically argued for. This is especially the case in hybrid bands, for in no analysis so far published has allowance been made for the slight reorientation of the inertial axes on excitation. The theory of such "axis-switching" has been developed by Hougen and Watson.[34] Sub-bands forbidden in first order are intensified. The consequences as seen in small-molecule spectra are considerable enough to merit exploratory consideration in work on aromatic molecules of low symmetry.

The last column gives the authors' assignment of the fine structure, though it seems certain that some of this structure might be more advantageously described in terms of the $2J - K_c$ mechanism which should be of wide generality: for example, Brown[15] so classifies the structure in phenol and benzothiadiazole, the C value used for indene was so derived,[35] and Innes and Parkin[36] for pyrazine arrived at the same principle by another, more approximate, route.

IV. APPLICATIONS

The results of high-resolution studies of aromatic molecules have been used in three ways: to determine polarizations of transitions, to cast light on excited-state geometries (if not fully to determine them), and to explore the elusive question of line widths in excited states. In addition, Stark splittings have yielded a new kind of information in the form of excited-state dipole moments.

A. Polarizations

The determination of the polarization of vibronic bands, and thereby of the symmetry of the excited electronic wave function, may be partly achieved through vibrational analysis alone, but the method is most useful when applied to forbidden transitions and even then is far from simple to carry through. It presupposes a detailed knowledge of the ground-state vibrations and usually requires that the fluorescence be observable as well, in order that these vibrations can be comprehensively seen in the electronic spectrum. The complete vibrational analysis of a large molecule is a formidable undertaking. The nearly settled list of vibrational fundamentals of naphthalene, for example, is the result of over twenty major studies (including band contour considerations[37]) spread over almost as many years. The vibrational analysis of anthracene is only now beginning to look as though it has more correct assignments than doubtful ones. And fluorescence, the other needed observation, has not been observed (or at least is intolerably weak) for many substances. The nitrogen heterocycles in general are especially grudging in this respect.[38]

For this reason much reliance has come to be put on polarized crystal spectra, either of pure crystals or of crystals in which the molecule to be studied is isomorphously incorporated in the lattice of a transparent host. Both techniques have considerable limitations. In pure crystals, the polarization ratios can be drastically altered by intermolecular coupling and the mixing of different free molecular transitions: the attempt to resolve free molecule problems frequently becomes one of contending with properties of exciton bands.[39] McClure[40] introduced the use of mixed crystals in an attempt to isolate absorbing molecules from each other, and thus more nearly to approximate the ideal of an oriented gas. The first systems studied yielded admirable results. Yet there are still polarization anomalies due to guest-host coupling, and if the guest and a strongly absorbing host level are close together the latter can call the tune entirely; with increasing use of the technique more and more anomalies are coming to be discovered. Two examples may be instructive:[41] azulene in naphthalene and biphenyl behaves tolerably like an oriented gas, yet in durene a type B transition (in rotational terminology) which should have polarization ratios 1 : 1 : 0 in the $a : b : c'$ crystal directions, is immaculately polarized except for the origin, which is exclusively b polarized—a result not obtainable from any combination of type A, B, and C transition moments if the guest molecules lie parallel to host molecules. The $\pi^* \leftarrow n$ systems of azanaphthalenes seen in naphthalene and durene as hosts likewise abound with anomalies: upward of four separate origins (different sites?), or in the one spectrum as many as six different polarization types.

By contrast, band contour analysis is unambiguous, and is also the *only* practical route to the secure determination of the direction of transition moments in molecules of symmetry C_{2h} or less. In molecules of symmetry C_{2v} and higher x, y, z axes can be established *a priori* such that the transition moment must lie along one of them. Polarized crystal spectra, despite their complications, can often identify the correct one of three orthogonal alternative directions, but if the transition can lie anywhere in a molecular plane, or anywhere at all, such spectra, because of the possibility of perturbation, are far too unreliable to stand alone. Band contour analyses at the present level of performance can determine the proportions of mixing in a twofold hybrid to about $\pm 5\%$, which means $\pm 13°$ for the direction of the transition moment of a band which is barely hybridized at all, or $\pm 3°$ for a 50-50 hybrid, since intensities are proportional to direction cosines squared. However, there is an inherent ambiguity about the actual direction which may leave the resultant transition direction considerably in doubt. For example, the hybrid character of the first transition of thionaphthen is given[42] as mostly B, plus $10 \pm 10\%$ A. The sense of rotation away from the B direction is not determined by the hybridization, the angle of inclination to B being thus $\pm 18°$ with an $18°$ uncertainty either way. That is, the polarization direction could lie anywhere within a $72°$ arc. It may perhaps be valid to use analogical arguments to resolve the phase uncertainty. In this case, related transitions in indene, indole, and benzofuran lie rather close to the long ($\sim A$) molecular axis. The A axis in thionaphthen is rotated 34°, towards the sulfur atom, away from a long axis defined as perpendicular to the bond between the two rings. The alternative polarization direction which brings the transition moment nearer this long axis is thus more likely, and the preferred direction is thus one 39° from the long axis, away from the sulfur atom—but plainly this is only a very tentative conclusion.

In 2-aminopyridine[43] on the other hand, the run of numbers is much more favorable and the transition moment is rather definitely $42 \pm 6°$ from the C—NH_2 direction, the direction of the rotation being away from the ring nitrogen atom.

These determinations are of considerable theoretical significance, for they impose new and needed constraints on the acceptibility of computed wave functions. Compounds of low symmetry have been relatively unpopular, as objects of theoretical interest, simply because it has not been possible to characterize the properties of their excited states by any property other than energy and the scalar quantity oscillator strength. Conversely, compounds of high symmetry are subject to such severe constraints on their wave functions that even rough calculations may give acceptable results. The two results first quoted, like others given in Table I, throw out a considerable challenge and opportunity in, respectively, the difficult theory of the d orbitals in sulfur, and the theory of perturbed benzene rings.

For these reasons, rotational band contour analyses can now, in the short time that the method has been pursued, be said to have settled (in concert with vibrational analysis) as many completely definite assignments of upper-state symmetries as all other methods put together. Of the list in Table I, comprising 36 transitions in 33 molecules, only three transitions had been conclusively assigned by other means, and the number of other transitions in aromatic molecules which are reassuringly well understood is less than a dozen.

A particularly notable achievement of band contour analysis was the assignment of a Rydberg transition in pyrazine,[44] the first assignment which had not been in part anticipated by other lines of evidence. The symmetry attributed to the Rydberg states of all monocyclic aromatics depends uniquely on this result.

A subtle application of the method has been made in interpreting a sequence band in the spectrum of benzoquinone.[45,22] Sequence bands derive from vibrational excited molecules, and a particular transition from the zero-point level (parent band) is invariably accompanied by a host of bands recording like transitions from thermally accessible higher vibrational states. Because frequencies change on excitation, sequence bands mostly stand clear of the parent, though near overlapping is a problem in rotational analysis (a particularly good example is seen in pyridazine;[31] the sequence interval between two sharp ^{q}Q peaks is only 1 cm^{-1} causing the central feature to appear split). Mostly, sequences tend to be treated as irrelevant complications in the way of identifying the parent band, much as the Alps were long regarded as so much obstructive rubbish swept from the desirable plains of Lombardy. In fact, of course, they contain, if only they can be identified, almost complete information about the lower frequencies of the excited state. Christoffersen and Hollas[22] have succeeded in identifying a sequence band in benzoquinone as arising from one of two possible non-totally symmetric vibrations, because a resulting reversal of nuclear statistical weights, relative to the levels involved in the 0-0 transition, radically redistributes intensity within the band envelope.

Differences in band contours corresponding to different polarizations are frequently so marked that differences in polarization (even if not specifically identified) can often be perceived at once even from spectra taken at quite low resolution. The case of naphthalene has already been cited; pyrazine and benzoquinone provide equally good examples. Detailed calculation can then demonstrate whether or not both contours are compatible with the same or only very slightly different rotational constants: this has proved to be so in all cases so far examined. Yet there is reason to be cautious in inferring that very different contours necessarily imply different polarizations, for enough is not yet known about the variation of the effective rotational constants with vibrational quantum states. Table I deliberately includes such data as are available in order to provide some guidelines, and this data may be profitably read in conjunction with Figure 4, which is a guide to the type A bands of azulene.[10]

The point of this figure is the extreme variability of band type resulting from changes in ΔA and ΔB which are small compared with A and $\bar{B}$ individually. A spectrum, as yet incompletely analyzed, in which this point may well be important is that of quinoxaline (1,4-diazanaphthalene).[41] It was noted some time ago that the first ($\pi^* \leftarrow n$) system contains two types of bands,[46] one type very broad, the other sharply peaked, and it would not be unnatural to take these bands as belonging to different transitions or different polarizations within the one transition. Closer inspection, however, reveals an almost continuous gradation of band types, and additionally even the sharp parent bands, which should have as many sequence bands built on them as does naphthalene (about 200), in fact have less than 10. A reasonable inference is that the great majority of sequence bands are too diffuse to be seen, and that the effect of vibrational excitation can readily

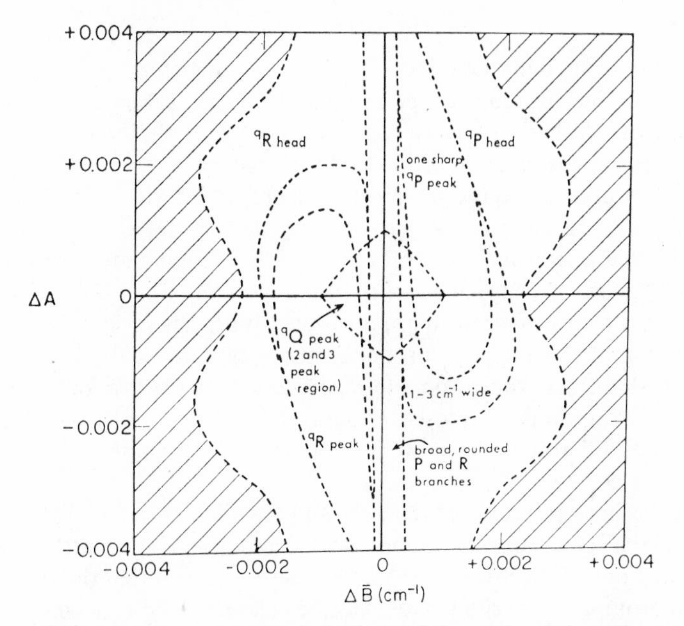

Fig. 4. Synoptic diagram of the form of type *A* bands in azulene. Shaded areas contain no peaks or heads. Reproduced from A. J. McHugh and I. G. Ross, *Spectrochimica Acta* **26A**, 441 (1970) by permission of the publishers.

move bands in and out of the structured area of diagrams like Figure 4 (which applies quite adequately to quinoxaline and as far as its boundaries are concerned, to type *B* and type *C* bands as well). If so, then the entire spectrum could well be a single transition with polarizations not necessarily following the sharp-broad classification (mixed crystal spectra have been interpreted as one transition,[47] but correlating these spectra with that of the vapor is not easy). Note, though, that this is an illustrative possibility, and *not* an undisputed interpretation.*

* *Added in proof.* Much progress has been made recently in the analysis of the first absorption region of quinoxaline as a single transition, and the origin has been positively identified.[48a, 41] The best apparent examples of a gradation of band contours are, in fact, due to coincidences of bands, one sharp (rotational type *A*) and the other broad (type *C*), associated with two almost identical vibrational frequencies of different symmetries. A third intermediate contour (type *B*) has also been discerned. Nevertheless, the contours are not uniform along the spectrum and the failure to see the sequence bands still needs explanation. In addition, it is of interest that among the azanaphthalenes, of which several are currently under study, $\pi^* \leftarrow \pi$ transitions show rotational fine structure (quinoline, isoquinoline), but where the first transition is $\pi^* \leftarrow n$ no such structure has been found.[69, 41, 48b]

B. Geometries

The complete determination of molecular structure from measured moments of inertia requires, quite apart from measurements, a resolute program of synthesis of isotopic variants, and in the compounds considered here deuterium substitution must be complemented by heavy atom substitution to secure sufficient independent information. The microwave work of Bak and co-workers is particularly notable in this area: for example their determination of a complete ground-state structure for pyridine[49] from measurements of pyridine-d_0, -2-d, -3-d, -4-d, -2-^{13}C, and -3-^{13}C. In the area of electronic spectra only Innes has seriously attempted to gather as much information about any single molecule, but even the work of Merer and Innes[5] on s-tetrazine (see Table I) invokes some simplifying assumptions. Proposed structures for other molecules draw on other kinds of argument as well. Proposed excited-state structures are here discussed.

First, because the topic is separate and concise, we refer to the matter of the planarity of aromatic molecules, in any electronic state. A reason commonly offered for believing that benzene is planar is that the vibrations, in both ground and excited states, can be classified as g or u, showing the presence of a center of symmetry. This determination does not set quantitative bounds to possible small distortions, since if benzene were only slightly nonplanar or its potential for small out-of-plane displacement nonquadratic, the planar selection rules would be only very slightly relaxed, and the consequences could be overlooked. X-ray crystallography cannot assist at this level or precision: in fact, in crystals, aromatic molecules appear to be slightly buckled. In other aromatics, proposals for nonplanar excited states have often been made, but have not survived further investigation, and the situation at present is that there is no evidence from vibrational structure that compels consideration of a nonplanar ring in any state of any aromatic molecule.

The requirement that the inertial defect Δ be zero for planar molecules would seem to offer a direct and precise settlement of the matter. Consider, however, the possibility that all the atoms in benzene and naphthalene are located 0.01 Å above or below the mean molecular plane. The resulting values of Δ are then just -0.016 and -0.026 amu-Å^2, respectively. The observed microwave figures for closely related dipolar molecules are $+0.03$ and -0.15, and these figures are attributable to vibrations. Thus the consequences of this slight degree of nonplanarity in the rigid molecule model are less than the vibrational contributions to Δ. A few cases appear in Table I in which Δ has been estimated for excited states; there are no great changes on excitation and the nonzero values obtained are quantitatively reasonable for vibrational perturbations of a planar nonrigid model.[21] However, since Δ increases as the squares of the out-of-plane displacements it seems at least safe to say that any nonplanarity cannot exceed the zero-point amplitudes of the out-of-plane vibrations—a few hundredths of an Ångström unit.

Of course, in certain of the molecules in Table I, Δ is necessarily nonzero (e.g., indene) and in substituted benzenes the disposition of polyatomic substituents has been a lively issue. Phenol and aniline are the notable cases. Phenol was shown to be planar in the ground state from microwave data ($\Delta = -0.03$), and all the evidence points to planarity in its first excited state as well (indeed, the barrier to rotation about the CO bond is greater).[50,51] The long-standing problem of aniline in the ground state was simultaneously resolved from the microwave spectrum and by Bist, Brand, and Cook[52] in a notable vibrational analysis of the ultraviolet spectrum; the HNH plane is inclined at about 36° to the ring plane in the ground state, and in the first excited state is either less inclined or else, in Brand's terminology, the NH_2 group is "quasi-planar" (flat-bottomed potential). Christoffersen, Hollas, and Kirby[53] subsequently analyzed the band contours, finding a substantial value of Δ consistent with a reduced NH_2-ring angle of 30°, though it must be conceded that there is as yet no theory of the inertial defect in quasi-planar molecules, and the rotational analysis cannot be said to have distinguished between the two alternatives. For this purpose the anharmonicities of the vibrational levels remain the most telling source of information.

We now consider bond lengths and angles. For hexagonal benzene, with only two structural parameters (CC and CH), the band contours of C_6H_6 and C_6D_6 yield interdependent values of ΔCC and ΔCH which agree with previously quoted estimates based on vibrational parameters (but are no more secure). Only for *s*-tetrazine has the excited-state geometry been as rigorously determined. More generally, attempts have been made to link the partial information given by rotational constants for a few isotopic variants with the intensities of vibrational progressions interpreted through the Franck-Condon principle, and with chemical intuition. In principle it should be possible to go a long way by combining these kinds of arguments; in practice it has proved difficult to reach consistent conclusions. Franck-Condon analysis depends on good measurements of band intensities (difficult because of overlapping) and accurate normal coordinate analyses (usually lacking).

Such Franck-Condon analyses as have been attempted are not particularly encouraging, and in any event lead to multiple solutions. That is, the intensities seen in a particular progression associated with a normal coordinate Q can almost always be reproduced by a shift along that coordinate of $\pm\Delta Q$, and this ambiguity applies to every progression. The failure of conscientious Franck-Condon analyses (e.g., on benzoquinone[54]) to match up with the inferences from rotational constants may be simply due to unreliable force fields and neglect of vibrations which, while individually inconspicuous, collectively contribute too much of the intensity to be ignored. But we also have reason to suspect,[41] in certain allowed transitions at least, that the orderly run of progressions (i.e., intensities proportional to squared vibrational overlap integrals) may be routinely distorted by simultaneous vibronic intensity stealing, even though the vibrations are totally symmetric.[55]

At times theoretical or intuitive considerations may be used to deduce results which, if not secure, are at least provocatively consistent. Two examples merit mention. Cvitas, Hollas, and Kirby[56] have considered collectively data on five monosubstituted and five *p*-disubstituted benzenes, and rationalize all the observed values of ΔA, ΔB, and ΔC through a model which assumes that the effects of two substituents are almost independent. They then conclude that excitation to the $^1B_{2u}$ benzene state expands all ring bonds, but contracts the ring-substituent bonds in the amounts N—C 0.08 Å, O—C, Br—C, Cl—C 0.04 Å, F—C 0.02 Å, the adjacent ring angle simultaneously opening by a few degrees. The slight interaction between substituents in *p*-distinguished benzenes is manifest as a slight additional contraction of the ring dimension along the long molecular axis.

The analysis just described is based on empirical proposals about bonds and angles which fit first the observed ground-state constants, and then the upper-state constants. A more overtly computational approach, but again including speculative assumptions, appears to be applicable to unsubstituted aromatic hydrocarbons, of which, however, there are few that yield suitable spectra. This method[57,58,32] uses the bond orders of molecular orbitals theory to predict bond lengths in excited states, and a purely heuristic assumption that the bond angles of aromatic rings are such as minimize the strain imposed by unequal bond lengths on rings which would otherwise be regular polygons. The method works acceptably for ground states, was used to predict in advance rotational constants for naphthalene which proved to be correct, and has been successful for azulene and azulene-d_8 as well. The predicted bond length changes in both of these molecules are mostly less than 0.02 Å, and are mostly extensions, except for the long central bond of azulene which is thought to contract on excitation by 0.03 Å. This contraction is responsible for the zero change in A in that molecule. Angles appear to change by no more than 1°. Even if these calculations are hardly a substitute for rigorous determinations, they do appear to explain the scale of values consistently observed for ΔA, ΔB, ΔC ,and have been used (see Section IVC) in a consideration of the general problem of the definition obtainable in polyatomic spectra.

Finally, we consider the changes of shape of N-heterocycles, consequent on $\pi^* \leftarrow n$ excitation.[59] (Quasi-)theoretical approaches are presently not useful, because the theory of n, π^* states now appears to be in a state of flux, if not actual disarray. The matter may be simplified for present purposes to a single issue: by how much does the ring angle at a nitrogen atom open or close on $\pi^* \leftarrow n$ excitation? The Franck-Condon activity of the ring-squashing mode indicates changes of the order of $\pm 3°$. The obvious objects of study, the $\pi^* \leftarrow n$ transitions of pyridine and *s*-triazine, are unsuitable as yet. In pyridine, only the -4-*d* isotope, of several that have been examined, has bands of well-formed contour.[60] The high-resolution spectrum of *s*-triazine is chaotic.[61] Innes and co-workers have, however, pursued the four other known azines with considerable tenacity, and have offered a complete structure for excited *s*-tetrazine, with only four independent structural parameters. However, even the six isotopes studied do not provide more than three independent structural quantities in the symmetric top approximation. In addition this approximation (i.e., $\bar{B} = 2C$) itself introduces uncertainties, and finally the analyses actually performed were not all on 0-0 bands. However, the coordinates of the hydrogen and carbon atoms are explicitly determined and the indeterminacy is in the nitrogen atoms, for which only the radial distance from the molecular center is determined. New information is added by acknowledging asymmetry and supplying, from the contour calculations, a value for $\Delta\kappa$. In this way, the problem is in principle determinate, and led to the following provisional figures:

$$\Delta \text{NCN} = -7.9° \text{ (whence } \Delta \text{NNC} = +4.0°\text{)},$$

$$\Delta \text{CN} = +0.05 \text{ Å}, \quad \Delta \text{NN} = -0.11 \text{ Å}, \quad \Delta \text{CH} \sim 0$$

More detailed "line" contour analysis has subsequently yielded values for ΔA, ΔB, and ΔC individually for the four isotopic species that have analyzable 0-0 bands. From these new figures come revised structural parameters as follow:[5b]

$$\Delta \text{NCN} = -11.9°, \ \Delta \text{CN} = +0.08 \text{ Å}, \ \Delta \text{NN} = -0.17 \text{ Å}, \ \Delta \text{CH} = -0.03 \text{ Å}$$

Qualitatively, these figures mean much the same as those given above, but the considerable quantitative change provides a further commentary on the difficulty of determining data of this kind for the general run of polyatomic molecules.

These results stand in sharp conflict with the naive version of $\pi^* \leftarrow n$ theory which treats the lone pair (n) orbitals of nitrogen as independent of the σ-bond system of the ring (but see recent calculations of Clementi[62]) and which predicts for the lowest transition in this molecule no significant change in bond lengths, and in the most likely alternative assignment of the transition, no significant change in the NN distance. However, the determination of the effect on the angles at the nitrogen is definite, and seems to be compatible with a less fully determined result in pyrazine,[63] in which the excitation brings the nitrogen atoms about 5% closer together, again most likely through opening of the angle. The argument here comes from a comparison of pyrazine and pyrazine-$^{15}N\,^{14}N$, which shows that increasing the mass at the nitrogen atom brings $\Delta\bar{B}$ (negative) nearer to zero; that is, assuming $\bar{B} = 2C$, the increase in I_c is diminished. However, it is instructive to remark that this conclusion (that CNC angles open up) reverses an apparently equally conclusive decision made on the basis of only slightly less accurate data.[36] In pyridazine the evidence once again (but now drawing extensively in Franck-Condon arguments) favors nitrogen-angle opening,[31] but in pyrimidine it may be that there is no change at all.[64]

C. Sharp and Diffuse Transitions

It has been suggested[46,65–69] that the diffuseness which characterizes nearly all the valence-shell transitions in molecules with more than about six atoms is associated with radiationless transitions, sufficiently rapid to cause uncertainty broadening. The matter has recently been the object of considerable theoretical attention.[70]

Vapor spectra are subject to various kinds of trivial causes of diffuseness, such as overcrowding of sequence bands,[67,69] but they may also appear diffuse if the rotational structure is broad and featureless. Calculations of upper-state geometries, of the kind described in Section IVC, have been applied to the known diffuse higher-energy transitions in three aromatic hydrocarbons.[10] They uniformly predict changes in inertial constants which would lead one to expect that the bands should be structured. That is, if the various calculated values of ΔA and $\Delta\bar{B}$ are plotted on diagrams such as Figure 4, the points all fall within the structured region.

An extensive survey of the high-resolution absorption spectra of a large number of larger polyatomic molecules[69] confirms the generality of diffuseness among excited states other than the first singlet, the first triplet, and Rydbergs. The calculations referred to in the previous paragraph show that diffuse rotational profiles cannot be the reason in certain specific instances, such as the third singlet transitions of azulene and naphthalene. Thus they force consideration of alternative explanations. The conclusion drawn, expressed in terms of time-dependent theory, is that these diffuse transitions owe their broadening to rapid nonradiative processes, which in certain cases must be predissociation or preisomerization, and in others is electronic relaxation. By electronic relaxation is meant radiationless transfer to the higher vibrational levels of lower electronic states; internal conversion and intersystem crossing are alternative names for the same process. The same explanation can be expressed alternatively in stationary state language as a coupling between individual levels of the electronic state in question and a continuum or quasi-continuum (dense vibrational manifold) associated with an electronic state of lower energy. The generalization concerning the incidence of diffuseness has a few exceptions that appear to be explicable in terms of a wide gap to the nearest lower state; the very sharp second excited state of azulene at 28757 cm^{-1} is the prime example. There are also more puzzling cases; the most notable is the five-membered ring aromatics (including, from very recent work,[71] fulvene). These compounds show no really sharp transitions, which explains their conspicuous absence from Table I, but even so their second transitions are, in most cases, defined much better than their lowest ones.

Considerations of sequence band crowding make it possible to predict that rotational envelopes, sufficiently well resolved for analysis, will not be observed for molecules significantly larger than anthracene,[67,69] unless the technology of multiple reflection cells can be so improved that these sparingly volatile large molecules can be studied at greatly reduced temperature. Flexible substituents also induce diffuseness; for example, rotational contours are just visible in isopropylbenzene, but not in *n*-propyl or higher alkylbenzenes.[69]

These conclusions about line widths are supported by low-temperature crystal spectra, and direct evidence for the effect of radiationless transitions appears to have been seen[72] in the line widths of the first electronic transition of azulene-d_8. The lines for this compound are narrower than those for azulene-d_0, whose excited-state lifetime, determined by the radiationless rate, is probably shorter than that of -d_8. These line widths were observed in naphthalene as host crystal, at 4°K, and because of environmental effects are considerably broader than the fine-structure features of second transition of the same molecule, observed as a vapor. It is not possible, however, to make a meaningful comparison with line widths in the vapor spectrum of the same transition, for the bands there are badly overlapped with sequences and no analyzable features are seen.[13]

Attention is also coming to be directed to the incidence of line broadening in the higher reaches of transitions whose origin and low-energy bands are sharp. Such broadening could be a measure of radiationless transition rates out of individual levels, or equally (or as well) of the rates of photoreactions. The likely operation of the latter effect was pointed out some years ago by Sponer,[73] with reference to the spectrum of benzene. Current very-high-resolution work on benzene's spectrum has revealed the onset of the expected line broadening.[74] There seem to be profound implications in this type of work for the detailed investigation of photochemical and other modes of electronic energy dissipation.

D. Dipole Moments in Excited States

Labhart[75] has recently reviewed a number of ways in which the dipole moments of excited molecules can be inferred from measurements of solution spectra. Of these methods and measurements it is probably fair to say, either that the actual figures obtained depend upon particular models of the liquid state, or else that they apply only to complex molecules with high dipole moments and which accordingly are beyond the reach of the more precise kinds of valence theory. There is thus still a need for direct measurements of excited-state dipole moments which do not depend on assumptions about the polarization of a medium, and which are applicable to simpler molecules. For this purpose the method of choice must be the direct observation of the splitting of absorption or emission lines by applied electric fields.

Hochstrasser and Noe[76] have succeeded in observing such a splitting in pure and in doped crystals. The latter case is simpler. The impurity enters randomly on host crystal sites, that is, head first or tail first. An applied electric field separates the energies of upfield and downfield dipoles and absorption lines are split proportionately to the local field strength and to the change in dipole moment, $\Delta\mu$. In practical fields the splittings may reach several cm^{-1}, compared with line widths of about 1 cm^{-1}, and the change in dipole moment has been measured in this way for the first and second excited states of azulene ($\Delta\mu = 1.2$, 1.1 in units of 10^{-18} esu-cm, used through this discussion). Errors in the method stem in part from problems of alignment of the field with crystal axes, in part from the actual measurement of the splitting, and in part from approximations made in calculating the cavity field; in all they should not exceed 10%. A similar measurement has been achieved for the first singlet state of benzophenone. If, however, intermolecular interactions are strong, so that it is necessary to use exciton functions as a basis for the perturbation, the analysis is more complex; Hochstrasser and Lin[77] present the appropriate Stark (and

Zeeman) theory for the triplet state of benzophenone, and from one of the most intricate experiments yet performed on molecular crystals obtain a value $\mu' = 1.75$ for this triplet state.

Meanwhile, direct observations of excited-state dipole moments had already been achieved in the vapor phase, from observations on the Stark effect on band structure. The first such measurement was on formaldehyde, by Freeman and Klemperer,[78] who used electric fields as high as dielectric breakdown will permit. Subsequent measurements by the same group have extended the technique first to larger carbonyl compounds, such as propynal,[79] and very recently to the cyclic molecules which are of concern here. Excited-state dipole moments of aromatic molecules can now be considered accessible, in favorable cases at least, following measurements by Lombardi[80,81] of $\Delta\mu$ for aniline, phenol, *p*-fluoroaniline, and *p*-fluorophenol.

The basic theory is still the conventional theory of the Stark effect in microwave spectroscopy. That is, symmetric rotor wavefunctions are now coupled, not only by the matrix elements attributable to asymmetry but also by matrix elements proportional to field strength and dipole moment, according to selection rules which are of course the same as those for electric dipole transitions between states classified by J, K, and M (M measures the components of J). They are, for a general orientation of μ, $\Delta J = 0, \pm 1$, $\Delta K = 0, \pm 1$, $\Delta M = 0, \pm 1$. The M degeneracy is lifted, and J is no longer a good quantum number. The energy matrix is infinite but for any practical calculation can still be safely truncated at J somewhat greater than the highest levels of interest.

The general theory as applied to electronic spectra has been set out in clear detail by Lombardi.[82] There are two limiting cases. Consider specifically an asymmetric rotor of the more nearly prolate kind (κ negative). Where the asymmetry splitting is small (cf. Fig. 1), the electric field perturbation acts on near-degenerate levels and displacements of levels are proportional to the applied field (first-order Stark effect). Specifically, the displacements are proportional to $\mu M K_a / J(J+1)$, and the consequences will be most evident for low J and relatively high K_a. (Figure 1 might be taken to suggest that low K_a levels, where the oblate rotor approximation holds, might also be favorable. It should be recalled, however, that Gora's approximations, with neglect of asymmetry splitting, apply best to high J levels. First-order splittings associated with levels of high J and very low K_a have yet to be seen.)

For lower K_a, where the asymmetry splitting may exceed the electric field perturbation, the latter may be treated through second-order perturbation theory, and the splitting is then proportional to the square of the field strength. While the information content of this kind of splitting is potentially high, the effect has not yet been detected in large molecules.

The analysis of large-molecule Stark-perturbed spectra is once again a matter of contour fitting, with assumed values of $|\Delta\mu|$, of those parts of the band envelope (if any) which are markedly changed by the field. Spectra are measured with light polarized parallel (selection rule $\Delta M = 0$) and perpendicular ($\Delta M = \pm 1$) to the field. In F_2CN_2[19]—the smallest of the ring molecules in Table I, and interesting as being a highly asymmetric top—the field observably affected only the ^rR fine structure at the high-wave-number end of the band. These peaks are sub-band heads, labeled by K_a, and their positions are calculable using the quasi-symmetric top equations. Since $J \geqslant K_a$, the fact that they are perturbed shows that the intensity in each peak is generated by lines with J not greatly in excess of K_a.

The Stark perturbation was most clearly seen in perpendicular polarization, for then the lines with $|M| = J$ are more intense than those with $M = 0$. Intensity is thus thrown out into the wings and the lines, though still unresolved, generate splittings of the original peaks. In F_2CN_2 these splittings, of irregular appearance because of the composite nature of the peaks, were fitted in the course of trial calculations with an uncertainty in $\Delta\mu$ of 0.2×10^{-18} esu-cm, to which should be added perhaps an additional uncertainty of about the same order due to the possible errors in the K_a numbering of the peaks. (The K values of the peaks run from 12 to 21, and could be in error by ± 2; the splitting is proportional to the product μK_a.) Dipole moments of ground states, obtained from traditional dielectric constant measurements, are hardly more reliable.

The analysis of the Stark spectra of aniline and phenol was similarly conducted from ^pP branch heads, with $K_a \sim 27$ and again $J \sim K_a$.

The sign of $\Delta\mu$ was not determined in these spectra, though in smaller molecules where splittings can be seen in more than one branch, the measurement is absolute.*[78] Nevertheless, the matter of sign can frequently be resolved by general chemical considerations, which are frequently secure enough to give strong weight to one or the other of two very different alternative values of μ'. Thus for phenol in the ground state μ'' is 1.45 with the negative end (from classical vector addition arguments applied to chlorophenols) on the ring. μ' is 1.25 or 1.65 in the same sense. The familiar valence-bond arguments of resonance with quinonoid structures, supported by the fairly secure conclusion that the CO bond shortens (Section IVB), provides a strong reason for preferring the latter. Likewise, in aniline

* *Added in proof.* An absolute determination of $\Delta\mu$, by the method referred to, has been reported recently for fluorobenzene. See Table I.

$\mu' = 2.4$ is preferred to $\mu' = 0.7$, and in *p*-fluoroaniline $\mu' = 3.3$ to $\mu' = 1.7$ (μ'' is 2.5).

For *p*-fluorophenol a direct physical argument was available to resolve the choice. Studies of fluorescence as a function of pH yield acid dissociation constants for excited states. For *p*-fluorophenol, excitation decreases the pK_a; that is, the excited molecule is the stronger acid. This points to an increased positive charge on the oxygen atom. Therefore $\mu' = 2.6$ rather than $\mu' = 1.8$ (μ'' is 2.2).[81]

The above discussion has been wholly concerned with Stark splittings. A magnetic field (Zeeman) splitting, described as large, has been observed for the triplet state of pyrazine vapor,[83] but details have not been published and there are no other reports of Zeeman effects in vapor spectra of large molecules. This is doubtless in part due to the difficulty of detecting the weak triplet absorption in most molecules; pyrazine and *p*-benzoquinone are rather exceptional in this regard. In crystals, however, the previous reference to benzophenone[77] will indicate the sophistication already achieved in Zeeman effect measurements, and theory, in the solid state.

V. EMISSION SPECTRA

There has been no work on emission spectra at resolutions comparable with the absorption studies which have been our concern up to this point, and hence no rotational interpretations. Nonetheless, the topic merits brief comment, to complete the experimental picture.

Emission spectra have received rather little attention, in part for the simple reason that they are hard to measure. Low-pressure vapors irradiated with available monochromatic light sources are weak emitters, and exposure times in typical systems with slow high-resolution spectrometers would run into days or weeks. Also, in the majority of larger molecules emission spectra are found to be diffuse unless irradiation has been applied specifically to the 0-0 band, a generalization supported by the spectra of naphthalene,[84] anthracene,[85] indazole,[38] and isoquinoline,[38] in which it has been possible to achieve such excitation (the experimenter is much restricted by the availability of strong arc lines, and of filters to remove unwanted lines). It is not clear why excitation elsewhere should generally lead to poorly defined spectra: an explanation often offered is that the vibrational energy is randomly redistributed before emission. This is possible, though there are grave objections in principle to the practice of characterizing the redistribution energy, even under no-collision conditions, by a "temperature." However, it is likely that part of the reason for the diffuseness is simply that monochromatic excitation at an arbitrary frequency will normally generate excited molecules in many vibrational levels. This is because of the dense packing of sequence structure and because, as the contours described here abundantly indicate, each individual absorption band extends over 30 cm^{-1} or more. Excitation in the 0-0 band will be effective if, as is often the case, it is at one end of the sequence structure, in which case it will be the least overlapped band in the spectrum. A significant experiment, not yet attempted, would be to excite fluorescence in a 1-0 band which likewise stands at one end of the sequence structure and which is well separated from the sequence envelope of any strong preceding band. If the emission is not sharp, the vibrational energy redistribution must have taken place.

Benzene and some of its simple substituted derivatives do not conform with this description. The emission spectrum of benzene is sharp when excited monochromatically at energies as high as 2400 cm^{-1} above the 0-0 band,[86,87] beyond which no fluorescence is seen. The narrow 2537 Å line group of a low-pressure mercury discharge is much narrower than the rotational envelope of the absorption bands, so that with 2537 Å irradiation only selected rotational states are excited. At very low pressures of benzene vapor, emission occurs without prior collisions and the emission spectrum is largely linelike, reflecting the excitation of a limited number of J states; with increasing pressure, the band contours develop and the changing profiles, at present recorded only at medium resolution, show the course of rotational relaxation, while the development of sequence bands records quantitatively the rate of vibrational relaxation.[88] Elegant work on this spectrum has been carried out recently by Parmenter,[89] using selective excitation in particular bands using a continuous light source and a monochromator. The practical limits of resolution in this type of irradiation experiment are of course far less than can be achieved in absorption, because the luminescence efficiencies of low-pressure vapors are so low: Parmenter's bandwidths for excitation were of the order of 35 cm^{-1}. Nevertheless, by careful avoidance of excitation in overlapping bands it has been possible to disentangle the previously unanalyzed low-pressure fluorescence, and to demonstrate that 2537 Å excites six different vibrational levels. The principles of analysis of the low-pressure emission having then been grasped—they prove to be reasonably conventional—excitation in regions where the absorption spectrum was unsatisfactorily analyzed has been used to unravel the absorption structure. It has been said that high resolution is as much an attitude of mind as a matter of instrumentation. Certaintly, this approach to the remaining unresolved complexities of benzene's spectrum is as delicately surgical as anything that has been done in absorption.

VI. SUMMARY

Band contour analysis of high-resolution spectra appears to be applicable to the first singlet and triplet transitions, and some Rydbergs, of many molecules at least as large as simply substituted naphthalenes, and may in favorable cases reach as far as tricyclic compounds. Present techniques of contour fitting permit the determination of changes in rotational constants to an accuracy of about 10^{-4} cm^{-1}, fairly routinely, and in favorable cases the presence of fine structure may permit accuracy to 10^{-5} cm^{-1} in some or all of the constants. These figures represent determinations of changes in the principal moments of inertia sensitive to, say, 0.5% to 0.05% of their ground state values. The next generation of high-resolution instrumentation could increase the routine precision to 10^{-5} cm^{-1} or slightly better, by detailed fitting of the contours of individual fine structure features formed by near-coincidences of strong lines. Nevertheless, there is a limit to what can be achieved with improved resolving power in conventional absorption spectroscopy, since an instrument with resolving power of 10^6 would be Doppler limited (at 300°K) for all molecules small enough to show sharp spectra.

A conventional objective of rotational analysis in small molecule spectroscopy is the determination of geometries in excited states. Among larger molecules, data obtained on suitable sets of isotopic species have in two cases yielded structures of excited states, which are about as precise as conventional X-ray structures of ground states. Bond lengths, for example, are determined to about 0.2 Å. However, this has been achieved only for benzene and *s*-tetrazine, each a symmetrical molecule with relatively few independent structural parameters. Structure determination from isotopic moments of inertia is a well-explored procedure in microwave spectroscopy where its application is limited so far to molecules of the degree of complexity of, for example, pyridine. Moments of inertia measured in electronic spectra are somewhat less accurate (compare Table I which includes data from both kinds of source). Also, certain needed isotopic species may not have analyzable 0-0 bands: pyridine and *s*-tetrazine are examples. For this reason, and because of the amount of work involved, it is unlikely that *complete* structure determinations in excited states will be other than occasional tours de force. Meanwhile, partial structural information is commonly sufficient to test theoretical proposals. Insights obtained so far confirm conventional theory for π, π^* states but not for n, π^* states.

Rotational analyses also determine polarization directions. A measure of the importance of this aspect of the analyses is that nearly all of the really secure symmetry assignments of the excited states of aromatic molecules now rest on the evidence of band contours. Polarizations are determined absolutely if restricted by symmetry to just three orthogonal directions. In molecules of lower symmetry, polarization directions can be determined in favorable cases to $\pm 5°$ relative to the inertial axes, although with an ambiguity of sign. The less favorable cases are those in which the polarization direction lies close to an inertial axis, when the uncertainty in angle is much greater. Still less favorable of course, for the purpose of theoretical comparisons, are molecules, notably flexible ones, in which the locations of the inertial axes are uncertain.

The fine structure seen in many contours has spacings as close as $0.1\ cm^{-1}$, implying that the individual fine structure half-widths can be less than $0.05\ cm^{-1}$. Hardly less significant than the ability to observe this fine structure is the possibility of observing the evolution in such lines, as observed in particular bands, of lifetime broadening due to photodecomposition or radiationless transitions. In lifetime terms, $0.05\ cm^{-1}$ is about 0.03 nsec. Such rapid processes have not been measured, in vapors, by direct decay rate measurements.

The first-order Stark effect, leading to measurement of excited-state dipole moments, has been detected through the perturbation of particular kinds of fine structure due to clustering of lines with $K \sim J$. Just how many spectra will prove to show measurable effects remains to be seen but the prospects are increasingly encouraging.

Acknowledgments

I would like to thank my colleagues in spectroscopy, whose work is cited in the references as "unpublished" or "in press," or with 1970 publication dates, for their generosity in permitting their results to be used in advance of publication.

References

1. A. Turkevich and M. Fred, *Rev. Mod. Phys.*, **14,** 246 (1942).
2. C. D. Cooper and H. Sponer, *J. Chem. Phys.*, **20,** 1248 (1952).
3. S. F. Mason, *J. Chem. Soc.*, **1959,** 1269.
4. H. C. Allen and P. C. Cross, *Molecular Vib-Rotors*, Wiley, New York, 1963.
5. (a) A. J. Merer and K. K. Innes, *Proc. Roy. Soc. (London) Ser. A*, **302,** 271 (1968), (b) A. Khan and K. K. Innes, unpublished work.
6. K. K. Innes, J. A. Merritt, W. C. Tincher, and S. C. Tilford, *Nature*, **187,** 500 (1960).
7. J. A. Merritt and K. K. Innes, *Spectrochim. Acta*, **16,** 945 (1960).
8. T. M. Dunn, in *Spectroscopy*, Institute of Petroleum, London, 1962, p. 303 and in *Studies on Chemical Structure and Reactivity*, J. H. Ridd, Ed., Methuen, London, 1966, p. 103.

9. D. P. Craig, J. M. Hollas, M. F. Redies, and S. C. Wait, *Phil. Trans. Roy. Soc. (London) Ser. A*, **253,** 543 (1961).
10. A. J. McHugh and I. G. Ross, *Spectrochim. Acta*, **26A,** 441 (1970).
11. J. L. Duncan and D. Ellis, *J. Mol. Spectrosc.*, **28,** 540 (1968).
12. G. Herzberg, *Infra-red and Raman Spectra of Polyatomic Molecules*, Van Nostrand, Princeton, N.J., 1945.
13. A. J. McHugh, D. A. Ramsay, and I. G. Ross, *Australian J. Chem.*, **21,** 2835 (1968).
14. E. K. Gora, *J. Mol. Spectrosc.*, **16,** 378 (1965).
15. J. M. Brown, *J. Mol. Spectrosc.*, **31,** 118 (1969).
16. J. T. Hougen, *Can. J. Phys.*, **42,** 433 (1964).
17. J. M. Hollas, *Spectrochim. Acta*, **22,** 81 (1966).
18. J. M. Brown, *Can. J. Phys.*, **47,** 233 (1969).
19. J. R. Lombardi, W. Klemperer, M. B. Robin, H. Basch, and N. A. Kuebler, *J. Chem. Phys.*, **51,** 33 (1969).
20. G. O. Sørensen, *J. Mol. Spectrosc.*, **22,** 325 (1967).
21. A. J. McHugh and I. G. Ross, *Australian J. Chem.*, **22,** 1 (1969).
22. J. Christoffersen and J. M. Hollas, *Mol. Phys.*, **17,** 655 (1969).
23. F. M. Garforth, C. K. Ingold, and H. G. Poole, *J. Chem. Soc.*, **1948,** 508.
24. D. P. Craig, *J. Chem. Soc.*, **1950,** 2146.
25. J. H. Callomon, T. M. Dunn, and I. M. Mills, *Phil. Trans. Roy. Soc. London Ser. A*, **259,** 499 (1966).
26. J. E. Parkin, *J. Mol. Spectrosc.*, **15,** 483 (1965).
27. L. Pierce, *J. Mol. Spectrosc.*, in press. See also J. D. Swalen and L. Pierce, *J. Math. Phys.*, **2**, 736 (1961).
28. J. M. Bennett, I. G. Ross, and E. J. Wells, *J. Mol. Spectrosc.*, **4,** 342 (1960).
29. F. W. Birss, S. D. Colson, and D. A. Ramsay, unpublished work.
30. J. H. Wilkinson, *Computer J.*, **1,** 90 (1958); *Numer. Math.*, **4,** 354 (1962).
31. K. K. Innes, W. C. Tincher, and E. F. Pearson, *J. Mol. Spectrosc.*, **36,** 114 (1970).
32. A. J. McHugh and I. G. Ross, *Australian J. Chem.*, **21,** 3055 (1968).
33. J. A. Merritt and K. K. Innes, *J. Mol. Spectrosc.*, **23,** 280 (1967).
34. J. T. Hougen and J. K. G. Watson, *Can. J. Phys.*, **43,** 298 (1965).
35. A. Hartford and J. R. Lombardi, *J. Mol. Spectrosc.*, **34,** 257 (1970); S. D. Colson and G. D. Johnson, unpublished work.
36. K. K. Innes and J. E. Parkin, *J. Mol. Spectrosc.*, **21,** 66 (1966).
37. S. C. Wait and F. E. Shaffer, *J. Mol. Spectrosc.*, **10,** 78 (1963).
38. L. M. Logan and I. G. Ross, *Acta Phys. Polon.*, **34,** 721 (1968).
39. D. P. Craig and S. H. Walmsley, *Excitons in Molecular Crystals*, W. A. Benjamin, New York, 1968.
40. D. S. McClure, *J. Chem. Phys.*, **22,** 1668 (1954).
41. A. D. Jordan, A. R. Lacey, and I. G. Ross, unpublished work.
42. A. Hartford, A. R. Muirhead and J. R. Lombardi *J. Mol. Spectrosc.*, **35,** 199 (1970). The present discussion of these results differs somewhat from this reference.
43. J. M. Hollas, G. H. Kirby, and R. A. Wright *Mol. Phys.*, **18,** 327 (1970). The figure quoted for the polarization direction is a corrected value supplied by J. M. Hollas.
44. J. E. Parkin and K. K. Innes *J. Mol. Spectrosc.* **15,** 407 (1965).
45. J. M. Hollas *Spectrochim. Acta* **20,** 1563 (1964).
46. L. M. Logan J. P. Byrne and I. G. Ross *Proc. Intern. Conf. on Luminescence* 1966 Hungarian Academy of Sciences Budapest 1968 p. 194.

47. R. H. Clarke R. M. Hochstrasser and C. J. Marzzacco *J. Chem. Phys.* **5,** 5016 (1969).
48. (a) R. W. Glass, L. C. Robertson, and J. A. Merritt, *J. Chem. Phys.*, **53,** 3857 (1970); (b) S. C. Wait, unpublished work.
49. B. Bak, L. Hansen-Nygaard, and J. Rastrup-Andersen, *J. Mol. Spectrosc.*, **2,** 54 (1958).
50. H. D. Bist, J. C. D. Brand, and D. R. Williams, *J. Mol. Spectrosc.*, **24,** 413 (1967).
51. J. Christoffersen, J. M. Hollas, and G. H. Kirby, *Proc. Roy. Soc. (London) Ser. A*, **307,** 97 (1968).
52. J. C. D. Brand, D. R. Williams, and T. J. Cook, *J. Mol. Spectrosc.*, **20,** 193, 359 (1966).
53. J. Christoffersen, J. M. Hollas, and G. H. Kirby, *Mol. Phys.*, **16,** 441 (1969).
54. T. Anno and A. Sadô, *J. Chem. Phys.*, **32,** 1611 (1960).
55. D. P. Craig and G. J. Small, *J. Chem. Phys.*, **50,** 3827 (1969).
56. T. Cvitas, J. M. Hollas, and G. H. Kirby, in press.
57. E. F. McCoy and I. G. Ross, *Australian J. Chem.*, **15,** 573 (1962).
58. K. K. Innes, J. E. Parkin, D. K. Ervin, J. M. Hollas, and I. G. Ross, *J. Mol. Spectrosc.*, **16,** 406 (1965).
59. K. K. Innes, J. P. Byrne, and I. G. Ross, *J. Mol. Spectrosc.*, **22,** 125 (1967).
60. F. W. Birss, S. D. Colson, J. P. Jesson, H. Kroto, and D. A. Ramsay, unpublished work.
61. G. Fischer, unpublished work.
62. E. Clementi, *Chem. Revs.*, **68,** 341 (1968).
63. K. K. Innes, *Proc. Intern. Conf. on Spectroscopy*, Bombay, 1967, p. 135.
64. K. K. Innes, H. D. McSwiney, J. D. Simmons, and S. G. Tilford, *J. Mol. Spectrosc.*, **31,** 76 (1969).
65. G. R. Hunt and I. G. Ross, *Proc. Chem. Soc.*, 11 (1961).
66. G. R. Hunt and I. G. Ross, *J. Mol. Spectrosc.*, **9,** 50 (1962).
67. J. P. Byrne and I. G. Ross, *Can. J. Chem.*, **43,** 3253 (1965).
68. R. M. Hochstrasser, *Accounts Chem. Res.*, **1,** 266 (1968).
69. J. P. Byrne and I. G. Ross, *Australian J. Chem.*, **24,** 1107 (1971).
70. M. Bixon and J. Jortner, *J. Chem. Phys.*, **48,** 715 (1968) and subsequent papers.
71. P. J. Domaille and J. E. Kent, unpublished work.
72. R. M. Hochstrasser, *Symposium on Molecular Spectroscopy and Molecular Structure*, Ohio State University, Columbus, Ohio, 1969.
73. H. Sponer, *Radiat. Res. Suppl.*, **1,** 658 (1959).
74. J. H. Callomon, unpublished work.
75. H. Labhart, *Experientia*, **22,** 65 (1966); *Adv. Chem. Phys.*, **13,** 179 (1967).
76. R. M. Hochstrasser and L. J. Noe, *J. Chem. Phys.*, **48,** 514 (1968); **50,** 1684 (1969).
77. R. M. Hochstrasser and T.-S. Lin, *J. Chem. Phys.*, **49,** 4929 (1968).
78. D. E. Freeman and W. Klemperer, *J. Chem. Phys.*, **40,** 604 (1964); **45,** 52 (1966).
79. D. E. Freeman, J. R. Lombardi, and W. Klemperer, *J. Chem. Phys.*, **45,** 58 (1966).
80. J. R. Lombardi, *J. Chem. Phys.*, **50,** 3780 (1969).
81. K.-T. Huang and J. R. Lombardi, *J. Chem. Phys.*, **51,** 1228 (1969).
82. J. R. Lombardi, *J. Chem. Phys.*, **48,** 348 (1968).
83. A. E. Douglas and E. R. V. Milton, *Discussions Faraday Soc.*, **35,** 235 (1963).
84. J. M. Hollas, *J. Mol. Spectrosc.*, **9,** 138 (1962).
85. J. E. Haebig, *J. Mol. Spectrosc.*, **25,** 117 (1968).

86. C. K. Ingold and C. L. Wilson, *J. Chem. Soc.*, **1936,** 955.
87. B. R. Cuthbertson and G. B. Kistiakowsky, *J. Chem. Phys.*, **4,** 9 (1936).
88. L. M. Logan, I. Buduls, and I. G. Ross, in *Molecular Luminescence*, E. C. Lim, Ed., W. A. Benjamin, New York, 1969, p. 53.
89. C. S. Parmenter and M. W. Schuyler, *J. Chim. Phys.*, **67,** 92 (1970).
90. K. K. Innes and R. M. Lucas, *J. Mol. Spectrosc.*, **24,** 247 (1967).
91. K. K. Innes, J. D. Simmons, and S. G. Tilford, *J. Mol. Spectrosc.*, **11,** 257 (1963).
92. K. K. Innes and L. E. Giddings, *Discussions Faraday Soc.*, **35,** 192 (1963).
93. L. A. Franks and K. K. Innes, *J. Chem. Phys.*, **47,** 863 (1967).
94. J. M. Hollas, E. Gregorek, and L. Goodman, *J. Chem. Phys.*, **49,** 1745 (1968).
95. G. W. King and S. P. So, *J. Mol. Spectrosc.*, **33,** 376 (1970).
96. J. C. D. Brand and P. D. Knight, *J. Mol. Spectrosc.*, **36,** 328 (1970).
97. A. R. Hartford and J. R. Lombardi, *J. Mol. Spectrosc.*, **35,** 413 (1970).
98. G. H. Kirby, *Mol. Phys.*, **19,** 289 (1970).
99. K. -T. Huang and J. R. Lombardi, *J. Chem. Phys.*, **52,** 5613 (1970).
100. T. Cvitas and J. M. Hollas, *Mol. Phys.*, **18,** 101 (1970).
101. T. Cvitas, *Mol. Phys.*, **19,** 297 (1970).
102. T. Cvitas and J. M. Hollas, *Mol. Phys.*, **18,** 793 (1970).
103. T. Cvitas and J. M. Hollas, *Mol. Phys.*, **18,** 261 (1970).
104. T. Cvitas and J. M. Hollas, *Mol. Phys.*, **18,** 801 (1970).
105. J. Christoffersen, J. M. Hollas, and G. H. Kirby, *Mol. Phys.*, **18,** 451 (1970).
106. H. E. Howard-Lock and G. W. King, *J. Mol. Spectrosc.*, **36,** 53 (1970).
107. A. Mani and J. R. Lombardi, *J. Mol. Spectrosc.*, **31,** 308 (1969).
108. J. Christoffersen, J. M. Hollas, and R. A. Wright, *Proc. Roy. Soc.* (*London*) *Ser. A*, **308,** 537 (1969).
109. J. M. Hollas and R. A. Wright, *Spectrochim. Acta*, **25A,** 1211 (1969).

AUTHOR INDEX

Numbers in parentheses are reference numbers and show that an author's work is referred to although his name is not mentioned in the text. Numbers in *italics* indicate the pages on which the full references appear.

SUBJECT INDEX